Lecture Notes in Statistics

Edited by P. Bickel, P. Diggle, S. Fienberg, U. Gather, I. Olkin, and S. Zeger

Viatcheslav B. Melas

Functional Approach to Optimal Experimental Design

 Springer

Viatcheslav B. Melas
Mathematics & Mechanics Faculty
St. Petersburg State University
28, University Avenue, Petrodvoretz
198504 St. Petersburg, Russia
v.melas@pobox.spbu.ru

Library of Congress Control Number: 2005930803

ISBN-10: 0-387-98741-X
ISBN-13: 978-0387-98741-5

Printed on acid-free paper.

Printed in the United States of America.   (MVY)

9 8 7 6 5 4 3 2 1

springeronline.com

Preface

The present book is devoted to studying optimal experimental designs for a wide class of linear and nonlinear regression models. This class includes polynomial, trigonometrical, rational, and exponential models as well as many particular models used in ecology and microbiology. As the criteria of optimality, the well known D-, E-, and c-criteria are implemented.

The main idea of the book is to study the dependence of optimal designs on values of unknown parameters and on the bounds of the design interval. Such a study can be performed on the base of the Implicit Function Theorem, the classical result of functional analysis. The idea was first introduced in the author's paper (Melas, 1978) for nonlinear in parameters exponential models. Recently, it was developed for other models in a number of works (Melas (1995, 2000, 2001, 2004, 2005), Dette, Melas (2002, 2003), Dette, Melas, Pepelyshev (2002, 2003, 2004b), and Dette, Melas, Biederman (2002)).

The purpose of the present book is to bring together the results obtained and to develop further underlying concepts and tools. The approach, mentioned above, will be called *the functional approach*. Its brief description can be found in the Introduction.

The book contains eight chapters. The first chapter introduces basic concepts and results of optimal design theory, initiated mainly by J.Kiefer. In the second chapter a general theory of the functional approach is developed. Particularly, it is proved that for the class of models considered in this book support points of optimal designs are real analytic functions of some values (the initial values of parameters for nonlinear models and the bounds of the design interval for linear models). This allows one to approximate the support points by the Taylor series. In Chapters 3 and 4 this approach is applied to polynomial and trigonometrical models, respectively. Chapters 5, 6, and 7 are devoted to rational and exponential models. In Chapter 8, a nonlinear model widely used in microbiology and called the Monod model is thoroughly studied.

I would like to thank Professor Sergey Ermakov, who attracted my attention to exponential models. A part of this book is based on works joint with Professor Holger Dette. I thank him for the permission to use our results here. Note that the computer calculations were performed under my guidance by my Ph.D. students Andrey Pepelyshev and Liudmila Krylova. I am grateful to several anonymous referees for helpful comments on an earlier version of the book and to Dr. John Kimmel for agreeing to prepare this book for publication. The work was performed partly under the financial support of Russian Foundation of Basic Research (Project Ns 00-01-00495 and 04-01-00519).

I thank my wife for her care of me and her help during this work.

Viatcheslav B. Melas
St.Petersburg

Contents

Introduction

The present book is devoted to studying optimal designs for linear and non-linear regression models possessing some Chebyshev properties – in particular, for polynomial, trigonometrical, rational and exponential models.

In the past, statistical procedures were applied to data, collected without a definite design. However, even in the 19th century many researchers felt the importance of rational choice of experimental designs. Interesting historical facts on the matter can be found in Stiegler (1986).

Fisher was the first to consider design problems systematically. His research on agrobiological experiments used the designs based on combinatorial principles, such as the Latin squares, to estimate the influence of some discrete factors. His popular book (Fisher, 1935) passed through many editions and affected the development and applications of experimental design. Fisher's approach is still developing (e.g., see Street and Street (1987)).

Another branch of experimental design goes back to the paper Box, Wilson (1951). This paper offers an approach to finding the conditions for some output variable to be of maximal value. The approach is based on applying fractional factorial experiments to estimate the gradient of the goal function. It is called *the response surface methodology*. The approach is outlined in the paper by Box (1996) (bibliography included) (see also Box and Draper (1987)).

The third branch, called *the optimal design*, was founded mainly by Kiefer (see the collection of papers Kiefer (1985)). In terms of this approach, the experimental design is a discrete probability measure, defined by the set of various experimental conditions and weight coefficients corresponding to them. These coefficients show how many experiments (with respect to their total amount) should be performed under the condition. Here, the optimality criteria are represented as various functionals defined on the set of information matrices and possessing some statistical sense. A design, at which such a functional attains its extremum, is called the optimal one.

This branch of the experimental design theory seems to be highly attractive from the viewpoint of calculations. The equivalence theorem, derived in Kiefer and Wolfowitz (1960) and its various analogs and expansions, which are reviewed in Kiefer (1974) and monograph Pukelsheim (1993), give the necessary and sufficient optimality conditions; they are the main

tools for developing both analytical and numerical methods of constructing the optimal designs. At the same time, it should be noted that the
classical optimality concept is quite severe, since constructing the design
requires a fixed regression model and a fixed range of experimental conditions to be prescribed. To a considerable extent, such a constraint can be
overcome by introducing multicriterial approach [e.g., see the introduction
to the monograph by Ermakov (1983)], considering systematic error [Box
and Draper (1987), Ermakov (1983), Ermakov and Melas (1995); Wiens
(1992, 1993)] and analyzing robust properties of optimal designs [see, e.g.,
Pronzato and Walter(1985)]. The brief review of the monographs, deal-
ing with the systematic statement of the optimal design theory, can be
found in the introduction to Fedorov and Hackl (1997). The monograph
by Schwabe (1996) considers constructing multidimensional designs on the
base of one-dimensional ones. The recent book by Müller (1998) examines
the optimal design for the random fields. Sometimes experimental designs
defined above cannot be applied in practice. In such a case, one should
consider so-called *replication free designs* (see Rasch (1996))).

For physical experiments, it proved very important to consider the design region as a functional space. The corresponding approach was developed by the Russian mathematician V. Kozlov (see the book of his selected
papers Kozlov (2000)).

Reviews of experimental design problems and results can be found in
Bock (1998)and Rasch (2003) as well as in a two-volumed handbook Rasch
et al. (1996, 1998) which is, unfortunately, in German only.

The present book considers optimal designs for a wide class of models
mentioned at the beginning of this Introduction.

Such models were considered by many authors and the corresponding
literature will be cited throughout this book. However, many problems
remained unsolved or were solved only numerically and we will demonstrate
that the approach developed here allows one to obtain almost exhaustive
solutions in many cases.

Note that under nonlinear models we include all of the models for which
optimal (in a usual sense) design depends on true values of parameters.
We concentrate here on problems of parameter estimating. But it should
be noted that a similar approach can be used for discrimination between
competing models and other problems for which the criteria considering
here are appropriate.

The main approach developed here is based on the following ideas. Let
us consider a nonlinear regression model given at a finite or infinite design
interval. Assume that our task is to construct a locally D-optimal design.
This is a discrete probability measure maximizing the determinant of the
information matrix under given initial values of nonlinear parameters. In
many cases, the number of support points of locally D-optimal designs is
equal to the number of parameters and the weights of all points are the
same. Note that the support points are functions of the initial values of

parameters. Sometimes these functions can be found explicitly. However, in general, these functions are given implicitly by an equation system generated by necessary conditions of optimality. This system can be received by equating the derivatives of the goal function by design points to zero. Let the regression function be real analytic, which is the case for all models considered in this book. Also, it can be proved that the Jacobi matrix of the equation system is invertible for any fixed value of the parameter vectors. Now, due to the Implicit Function Theorem (see, e.g., Gunning and Rossi (1965)), the support points appear to be also real analytic functions of the initial values. This allows one to approximate these functions by segments of their Taylor series. In Chapter 2 of this book we introduce very convenient recurrent formulas for calculating the Taylor coefficients. Note that the zero coefficients can be calculated analytically by constructing a locally D-optimal design for a special value of the vector parameter. To this end, a method of asymptotic analysis is introduced.

A similar approach is introduced for studying the dependence of support points of optimal designs for linear models on the bounds of the design interval. These ideas go back to papers Melas (1978, 1995, 2000, 2004, 2005) and Dette, Melas and Pepelyshev (2004b).

Numerical studies show that the approach allows one to calculate optimal designs with a high precision. Note that it can be done simply by hand, using the tables of coefficients given in this book.

The book contains eight chapters and an Appendix. The dependence among chapters is represented in Figure 1.

In the first chapter, the basic concepts and results of the optimal design theory are briefly described. This chapter also introduces the Implicit Function Theorem and basic properties of Chebyshev systems, providing important tools for the functional approach. A general theory of this approach is developed in Chapter 2. This chapter is devoted to studying locally D-optimal and maximin efficient designs for nonlinear regression models of a Chebyshev type. Basic results of this chapter were briefly described earlier. It is also proved here that under some conditions, support points of the optimal designs are monotonic functions of initial values of nonlinear parameters. Similar results for c- and E-optimal designs are obtained in the following chapters for more special types of model.

Chapter 3 is devoted to the implementation of the approach to polynomial models on an arbitrary interval. It considers E-optimal designs and designs optimal for estimating individual coefficients (e_k-optimal designs). Such designs can be found explicitly only for some types of interval. Also, the functional approach allows one to study e_k-optimal designs for arbitrary intervals and E-optimal designs for symmetrical intervals of arbitrary length. In Chapter 4, D- and E-optimal designs are constructed for trigonometrical models on arbitrary design intervals. Locally D-optimal designs for rational and exponential models are studied in Chapter 5 and 6, respectively. Chapter 7 is about locally E- and c-optimal designs for a

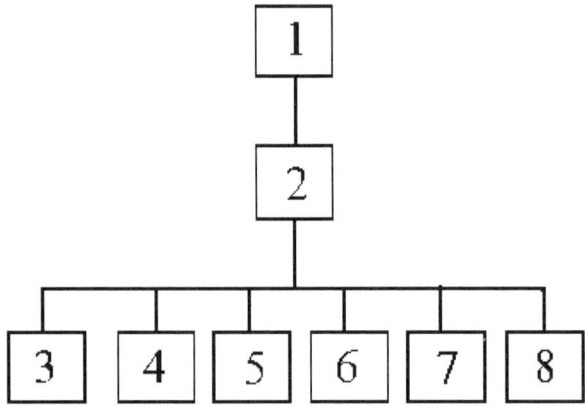

Figure 1: Logical dependence between chapters

wider class of nonlinear models.

The last chapter considers a nonlinear model used in microbiology and called the Monod model. For this model, locally D-, E- and c-optimal designs are studied. Here maximin efficient designs defined in Müller (1995) are also investigated. In the Appendix some remarks on computer calculating the Taylor coefficients are given.

Note that due to a great variety of models, the notations in different chapters are slightly different. However, the basic notations are the same.

It is worth mentioning that the models considered in this book depend on a single variable. However, the recurrent formulas and the general scheme of the approach are appropriate for multivariate models as well. Such a model was considered in Melas, Pepelyshev and Cheng (2003), devoted to studying locally optimal designs for estimating an extremum point of quadratic regression on a hyperball. In that paper, the designs were constructed explicitly and that is why they were not included in the present book. The development of the approach to multivariate models is a matter of a future work.

Chapter 1

Fundamentals of the Optimal Experimental Design

The present chapter is to recall some basic statements and definitions of the theory of optimal design needed to develop the functional approach.

Here, the results will be outlined only. More detailed layout for Sections 1.1–1.7 can be found in the introductory chapters of Fedorov (1972), Pukelsheim (1993), Fedorov and Hackl (1997).

The last two sections of the chapter are devoted to the Implicit Function Theorem and properties of Chebyshev systems, respectively. The Theorem is the corner stone of the functional approach. Also, regression models considered in this book are closely connected with Chebyshev systems.

1.1 The Regression Equation

The following equation appears to be the basic one in regression theory:

$$y_j = \eta(x_j, \Theta) + \epsilon_j, \quad j = 1, \ldots, N, \qquad (1.1)$$

where $y_1, \ldots, y_N$ are experimental results, $\eta(x, \Theta)$ is a given function with unknown parameter vector $\Theta = (\theta_1, \ldots, \theta_m)^T$, $\epsilon_1, \ldots, \epsilon_N$ are random variables, corresponding to the observation error and $x_1, \ldots, v_N$ are experimental conditions, belonging to the set $\mathfrak{X}$ usually called the *design region*.

The opportunity to represent the results of real experiments in form (1.1) has been shown in many examples in Rao (1973), Fedorov (1972), and Pukelsheim (1993).

Let us recall some basic assumptions of the classical research.

(a) Unbiasedness: $E\epsilon_j = 0$ $(j = 1, \ldots, N)$. It means that $Ey_j = \eta(x_j, \theta)$ (i.e., the model is free of a systematic error).

(b) Uncorrelatedness: $E\epsilon_i\epsilon_j = 0$ $(i \neq j)$.

(c) Variance homogeneity: $E\epsilon_j^2 \equiv \sigma^2 > 0$ $(j = 1, \ldots, N)$.

(d) Linearity of parametrization: $\eta(x, \theta) = \Theta^T f(x)$, where $f(x) = (f_1(x), \ldots, f_m(x))^T$, $f_i(x)$, $i = 1, \ldots, m$, are given basic functions.

Furthermore, the following assumptions are usual:

(e) Functions $\{f_i(x)\}_{i=1}^m$ are continuous and linearly independent on $\mathcal{X}$.

(f) $\mathcal{X}$ is a fixed set, which can be considered as a compact topological space.

As is usual in mathematical theory, these assumptions provide observable results to be obtained and correspond to some extent to the features of real experiments. Note that any assumption can be weakened (e.g., see Rao, 1973; Ermakov, 1983). However, it will imply some substantial difficulties. The present book deals with weakened assumptions (d).

The main purpose in an experiment is either to estimate a vector of unknown parameters, or to test a hypothesis on values of the parameters. Here, the accuracy of statistical conclusions depends on both the method of the statistical inference and the choice of the experimental conditions $(x_1, x_2, \ldots, x_N)$.

If (a)–(f) are assumed, then there exists the method (the least squares technique), which provides the most accurate, in a well-defined sense, estimation of the vector Θ of parameters under any fixed experimental conditions. Thus, the general problem of estimating and selecting experimental conditions is split into two independent problems. The following section considers the first of them.

1.2   Gauss–Markov Theorem

Set $X = (f_i(x_j))_{j,i=1}^{N\,m}$, X is a matrix of order $N \times m$.

Model (1.1) under assumptions (a)–(d) can be represented as

$$Y = X\Theta + \epsilon, \tag{1.2}$$

where $Y = (y_1, \ldots, y_N)^T$, $\epsilon = (\epsilon_1, \ldots, \epsilon_N)^T$,

$$E\epsilon = 0, \quad V_\epsilon = \sigma^2 I \tag{1.3}$$

(I is the identity matrix of order $N \times N$ and V_ϵ is the variance matrix).

As usual, we call *the estimator* $\tilde{\Theta}$ of the vector Θ unbiased if $E\tilde{\Theta} = \Theta$ for any vector $\Theta \in \mathbf{R}^m$. An estimator is called *linear* if it can be represented in the form $\Theta = SY$, where S is some matrix of order $m \times N$ independent of Y.

A linear *unbiased estimator* $\hat{\Theta}$ is called the best one if the matrix

$$V_{\hat{\Theta}} - V_{\tilde{\Theta}}$$

is nonpositive definite for any unbiased estimator $\tilde{\Theta}$; that is,

$$V(\hat{\Theta}z) = z^T V_{\hat{\Theta}} z \leq z^T V_{\tilde{\Theta}} z = V(\tilde{\Theta}^T z)$$

for any vector $z \in \mathbf{R}^m$. Here, the variance of scalar $\hat{\Theta}^T z$ is denoted by the same letter as the variance matrix, so do not mix them up.

The procedure of the least squares technique is to select some $\hat{\Theta}$ such that

$$\hat{\Theta} = \arg \inf_{\tilde{\Theta} \in \mathbf{R}^m} (Y - X\tilde{\Theta})^T (Y - X\tilde{\Theta}).$$

To simplify the statement, let us assume that the condition

(g) $\det X^T X \neq 0$

is satisfied.

The following theorem is well known.

Theorem 1.2.1 (Gauss–Markov theorem) *The estimator of the least squares method for model (1.2)–(1.3) under condition (g) is uniquely determined, it is of the form*

$$\hat{\Theta} = (X^T X)^{-1} X^T Y,$$

and it is the best linear unbiased estimator. Moreover,

$$V_{\hat{\Theta}} = \sigma^2 (X^T X)^{-1}.$$

The proof of the theorem and its expansions for the case of $E \epsilon \epsilon^T = W$, where W is a given nonnegatively definite matrix, and for the case of estimating the vector $K\theta$, where K is a given matrix (here condition (g) can be omitted) can be found, for instance, in Rao (1973) and Pukelsheim (1993).

The precision of the estimator can be improved by the optimal selection of experimental conditions. This is the subject of the theory of optimal design.

1.3   Experimental Designs and Information Matrices

The set $\{\hat{x}_1, \ldots, \hat{x}_N\}$ of elements in $\mathfrak{X}$ (although some of the elements may coincide with one another) is called an exact (or discrete) design of experiment of size N. Taking into account the possibility of coinciding elements, we can present such a design in different form.

Let us have only $n < N$ distinct point.   We can rename them $x_1, \ldots, x_n$. Suppose that x_i occurs r_i times among the points $\{\hat{x}_1, \ldots, \hat{x}_N\}$, $i = 1, \ldots, n$.

Let us associate weight coefficients $\omega_i = r_i/N$ with each of points x_i, $i = 1, \ldots, n$. A discrete probability measure, given by the table

$$\xi = \left(\begin{array}{ccc} x_1 & \cdots & x_n \\ \omega_1 & \cdots & \omega_n \end{array} \right), \tag{1.4}$$

will be called the normed exact (discrete) design or n-points design of size N.

The matrix

$$M(\xi) = \sum_{i=1}^{n} f(x_i) f^T(x_i) \omega_i \tag{1.5}$$

is called the information matrix of design ξ. By the Gauss–Markov theorem we have

$$V_{\hat{\Theta}} = \frac{\sigma^2}{N} M^{-1}(\xi)$$

for the variance matrix of the least squares estimate.

The design ξ is a discrete probability measure, defined by table (1.4), which includes the points of the set $\mathfrak{X}$ and the weight coefficients under some additional restrictions. If these restrictions are omitted, the study of designs becomes substantially more easy.

Introduce σ–algebra of subsets $\mathcal{B}$ on $\mathfrak{X}$, including all subsets of one point. The probability measure on $(\mathfrak{X}, \mathcal{B})$ is called *the approximate (continuous) design*.

In many practical situations it is impossible to realize these designs and such designs should be considered as approximation of some discrete designs.

Let us write a design, concentrated at a finite number of points, in the form (1.4), where coefficients ω_i are arbitrary positive numbers such that $\sum \omega_i = 1$. As for the general case, let $\xi(dx)$ stand for the corresponding design. The matrix

$$M(\xi) = \int f(x) f^T(x) \xi(dx)$$

is called *the information matrix* of the approximate design. This matrix assumes the form (1.7) for designs, concentrated in a finite number of points.

Let Ξ be the set of all approximate designs and $\mathcal{M}$ be the set of information matrices corresponding to them:

$$\mathcal{M} = \{M; \quad M = M(\xi) \quad \text{for some} \quad \xi \in \Xi\}.$$

Let Ξ_n be the set of approximate designs, concentrated at n points (with nonzero weights).

The basic properties of information matrices can be stated as a theorem.

Theorem 1.3.1 *((Properties of information matrices)*

   (i) Any information matrix is non-negative definite.

   (ii) If $n < m$, then $\det M(\xi) = 0$ for $\xi \in \Xi_n$.

   (iii) The set $\mathcal{M}$ is convex.

   (iv) If conditions (a)–(e) are satisfied, set $\mathcal{M}$, considered as a collection of vectors composed of diagonal and off-diagonal elements of the matrices, is a bounded and closed subset of $\mathbf{R}^s$, $s = (m+1)m/2$.

   (v) For any design $\xi \in \Xi$, there exists a design $\tilde{\xi} \in \Xi_n$, where $n \leq (m+1)m/2 + 1$, such that

$$M(\tilde{\xi}) = M(\xi).$$

Proof of the theorem can be found in Karlin and Studden (1966, Chap. X).

According to property (v), it is sufficient to consider only approximate designs with a finite support. Thus, they will be considered as experimental designs if not stated otherwise.

1.4 Optimality Criteria

Let us call the design ξ nonsingular if $\det M(\xi) \neq 0$. Such designs exist by assumption (e). Let us consider only the case of estimating the whole set of parameters $\Theta_1, \ldots, \Theta_m$. Here, only nonsingular designs are of interest. The version of Gauss–Markov theorem adduced in Section 1.2 is valid for them.

Typically, there is no design $\hat{\xi}$ such that the matrix

$$M^{-1}(\hat{\xi}) - M^{-1}(\xi),$$

is non-positive definite, where ξ is an arbitrary design. Therefore, some functions of information matrices, having strict statistical sense, are used as the optimality criteria.

Let us consider some commonly used optimality criteria.

1.4.1 *D*-Criterion

The D-criterion is of the form

$$\det M(\xi) \to \sup_{\xi \in \Xi},$$

(here and further, the extremum is taken over all approximate designs).

If the errors are normally distributed, this criterion corresponds to the requirement to minimize the volume of the confidence ellipsoid with an arbitrary fixed confidence level for the least squares estimators. This ellipsoid is of the form

$$\{\tilde{\theta};\quad (\tilde{\theta} - \hat{\theta})^T M^{-1}(\tilde{\theta} - \hat{\theta}) \le c\}, \tag{1.6}$$

where c is a constant (depending only on the confidence level).

1.4.2   *G*-Criterion

Set $d(x, \xi) = f^T(x) M^{-1}(\xi) f(x)$.

The G-optimality criterion is of the form

$$\max_{x \in \mathfrak{X}} d(x, \xi) \to \inf_{\xi}.$$

Note that for normed discrete design ξ,

$$d(x, \xi) = \frac{\sigma^2}{N} V(\hat{\theta}^T f(x));$$

that is $d(x, \xi)$ is equal (to constant precision) to the variance of a value, predicted by the model at point x. The G-criterion has the minimax sense, it means the minimization of the maximum of prediction variance.

1.4.3   *MV*-Criterion

$$\operatorname{tr} M^{-1}(\xi) \to \inf_{\xi}.$$

This criterion is to minimize the sum of the variances of the least squares estimator $\hat{\Theta}$.

1.4.4   *c*-Criterion

Let us introduce the value

$$\Phi_c(\xi) = \begin{cases} c^T M^-(\xi) c & \text{if } c \in \text{range } M(\xi) \\ \infty & \text{overwise,} \end{cases}$$

where c is a given vector, M^- denotes a generalized inverse for M, and the notation $c \in \text{range } M$ means that c is a linear combination of rows of the matrix M.

Let us note that the generalized inverse to a given matrix A can be defined as an arbitrary matrix with the property $A A^- A = A$, and if an equation system $Ax = y$ has a solution, say $\hat{x}$, this solution is of the form $\hat{x} = A^- y$.

A design minimizing $\Phi_c(\xi)$ will be called *c-optimal*. The statistical sense of this criterion consists of the minimization of the variance of the best linear unbiased estimate for a given linear combination of the model parameters $\tau = c^T \theta$.

1.4.5 *E*-Criterion

$$\lambda_{\min}(M(\xi)) \to \sup_{\xi},$$

where $\lambda_{\min}(M)$ is the minimal eigenvalue of the matrix $M = M(\xi)$.

The *E*-optimality criterion secures the minimization of the maximum axis of the confidence ellipsoid (1.6). This criterion was introduced in Ehrenfeld (1955).

Note that since

$$\lambda_{\min}(M) = \min_{c^T c = 1} c^T M c,$$

the *E*-criterion secures minimization of the maximum of variances of linear combinations $c^T \theta$ under the restriction $c^T c = 1$.

Sometimes it is useful to consider classes of criteria. The class of linear criteria is a class of criteria of the form

$$\operatorname{tr} L M^{-1}(\xi) \to \inf_{\xi},$$

where L is some given non-negative definite matrix. Particularly, for $L = I$ and $L = cc^T$, we have *MV*-criterion and *c*-criterion, respectively.

The class of Φ_p-criteria, introduced in Kiefer (1974), is of the form

$$\left(\operatorname{tr} M^{-p}(\xi)\right)^{1/p} \to \inf_{\xi},$$

where $0 \le p \le \infty$. For $p = \infty$ we have *E*-criterion, and for $p = 1$ we have *MV*-criterion.

Note that all the criteria above can be represented as

$$\Phi(M(\xi)) \to \sup_{\xi}$$

or as

$$\Psi(V(\xi)) \to \inf_{\xi},$$

where $V(\xi) = M^{-1}(\xi)$, $\Phi(M)$ is a concave function of the matrix M, and $\Psi(V)$ is a convex function of the matrix V. So, the methods of solving corresponding extremum problems can be unified.

If more optimality criteria are needed, especially for estimating the vector $K\Theta$, where K is a given matrix of order $s \times m$, $s \le m$, see Pukelsheim (1993).

1.5 Equivalence Theorems

The following result from Kiefer and Wolfowitz (1960) is of great importance in the theory of optimal experimental design.

Theorem 1.5.1 (Kiefer–Wolfowitz equivalence theorem) *For model (1.1), a D-optimal design exists under assumptions (a)–(f) and the following conditions are equivalent:*

(i) ξ^* *is a D-optimal design.*

(ii) ξ^* *is a G-optimal design.*

(iii) $\max_{x \in \mathfrak{X}} d(x, \xi^*) = m$.

Moreover, all D-optimal designs have the same information matrix, and the prediction variance function $d(x, \xi^*)$ attains its maximum at the points of any D-optimal design with finite support.

It is worth stressing that the theorem is true for designs to be D-optimal in the class of approximate designs.

This theorem not only states equivalency between D- and G-criteria but also gives the important necessary and sufficient condition of D-optimality: Design ξ^* is D-optimal if and only if $\max_{x \in \mathfrak{X}} d(x, \xi^*) = m$.

The proof of the theorem can be found in Kiefer and Wolfowitz (1960). Note that the problem $\det M(\xi) \to \sup_\xi$ is equivalent to $\ln \det M(\xi) \to \sup_\xi$. In fact, the proof is based on the concavity of the function $\ln \det M$ and the ability to evaluate its derivative in an explicit form.

It is worth mentioning that the D-efficiency of a given design with respect to a D-optimal design can be evaluated by Kiefer's inequality without an explicit construction of a D-optimal design. This inequality yields for the D-efficiency,

$$\left(\frac{\det M(\xi)}{\max_\xi \det M(\xi)} \right)^{1/m} \geq e^{1-v/m}, \tag{1.7}$$

where the constant v is defined by

$$v = \max_{t \in [0,T]} f^T(t) M^{-1}(\xi) f(t)$$

(see Pukelsheim (1993)).

Many analogs of the Kiefer–Wolfowitz theorem can be found in Kiefer (1974). The equivalence theorem seems to be most general one is given in Whittle (1973).

1.6   Iterative Numerical Techniques

The Kiefer–Wolfowitz duality theorem and its analogs are still the main tool of constructing optimal design. For some quite slender class of standard models and design regions (an interval, a circle, a ball and a hyperball, a parallelepiped, and a hyperparallelepiped) the optimal designs (mainly for D-criterion) have been found in explicit form (e. g., see Fedorov (1972), Kiefer (1985), Pukelsheim (1993), Ermakov et al. (1983)).

If it proves impossible to find an optimal design in an explicit form, it can be found by numerical techniques.

The equivalence theorems give the basis for the special numerical techniques to be constructed. The special iterative methods of constructing D-optimal designs, similar to one another and based on the Kiefer–Wolfowitz equivalence theorem, were originally offered by Fedorov (1972) and Wynn (1970). Let us outline Fedorov's version.

Set $\xi_x = \{x; 1\}$. Let ξ_0 be some nonsingular design (i. e., $\det M(\xi_0) \neq 0$),

$$\xi_0 = \{x_1, \ldots, x_{n_0}; \mu_1, \ldots, \mu_{n_0}\}.$$

For $s = 0, 1, \ldots$, find

$$x_{n_0+s+1} = \arg\max_{x \in \mathfrak{X}} d(x, \xi_s),$$

$$\alpha_s = \arg\max_{\alpha \in [0,1]} \det M(\xi_{s+1}(\alpha)),$$

where

$$\xi_{s+1}(\alpha) = (1 - \alpha)\xi_s + \alpha\xi_{x_{n_0+s+1}}; \text{ that is,}$$

$$\xi_{s+1}(\alpha) = \{x_1, \ldots, x_{n_0+s+1}; (1 - \alpha)\mu_{1(s)}, \ldots, (1 - \alpha)\mu_{n_0+s(s)}, \alpha\}.$$

It may be proved that α_s has the explicit form

$$\alpha_s = \frac{d_s - m}{(d_s - 1)m}, \quad d_s = d(x_{n_0+s+1}, \xi_s).$$

If $s \to \infty$, the sequence of designs ξ_s under the assumptions of the Kiefer–Wolfowitz theorem converges to some D-optimal design (in the sense of weak convergence of probability measures).

A similar algorithm for optimality criteria of a general form also can be designed (see Fedorov and Hackl (1997)).

The main advantage of such algorithms is that only one-point designs are to be sought at each step, so dimensionality of the experimental problem can be sufficiently reduced. In this view, currently they are the main tool of numerical evaluation of optimal designs.

1.7 Nonlinear Regression Models

In the present book, we will consider the regression function $\eta(x, \Theta)$, which can not be represented in the form $\Theta^T f(x)$. Other usual assumptions will be reserved. Let us describe our basic model in more detail.

Let Ω be a compact in $\mathbf{R}^m$, and $\mathfrak{X}$ be a compact in $\mathbf{R}^k$. We will assume that experimental results $y_1, \ldots, y_N \in \mathbf{R}^l$ can be represented in the form

$$y_j = \eta(v_j, \Theta) + \varepsilon_j, \tag{1.8}$$

where $\{\varepsilon_j\}$ are independent and identically distributed random values, such that $E\varepsilon_j = 0$, $E\varepsilon_j^2 = \sigma^2 > 0$, $\eta(v, \Theta)$ is a known function of unknown parameters, $\Theta^T = (\theta_1, \ldots, \theta_m)$, $v_j \in \mathfrak{X}$, a $\Theta \in \Omega$, and σ^2 is unknown.

Introduce also the following assumptions:

(a) The function $\eta(x, \Theta)$ is continuous on $\mathfrak{X} \times \Omega$.

(b) The series of designs $\{\xi_N\}$ weakly converges to design ξ that is, the following relation is valid for any continuous function $g(x)$ on $\mathfrak{X}$:

$$\int_{\mathfrak{X}} g(x)\xi_N(dx) \to \int_{\mathfrak{X}} g(x)\xi(dx)$$

while $N \to \infty$.

(c) The value of

$$\int_{\mathfrak{X}} [\eta(x, \Theta) - \eta(x, \bar{\Theta})]^2 \xi(dx)$$

for $\bar{\Theta}$, $\Theta \subset \Omega$ vanishes if and only if $\Theta = \bar{\Theta}$.

(d) The derivatives

$$\partial\eta/\partial\theta_i, \quad \partial^2\eta/\partial\theta_i\partial\theta_j, \quad i, j = 1, \ldots, m,$$

exist and are continuous on $\mathfrak{X} \times \Omega$.

(e) Θ_{tr}, the true value of parameter vector, is an internal point of Ω and the matrix

$$M(\xi, \Theta) = \int_{\mathfrak{X}} f(x, \Theta)f^T(x, \Theta)\xi(dx),$$

where

$$f^T(x, \Theta) = \left(\frac{\partial\eta(x, \Theta)}{\partial\theta_1}, \ldots, \frac{\partial\eta(x, \Theta)}{\partial\theta_m}\right),$$

is nonsingular at $\Theta = \Theta_{\text{tr}}$.

Let ξ_N be of the form

$$\xi_N = \begin{pmatrix} v_1 & \cdots & v_n \\ 1/N & \cdots & 1/N \end{pmatrix},$$

where some points of v_i may coincide with each other:

$$\hat{\Theta}_N = \arg\min_{\Theta \in \Omega} \sum_{i=1}^{N} (\eta(v_i, \Theta) - y_i)^2. \tag{1.9}$$

Theorem 1.7.1 *If the random errors obey the above assumptions and assumptions (a)–(c) are satisfied, then*

$$\hat{\Theta}_{(N)} \to \Theta_{tr}$$

with probability 1 for $N \to \infty$, where $\hat{\Theta}_N$ is defined by formula (1.9). If, in addition, assumptions (d) and (e) are satisfied, then for $N \to \infty$, the distribution of the random vector $\sqrt{N}(\hat{\Theta}_N - \Theta_{tr})$ converges to the normal distribution with zero vector mean value and variance matrix $\sigma^2 M^{-1}(\xi, \Theta_{tr})$.

A proof of this theorem can be found in Jennrich (1969).

Simulation studies (see, e.g., Section 8.4) show that the sampling covariance matrix become rather close to the asymptotic one, given by Theorem 1.7.1, under moderate values of N. Therefore, the information matrix $M(\xi, \theta)$ can be used for constructing efficient experimental designs. The majority of papers on design for nonlinear models are based on this matrix. Note that for very small N an alternative approach developed in Vila (1990) and Pazman and Pronzato (1992) could be more appropriate.

The dependence of the information matrix

$$M(\xi, \Theta) = \left(\sum_{s=1}^{n} \frac{\partial \eta(x_{(s)}, \Theta)}{\partial \Theta_i} \frac{\partial \eta(x_{(s)}, \Theta)}{\partial \Theta_j} \mu_s \right)_{i,j=1}^{m}$$

on at least one parameter is the basic fact for the models to be nonlinear in the parameters.

Due to the theorem, the same criteria as in the linear case (e.g., $\det M(\xi, \Theta)$), can be selected as the optimality criteria, but, here, the optimal design depends on the vector of true values of parameters.

To overcome this difficulty, we can apply one of the standard statistical approaches: locally optimal, sequential, minimax, or Bayesian.

The concept of locally optimal designs was introduced in Chernoff (1953). A locally optimal design maximizes a certain functional of the information matrix in which an initial value of the parameter vector is used instead of the unknown proper value. The same functionals as that for linear models can be implemented (e.g., $\det M(\xi, \theta)$).

For some models with one nonlinear parameter locally optimal designs were found explicitly in a closed form (see the pioneer paper by Box and Lucas (1959) or the recent paper by Han and Chaloner (2003) and references in it). In Melas (1978), support points of locally D-optimal designs with an arbitrary number of parameters were studied as implicitly given functions of nonlinear parameters. This approach was developed in Melas (2001, 2004, 2005) for a wide class of nonlinear models. It can be called *a functional approach* and will be further elaborated in the present book.

The idea of the sequential approach consists of partitioning the whole set of experiments into a number of series. The estimates of parameters received on the basis of results of previous series are used for constructing an optimal design for the current series. This approach is thoroughly described in Fedorov (1972) and Silvey (1980). The design received in this way tends in the limit (if the number of series tends to the infinity) to the true locally optimal design (i.e., the design locally optimal for initial values equal to the proper ones). Thus, studying locally optimal designs is important for the sequential approach.

A minimax approach was implemented by Melas (1978). The idea of this approach is to find designs optimal for the least favorable values of parameters inside a given set of possible values. Let us describe an advanced version of this approach developed in Müller (1995) and based on the notion of designs efficiency.

Let $\Phi(M(\xi, \theta))$ be a certain optimality criterion. In the present book, we will consider criteria of D-, E-, and c-optimality:

$$\Phi_D(M(\xi, \theta)) = (\det M(\xi, \theta))^{1/m},$$

$$\Phi_E(M(\xi, \theta)) = \lambda_{\min}(M\xi, \theta)),$$

$$\Phi_c(M(\xi, \theta)) = \begin{cases} \left(c^T M^-(\xi, \theta)c\right)^{-1}, & \text{if } c \in \text{range} M(\xi, \theta) \\ 0 & \text{otherwise}, \end{cases}$$

where m is the number of parameters of the model, $\lambda_{\min}(A)$ denotes the minimal eigenvalue of the matrix A, A^- denotes a generalized inverse for the matrix A, and c is a given vector.

A design $\xi^*(\theta)$ will be called *locally Φ-optimal* if it maximizes

$$\Phi(M(\xi, \theta))$$

for a given θ.

A design will be called *maximin efficient Φ-optimal* (or, briefly, maximin efficient) if it maximizes

$$\Psi_\Omega(\xi) = \inf_{\theta \in \Omega} \frac{\Phi(M(\xi, \theta))}{\Phi(M(\xi^*(\theta), \theta))},$$

where Ω is a given set of possible values of the vector parameter.

Note that $\Psi_\Omega(\xi)$ is the efficiency of the design ξ with respect to a locally Φ-optimal design for a least favorable value of θ inside Ω. This value indicates how many more experiments we will need under the design ξ with respect to an "ideal" design to receive the same accuracy of estimating in the worst case. This is the reason of the title "maximin efficient".

Note that the construction of maximin efficient designs includes that of locally optimal ones. Maximin efficient designs were found numerically for different models and criteria in Dette, Melas, and Pepelyshev (2003), Dette, Melas, and Wong (2004a) and other papers. Equivalence theorems for such designs were obtained in Müller and Pazman (1998), Dette, Haines, and Imhof (2003), and Dette, Melas, and Pepelyshev (2003).

Bayesian approach to constructing optimal designs for nonlinear model consists of maximization of functionals of the form

$$\int \Phi(M(\xi, \theta))p(d\theta) \tag{1.10}$$

or of the form

$$\int \frac{\Phi(M(\xi, \theta))}{\Phi(M(\xi^*(\theta), \theta)} p(d\theta),$$

where $p(d\theta)$ is given prior probability measure of θ's possible values. This approach was considered in a number of papers (see Pronzato and Walter (1985) or Chaloner and Larntz (1989) among many others).

It proves that Bayesian designs can be constructed in a close form only for some simple models with one nonlinear parameter. Studying locally optimal designs seems to be important in the frame of this approach even if we use criterion (1.10). The locally optimal approach can be also considered as a special case of the Bayesian approach with $p(d\theta)$ equal to a probability measure concentrated in one point.

Thus, in all cases, constructing locally optimal design remains an important intermediate problem. If one would like to use such designs in practice, it seems important to study the sensitivity of these designs to the initial value. Thus the problem of dependence of the designs on these values is actual.

In this book, we will study such a dependence with the help of the functional approach. This approach is based on the Implicit Function Theorem considering in the next section.

1.8 The Implicit Function Theorem

As will be shown in the next chapter, the necessary conditions of a design optimality (D-criterion will be considered) can be transformed to a relation of the form

$$q(\tau, z) = 0, \tag{1.11}$$

where τ is the vector containing the design support points and weights, z is the vector of the auxiliary parameters, and $q = (q_1, \ldots, q_s)^T$ is a vector function.

In such a way, points and weights of a design to be locally D-optimal in the class of designs with a fixed number of points can be considered as functions of auxiliary parameters implicitly given by (1.11). Studying this equation can be performed on the base of the well-known Implicit Function Theorem. Let us formulate the version of this theorem to be used in the following chapters.

Remember the following well-known concept of analysis: Let U be an open set in $\mathbf{R}^t$, $t \geq 1$.

Definition 1.8.1 A real function of vector variable $u \in U$ will be called *a real analytic function* if in a vicinity of any point $u_{(0)} \in U$, it can be expanded into a (convergent) Taylor series.

Theorem 1.8.1 (Implicit Function Theorem).
Let $q(u) = (q_1(u), \ldots, q_s(u))^T$, $u = (\tau, z)$, $\tau \in \mathbf{R}^s$, $z \in \mathbf{R}^k$ be a real continuously differentiable vector function defined in a vicinity U of a point $(\tau_{(0)}, z_{(0)})$ and

$$q(\tau_{(0)}, z_{(0)}) = 0, \quad \det \left(\frac{\partial}{\partial \tau_j} q_i(\tau_{(0)}, z_{(0)}) \right)_{i,j=1}^s \neq 0.$$

Then there exists a vicinity V of the point $\tau_{(0)}$ such that at this vicinity, a unique continuous vector function $\tau(z)$ with the properties $\tau(z_{(0)}) = \tau_{(0)}$, and $(\tau(z), z) \in U$, $q(\tau(z), z) = 0$ is determined.

Moreover, the function $\tau(z)$ for $z \in V$ satisfies the following differential equations:

$$J(\tau(z), z) \frac{\partial \tau(z)}{\partial z_j} = -L_j(\tau(z), z), \; j = 1, \ldots, k,$$

where

$$J(\tau, z) = \left(\frac{\partial}{\partial \tau_j} q_i(\tau, z) \right)_{i,j=1}^s, \quad L_j(\tau, z) = \left(\frac{\partial}{\partial z_j} q_i(\tau, z) \right)_{i=1}^s.$$

If $q(u)$ is a real analytic vector function, then $\tau(z)$ is also real analytic vector function.

The proof of this theorem can be found in Gunning and Rossi (1965).

1.9 Chebyshev Models

For applying the Implicit Function Theorem to studying optimal designs, it is necessary to verify the invertibility of the Jacobi matrix of a corresponding equation system. It can be done usually by numerical methods. However, there is a wide class of regression models such that it can be done in a strong theoretical way, which is certainly very important for analytical studies. This class includes linear models with basis functions generating a Chebyshev system. It includes also nonlinear models such that the basis functions of corresponding linearized models generates such a system. For brevity, we will call all of these models Chebyshev ones.

Remember that a function system $f_1(t), \ldots, f_m(t)$ is call a *Chebyshev system* on an interval $\mathfrak{X} = [a, b]$ if for any $x_1, \ldots, x_m$ such that

$$a \leq x_1 < \cdots < x_m \leq b,$$

$$\det \left(f_i(x_j) \right)_{i,j=1}^{m} \neq 0.$$

A number of following concepts and results are taken from Karlin and Studden (1966, Chap. 1).

A set of functions $f_1, \ldots, f_m : I \to \mathbb{R}$ is called *a weak Chebyshev system* (on the interval I) if there exists an $\varepsilon \in \{-1, 1\}$ such that

$$\varepsilon \cdot \begin{vmatrix} f_1(x_1) & \cdots & f_1(x_m) \\ \vdots & \ddots & \vdots \\ f_m(x_1) & \cdots & f_m(x_m) \end{vmatrix} \geq 0 \qquad (1.12)$$

for all $x_1, \ldots, x_m \in I$ with

$$x_1 < x_2 < \cdots < x_m.$$

If the inequality in (1.12) is strict, then $\{f_1, \ldots, f_m\}$ is called *a Chebyshev system*. It is well known (see Karlin and Studden (1966, Theorem II 10.2)) that if $\{f_1, \ldots, f_m\}$ is a weak Chebyshev system, then there exists a unique function

$$\sum_{i=1}^{m} c_i^* f_i(t) = c^{*T} f(t), \qquad (1.13)$$

with the following properties:

(i) $|c^{*T} f(t)| \leq 1 \quad \forall \, t \in I$,

(ii) There exist m points $s_1 < \cdots < s_m$ such that $\qquad (1.14)$

$$c^{*T} f(s_i) = (-1)^i, \quad i = 1, \ldots, m.$$

The function $c^{*T} f(t)$ is called *a Chebyshev polynomial*, the points $s_1, \ldots, s_m$ are called Chebyshev points and need not to be unique. They are unique if $1 \in \text{span}\{f_1, \ldots, f_m\}, m \geq 1$ and I is a bounded and closed interval, where, in this case, $s_1 = \min_{x \in I} x$, $s_m = \max_{x \in I} x$.

Let us also define the generalized Chebyshev system of order p, originally introduced by Karlin and Studden (1966).

Let $u_0, u_1, \ldots, u_m$ be continuous real functions, defined on the closed finite interval $[c, d]$. Let us assume that these functions are p times continuously differentiable. Set

$$U \begin{pmatrix} 1, & 2, & \dots, & m \\ t_1, & t_2 & \dots, & t_m \end{pmatrix} = \det \begin{pmatrix} u_1(t_1) & \dots & u_1(t_m) \\ u_2(t_1) & \dots & u_2(t_m) \\ \dots & \dots & \dots \\ u_m(t_1) & \dots & u_m(t_m) \end{pmatrix},$$

$$F \begin{pmatrix} 1, & 2, & \dots, & m \\ t_1, & t_2 & \dots, & t_m \end{pmatrix} = \det \begin{pmatrix} 1 & 1 & \dots & 1 \\ t_1 & t_2 & \dots & t_m \\ \dots & \dots & \dots & \dots \\ t_1^{m-1} & t_2^{m-1} & \dots & t_m^{m-1} \end{pmatrix},$$

$$\Delta_m = \{\bar{t} = (t_1, \dots, t_m) | c \le t_1 < t_2 < \dots < t_m \le d\},$$

$$\bar{\Delta}_m = \{\bar{t} = (t_1, \dots, t_m) | c \le t_1 \le t_2 \le \dots \le t_m \le d\}$$

Definition. Let us call $\{u_i\}_i^m$ the *ET-system* of order p, if $u_i \in C^{p-1}[c, d]$, $i = 0, 1, \dots, n$, and

$$\lim_{\bar{s} \to \bar{t}} \frac{U \begin{pmatrix} 1, & 2, & \dots, & m \\ s_1, & s_2, & \dots, & s_m \end{pmatrix}}{F \begin{pmatrix} 1, & 2, & \dots, & m \\ t_1, & t_2, & \dots, & t_m \end{pmatrix}} > 0,$$

where $\bar{s} \in \Delta_m$, $t \in \bar{\Delta}_m$, and not more than p successive components of $\bar{t}$ are equal to each other.

Many models are Chebyshev ones. In this book, we will consider polynomial, trigonometrical, rational, and exponential model as well some models used in microbiology.

Polynomial models are of the form

$$\eta(t, \theta) = \sum_{i=1}^{m} \theta_i f_i(t), \tag{1.15}$$

where $f_i(t) = t^{i-1}$, $i = 1, \dots, m$.

These models as well as their modification with $f_i(t) = e^{\lambda}t^{i-1}$ are Chebyshev ones for any $\mathfrak{X} = [a, b]$, $-\infty < a < b < \infty$, and any real λ since the Wandermonde determinant does not vanish.

With $f_i(t) = t^i/(t + \alpha)$, $i = 1, \dots, m$, the model (1.15) is Chebyshev for $\mathfrak{X} = [a, b]$, $-\alpha < a < b \le \infty$, by the same reason.

Trigonometrical models have also form (1.15), where $m = 2k + 1$,

$$f_1(t) \equiv 1, \quad f_{2j}(t) = \sin(jt), \quad f_{2j+1}(t) = \cos(jt), \tag{1.16}$$

$j = 1, \dots, k$. As shown in Karlin and Studden (1966, Chap. 1) the function system (1.16) is Chebyshev on any interval $\mathfrak{X} \subset (-\pi, \pi)$.

Consider now rational and exponential models

$$\eta(t, \theta) = \sum_{i=1}^{k} \theta_i \varphi(t, \theta_{i+k}), \ t \in [0, \infty),$$

where $\varphi(t, \theta_{i+k}) = 1/[t + \theta_{i+k}]$ for rational models and $\varphi(t, \theta_{i+k}) = \exp(-\theta_{i+k}t)$ for exponential ones. Corresponding linearized models have the form

$$\beta^T f(t, \theta),$$

where $\beta = (\beta_1, \ldots, \beta_{2k})^T$ are the parameters to be estimated

$$f(t, \theta) = \frac{\partial}{\partial \theta} \eta(t, \theta),$$

$$f(t, \theta) = (f_1(t, \theta), \ldots, f_m(t, \theta))^T,$$

$$f_{2j-1}(t, \theta) = \frac{1}{t + \theta_{j+k}}, \ f_{2j}(t, \theta) = \frac{\theta_j}{(t + \theta_{j+k})^2}, \tag{1.17}$$

$j = 1, \ldots, k$ for rational models, and

$$f_{2j-1}(t, \theta) = e^{-\theta_{j+k}t}, \ f_{2j}(t, \theta) = -\theta_j t e^{-\theta_{j+k}t}, \tag{1.18}$$

$j = 1, \ldots, k$ for exponential models.

The Chebyshev property of the function system (1.18) on arbitrary intervals $[a, b]$, $-\infty \le a < b \le \infty$ is proved in Karlin and Studden (1966). This property for (1.17) on interval $[0, \infty)$ for $\theta_1, \ldots, \theta_k > 0$ will be proved in Chapter 6.

In Chapter 8, we will prove that a nonlinear model called the Monod model and widely used in microbiology is also a Chebyshev model.

Chapter 2

The Functional Approach

This chapter is devoted to studying optimal designs for a wide class of nonlinear regression models on the basis of a functional approach. This class includes exponential and rational models as well as many particular models of the Chebyshev type used in microbiology and other fields of experimental research.

We consider designs that are locally D-optimal or maximin efficient D-optimal among designs with the number of points equal to the number of parameters. In many cases, such designs prove to be optimal or maximin efficient among all approximate designs.

Support points of such designs are considered here as implicit functions on the initial value of the nonlinear parameters or on characteristics of sets containing, by the assumption, the true parameter value. A corresponding equation system is derived and is called the basic equation system or the basic (vector) equation. Studying this system allows one to prove that the functions are real analytic and therefore can be represented by a Taylor series under natural conditions. Recurrent formulas for computer-calculating the Taylor coefficients are introduced.

2.1 Introduction

Most results in the modern regression design theory were obtained for linear models with a fixed design region (see Fedorov, 1972; Silvey, 1980; Kiefer, 1985; Pukelsheim, 1993). However, many models of practical importance are nonlinear models (see, e.g., Seber and Wild, 1989). The commonly used approach for experimental design in such models consists of their linearization in a vicinity of some initial values of the nonlinear parameters and application of locally optimal designs, briefly discussed in the previous chapter. In spite of such designs are usually depending on the initial values, they can be used if a reliable knowledge about the parameters is available. These designs are also used in studying more complicated approaches: maximin and Bayesian ones (see Section 1.7).

Even if linear models are implemented, the design region often cannot be considered as fixed. For example, in many microbiological studies (see Pirt, 1984; Dette, Melas, and Strigul, 2005), the design region is a time interval and can be chosen by an experimentator in different ways. The introduction of design intervals with variable bounds can be considered also as an artificial method for investigations of the structure of optimal designs.

In the present chapter, we will consider nonlinear models given at a design interval. Our basic method here is the functional approach introduced in Melas (1978) for studying exponential nonlinear models and our aim is to apply it to a wider class of models.

The main idea of this approach consists of studying optimal design points and weights as implicitly given functions of the bound of the design interval and/or nonlinear parameters of the model. These functions can be investigated on the basis of the Implicit Functional Theorem formulated in Section 1.8 (see also Gunning and Rossi (1965)). In particular, in many cases these functions prove to be real analytic which enables one to present them by segments of the Taylor series. We will introduce here general recurrent formulas for constructing such series and discuss their applications for studying properties of optimal designs.

The functional approach seems to be useful when an explicit analytical form of optimal designs is not available. It can be considered as an alternative or useful addition to merely numerical methods. It is worth mentioning that similar approaches are well known in many fields of mathematics and its application. For example, representing indefinite integrals by a power series is the recognized technique of their calculation, and coefficients of such series are tabulated and given in textbooks. However, in the field of experimental design the functional approach is relatively new. References to existing literature will be given throughout the book.

In Section 2.2*, we will introduce the basic ideas of the functional approach using exponential models (nonlinear by parameters) as an example. Section 2.3 contains a list of assumptions justifying the implementation of the functional approach and formulates without proofs the main theoretical results. Section 2.4 is devoted to studying the basic equation. It is also introduces general recurrent formulas for calculating the Taylor coefficients. The application of the theory to the three-parameter logistic model is given in Section 2.5. All lengthy proofs are deferred to Section 2.6.

---

*Note that in Section 2.2 and in Sections 2.3–2.6 a part of materials are taken from Melas, V.B. (2005). On the functional approach to optimal designs for nonlinear models. *J. Statist. Plan. and Inference*, **132**, 93–116. ©2004 Elsevier B.V. with permission of Elsevier Publisher.

2.2   Basic Ideas of the Functional Approach

In this section we will introduce some basic ideas of the approach. We will consider regression models given by linear combinations of unknown exponentials as a typical example of nonlinear models.

In order to make the explanation more apparent, all technically difficult mathematical results will be only formulated and their proofs will be given in further sections.

Let us restrict our attention by the D-criterion and study locally D-optimal designs and maximin efficient D-optimal designs.

As it was discussed in Section 1.7, the first of the problems has some independent interest. It is also a necessary step for investigating the second problem.

2.2.1   Exponential regression models

Let us consider the models given by relations

$$Y_j = \sum_{i=1}^{k} a_i e^{-\lambda_i x_j} + \varepsilon_j, \ j = 1, \ldots, N, \tag{2.1}$$

where $Y_1, \ldots, Y_N$ are experimental results, $a_1, \ldots, a_k$ and $\lambda_1, \ldots, \lambda_k$ are the parameters to be estimated; and

$$a_i \neq 0, \ \lambda_i > 0, \ i = 1, \ldots, k, \ \lambda_i \neq \lambda_j \ (i \neq j), \tag{2.2}$$

$\varepsilon_1, \ldots, \varepsilon_N$ are independent and identically distributed random values (experimental errors) with zero mean ($E\varepsilon_i = 0$) and the variance $E\varepsilon_i^2 = \sigma^2 > 0$, and $x_1, \ldots, x_N \in [0, \infty)$ are observation points.

Let us assume that k is known and the problem consists of an optimal choice of observation points in order to estimate the parameters as accurately as possible for a given number of possible observations at the interval $[0, \infty)$.

The model (2.1) is of a great theoretical and practical interest. It is often used in chemical and biological investigations (see, e.g., Becka and Urfer (1996) and Han and Chaloner (2003)).

A discrete probability measure

$$\xi = \left(\begin{array}{ccc} x_1 & \cdots & x_n \\ \omega_1 & \cdots & \omega_n \end{array} \right), \tag{2.3}$$

where $0 < x_1 < \cdots < x_n$ are support points and $\omega_i > 0$, $i = 1, \ldots, n$, and $\sum \omega_i = 1$ are weight coefficients, will be called the (approximate) experimental design.

Let we have an opportunity to realize N experiments. We will say that the experiments are performed in accordance with the design (2.3) if r_i

observations are performed in points x_i $(i = 1, \ldots, n)$, where

$$r_i = \lfloor \omega_i N \rfloor \text{ or } \lfloor \omega_i N \rfloor + 1,$$

and $\lfloor a \rfloor$ denotes the integer part of a, in such a way that $\sum r_i = N$.

Set $\theta = (a_1, \lambda_1, \ldots, a_k, \lambda_k)^T$. Denote by $\hat{\theta}(N)$ the least squares estimator for θ obtained from the results of N experiments in accordance with a design of the form (2.3); that is,

$$\hat{\theta} = \hat{\theta}(N) = \arg \min_{\theta \in R^{2k}} \sum_{i=1}^{n} \sum_{j=1}^{r_i} [Y_{ij} - \eta(x_i, \theta)]^2,$$

where

$$\eta(x, \theta) = \sum_{i=1}^{k} a_i e^{-\lambda_i x}$$

and Y_{ij} is the result of the j-th experiment in the point x_i.

Let θ^* denote the true value of θ in the model (2.1). It can be shown by verification of regularity conditions of the Jennrich theorem (Jenrich, 1969) that with $n \geq 3$ and $N \to \infty$, the covariance matrix of the vector $(\hat{\theta}(N) - \theta^*)/\sqrt{N}$ tends to the matrix

$$\sigma^2 \left[\int f(x, \theta) f^T(x, \theta) \xi(dx) \right]^{-1}, \tag{2.4}$$

where

$$f(x, \theta) = \frac{\partial}{\partial \theta} \eta(x, \theta),$$

$$\int g(x) \xi(dx) = \sum_{i=1}^{n} g(x_i) \omega_i,$$

$$\theta = \theta^*.$$

The matrix

$$\left(\sum_{s=1}^{n} \frac{\partial \eta(x_s, \theta)}{\partial \theta_i} \frac{\partial \eta(x_s, \theta)}{\partial \theta_j} \omega_s \right)_{i,j=1}^{2k}$$

is usually called the Fisher information matrix.

By immediate application of Binet–Cauchy's formula to the determinant of this matrix, we obtain

$$\det \left(\int f(x, \theta) f^T(x, \theta) \xi(dx) \right)$$

$$= a_1^2 \ldots a_k^2 \sum_{1 \leq i < \cdots < i_{2k} \leq n} \left(\prod_{s=1}^{2k} \omega_{i_s} \right) \det^2 \left(\psi_l(x_{i_j}) \right)_{l,j=1}^{2k}, \tag{2.5}$$

where

$$\psi_1(x) = \frac{\partial}{\partial a_1} \eta(x, \theta) = e^{-\lambda_1 x},$$

$$\psi_2(x) = \frac{\partial}{\partial \lambda_1} \eta(x, \theta) = -x e^{-\lambda_1 x},$$

$$\vdots$$

$$\psi_{2k-1}(x) = \frac{\partial}{\partial a_k} \eta(x, \theta) = e^{-\lambda_k x},$$

$$\psi_{2k}(x) = \frac{\partial}{\partial \lambda_k} \eta(x, \theta) = -x e^{-\lambda_k x}.$$

Let us restrict ourselves by the D-criterion of optimality (for other criteria, we can proceed in a similar way). A design is called D-optimal if it maximizes the determinant of the information matrix. The problem is to find a design maximizing the determinant among all possible (approximate) designs. Note that with $a_i \neq 0$, $i = 1, \ldots, k$, values of $a_1, \ldots, a_k$ do not influence the solution of this problem (since they involve only in the multipliers $a_1^2, \ldots, a_k^2$). Therefore, we can assume in the following that $a_1^2 = \cdots = a_k^2 = 1$.

However, the design maximizing the value (2.5) depends, generally speaking, on the value $\Lambda = (\lambda_1, \ldots, \lambda_k) = \Lambda^* = (\lambda_1^*, \ldots, \lambda_k^*)$. Such a dependence is the main feature of all nonlinear models.

There are several ways to overcome this difficulty. Let us begin with the locally optimal approach (introduced by Chernoff (1953)). This approach consists of the replacement of the unknown value Λ^* by a known approximation for it (an initial guess).

A design will be called locally D-optimal if it maximizes the determinant (2.5) with $a_1 = \cdots = a_k = 1$ and $\Lambda = (\lambda_1, \ldots, \lambda_k) = \Lambda^{(0)} = (\lambda_1^{(0)}, \ldots, \lambda_k^{(0)})$.

Let us set

$$M(\xi, \Lambda) = \int f(x, \theta) f(x, \theta)^T \xi(dx),$$

where $\theta = (1, \lambda_1, \ldots, 1, \lambda_k)$ and $\Lambda = (\lambda_1, \ldots, \lambda_k)^T$.

Note that this matrix coincides with the information matrix for the corresponding linear model

$$Y = \beta_1 e^{-\lambda_1 x} + \beta_2 x e^{-\lambda_1 x} + \cdots + \beta_{2k-1} e^{-\lambda_k x} + \beta_{2k} e^{-\lambda_k x} + \varepsilon, \quad (2.6)$$

where $\beta_1, \ldots, \beta_{2k}$ are parameters to be estimated and $\lambda_1, \ldots, \lambda_k$ are assumed to be known.

Now, we should make a very important remark. Note that we assumed $\lambda_i \neq \lambda_j$ $i \neq j$, when we formulated our model. In fact, if $\lambda_i = \lambda_j$ for some $i \neq j$, then the model (2.1) contains no more than $k - 1$ terms of the form

$$a_i e^{-\lambda_i x}.$$

However, we restrict ourselves by the models with k terms.

It should be noted that if $\lambda_i = \lambda_j$ for some i and j ($i \neq j$), then for each of the determinants in the right-hand side of (2.5) the two corresponding columns coincide. Thus, in this case, $\det M(\xi, \Lambda) = 0$ for any design ξ.

However, from a mathematical point of view, it is useful to admit that the value

$$\min_{i \neq j} |\lambda_i - \lambda_j|$$

can be as small as we like. Moreover, it can be verified [see Melas (1978)] that the function

$$V(\xi, \Lambda) = (\det M(\xi, \Lambda)) / \prod_{i<j} (\lambda_i - \lambda_j)^8 \qquad (2.7)$$

can be codefined with preserving continuity at the set of all positive values $\lambda_1, \ldots, \lambda_k$.

Now, we are prepared to introduce a more convenient definition.

Definition 2.2.1 A design, maximizing the value (2.7) among all (approximate) designs for an arbitrary fixed vector Λ with positive coordinates will be called *a locally D-optimal design*.

For arbitrary Λ such that $\lambda_i \neq \lambda_j$ ($i \neq j$), this definition corresponds to the usual definition of locally D-optimal designs.

Note that designs that maximize the limit of (2.7) with $\Lambda \to \Lambda_\gamma = \gamma(1, \ldots, 1)$ that is locally D-optimal designs for points $\Lambda = \Lambda_\gamma$ will play an important role in the following consideration. Due to the continuity arguments these designs will be nearly optimal for all vectors Λ whose coordinates are close enough to each other.

We will construct and study locally D-optimal designs in the next subsection.

It should be noted that locally D-optimal (LD)designs depend on the initial vector $\Lambda = \Lambda^{(0)}$ and could be not very efficient if this vector is far from the vector of true parameter values Λ^*. However, the design could be implemented in a sequential manner. One can take $\Lambda = \Lambda^{(0)}$, construct an LD design for this vector, and realize N_1 experiments in accordance with this design. Then one can construct the LS (least squares) estimator $\hat{\theta} = \hat{\theta}(N_1)$ and take the parameter vector $\hat{\Lambda}^{(1)} = \hat{\Lambda}(N_1)$ in order to construct the new LD design. By repeating this procedure several times, we will obtain a design close to the LD design with $\Lambda = \Lambda^*$.

The described procedure (see, e.g., Silvey (1980) for more accurate explanation) cannot be appropriate if we need to have a design for all experiments in advance. An alternative to such a sequential implementation of LD design consists of using a minimax approach (see Section 1.7 for a more detailed discussion).

Let us consider a reasonable version of the minimax approach.

Assume that for the vector Λ^*, a set Ω of its possible values is given. In particular, such a set can be obtained from preliminary experiments or by theoretical consideration of the underlying real problem. From a practical point of view, the following type of set seems to be of a great interest:

$$\Omega = \Omega(\delta) = \{\Lambda; (1-\delta)x_i \leq \lambda_i \leq (1+\delta)c_i, \ i = 1, \ldots, k\}, \qquad (2.8)$$

where c_i is an approximation to λ_i^*, $i = 1, \ldots, k$, and the value $\delta \in (0,1)$ can be interpreted as a relative error of this approximation.

Note that the intervals $[(1-\delta)c_i, (1+\delta)c_i]$ can be overlapped and even can coincide with each other.

From a methodical point of view, it is very convenient that the set (2.8) under fixed $c_1, \ldots, c_k$ is determined by a single parameter δ.

Let us call a design a maximin efficient D-optimal design if it maximizes the value

$$\min_{\Lambda \in \Omega} \left[\frac{V(\xi, \Lambda)}{V(\xi(\Lambda), \Lambda)} \right]^{1/m}, \ m = 2k, \qquad (2.9)$$

where $\xi(\Lambda)$ is a LD design, $\Omega = \Omega(\delta)$ is determined by (2.8).

Note that the minimum here is achieved at some values $\bar{\Lambda} \in \Omega$ since Ω is a bounded and closed set.

The value (2.9) for a given design will be called the minimal efficiency. Note that

$$\left[\frac{V\xi, \Lambda)}{V(\xi(\Lambda), \Lambda)} \right]^{1/m} = \left[\frac{\det M(\xi, \Lambda)}{\det M(\xi(\Lambda), \Lambda)} \right]^{1/m}$$

if Λ satisfies the restriction $\lambda_i \neq \lambda_j$ $(i \neq j)$.

If we perform N experiments in accordance with a design ξ, then the volume of a confidence ellipsoid for LS estimates will be proportional to

$$\left(\frac{1}{\sqrt{N}} \right)^m \sqrt{\det M(\xi, \Lambda)}$$

(see, e.g., Pukelsheim (1993)).

Thus, the minimal efficiency of a given design is equal to the ratio N/N^*, where N is the number of experiments along the design ξ needed for obtaining estimates with a given accuracy and N^* is the similar number for a LD design.

In the following subsections we will demonstrate opportunities of the functional approach to constructing and studying LD and maximin efficient D-optimal designs.

2.2.2 Locally D-optimal designs

It is easy to check that if the number of support points of a design ξ is less than the number of parameters to be estimated ($n < 2k$), then $\det M(\xi, \Lambda) = 0$. By this reason, the designs with $n = 2k$ is usually called designs with minimal support. In the following we restrict our attention by such designs, and in Chapter 6, it will be shown that LD designs for the exponential model (2.1) usually belong to this class of designs. Designs that are LD in the class of designs with minimal support will be called, for brevity, LDMS designs.

An immediate calculation shows that with $n = 2k$,

$$M(\xi, \Lambda) = F^T W F,$$

where $W = \text{diag}\{\omega_1, \ldots, \omega_{2k}\}$, $F = (\psi_l(x_j)_{l,j=1}^{2k}$, and $\psi_l(x)$ are defined in (2.8). Therefore,

$$\det M(\xi, \Lambda) = \prod_{i=1}^{2k} \omega_i \det^2 F$$

$$\leq \left(\frac{\sum \omega_i}{2k}\right)^{2k} \det^2 F = \left(\frac{1}{2k}\right)^{2k} \det^2 F,$$

whereas the equality takes place if and only if $\omega_i = \frac{1}{2k}$, $i = 1, \ldots, 2k$. Thus LDMS designs have the form

$$\xi = \begin{pmatrix} x_1 & \cdots & x_m \\ \frac{1}{m} & \cdots & \frac{1}{m} \end{pmatrix}, \ 0 \leq x_1 < \cdots < x_m, \ m = 2k,$$

that is, all weight coefficients in such designs are the same.

Let us prove that in each of LDMS designs $x_1 = 0$. Set

$$\xi_\Delta = \begin{pmatrix} x_1 + \Delta & \cdots & x_m + \Delta \\ \frac{1}{m} & \cdots & \frac{1}{m} \end{pmatrix}, \ F_\Delta = (\psi_l(x_j + \Delta))_{l,j=1}^m.$$

Consider the determinant

$$\det F_\Delta = \det \begin{pmatrix} e^{-\lambda_1(x_1+\Delta)} \cdots e^{-\lambda_1(x_m+\Delta)} \\ -(x_1 + \Delta)e^{-\lambda_1(x_1+\Delta)} \cdots - (x_m + \Delta)e^{-\lambda_1(x_m+\Delta)} \\ e^{-\lambda_k(x_1+\Delta)} \cdots e^{-\lambda_k(x_m+\Delta)} \\ -(x_1 + \Delta)e^{-\lambda_k(x_1+\Delta)} \cdots - (x_m + \Delta)e^{-\lambda_k(x_m+\Delta)} \end{pmatrix}.$$

Let us add the first line multiplied by Δ to the second line, $\ldots$, and the $(2k-1)$-st line multiplied by Δ to the $(2k)$-th line. Then let us extract from each of the lines the multiplies of the form $e^{-\lambda_i \Delta}$, $i = 1, \ldots, k$. In this way, we obtain

$$\det F_\Delta = e^{-2(\sum_{i=1}^k \lambda_i)\Delta} \det F,$$

and with $\Delta < 0$,

$$\det{}^2 F_\Delta > \det{}^2 F.$$

Thus, with $x_1 > 0$, a design ξ cannot be LDMS since with $\Delta = -x_1$,

$$\det M(\xi_\Delta, \Lambda) = \left(\frac{1}{m}\right)^m \det{}^2 F_\Delta > \det M(\xi, \Lambda).$$

Therefore, for any LDMS design, we have $x_1 = 0$.

Let us introduce the following notation:

$$\tau = (\tau_1, \ldots, \tau_{m-1}) = (x_2, \ldots, x_m),$$

$$\xi_\tau = \begin{pmatrix} 0 & x_2 & \cdots & x_m \\ \frac{1}{m} & \frac{1}{m} & \cdots & \frac{1}{m} \end{pmatrix},$$

$$\varphi(\tau, \Lambda) = (V(\xi_\tau, \Lambda))^{1/m},$$

$$R_+^s = \{u : u \in R^s, u = (u_1, \ldots, u_s); u_i > 0, i = 1, \ldots, s\}.$$

Note that there exists a one-to-one correspondence between vectors $\tau \in R_+^{m-1}$ and designs of the form

$$\xi = \xi_\tau = \begin{pmatrix} 0 & x_2 & \cdots & x_m \\ \frac{1}{m} & \frac{1}{m} & \cdots & \frac{1}{m} \end{pmatrix}.$$

The problem of LDMS designs is now reduced to the maximization of the function $\varphi(\tau, \Lambda)$ by $\tau \in R_+^{m-1}$ under a fixed Λ, where $\Lambda = (\lambda_1, \ldots, \lambda_k)$, $\lambda_i > 0$, and $i = 1, \ldots, 1$.

Since

$$\varphi(\tau, \Lambda) = C(\Lambda)(\det F)^{2/m},$$

$F = (\psi_l(x_j))_{l,j=1}^{2m}$, where $C(\Lambda)$ does not depend on τ and each of elements of F tends to zero with $x_m \to \infty$, then the maximum of $\varphi(\tau, \Lambda)$ by $\tau \in R_+^{m-1}$ is achieved in an inner point of R_+^{m-1} for which $0 < \tau_1 < \cdots < \tau_m$. Due to the known necessary conditions for extremum points in order for a design $\xi = \xi_{\tau*}$ to be an LDMS design, it is necessary that with $\tau = \tau*$, the following equalities be satisfied,

$$\frac{\partial}{\partial \tau_i} \varphi(\tau, \Lambda) = 0, \quad i = 1, \ldots, m - 1. \tag{2.10}$$

Consider the case $k = 1$. In this case,

$$\det M(\xi_\tau, \Lambda) = \left[\frac{1}{2} \det \begin{pmatrix} 1 & e^{-\lambda_1 x_2} \\ 0 & -x_2 e^{-\lambda_1 x_2} \end{pmatrix} \right]^2$$

$$= \frac{1}{4} x_2^2 e^{-2\lambda_1 x_2},$$

$$\varphi(\tau, \Lambda) = [\det M(\xi_\tau, \Lambda)]^{1/2}$$

$$= \frac{1}{2} x_2 e^{-\lambda_1 x_2} = \frac{1}{2} \tau_1 e^{-\lambda_1 \tau_1}.$$

Equalities (2.10) assume the form of the single equation

$$\frac{\partial}{\partial \tau_1} (\tau_1 e^{-\lambda_1 \tau_1}) = e^{-\lambda_1 \tau_1} (1 - \lambda_1 \tau_1) = 0.$$

The unique solution of this equation under fixed λ_1 is

$$\tau_1 = 1/\lambda_1.$$

Thus, in the case $k = 1$, there exists the unique LDMS design

$$\xi_* = \xi_{\tau_*} = \begin{pmatrix} 0 & 1/\lambda \\ 1/2 & 1/2 \end{pmatrix}.$$

It can be proved (see Chapter 6) that this design is a LD design among all approximate designs.

In the case $k > 1$, it seems impossible to find such an explicit solution of the problem for arbitrary vectors Λ. However, we can find an explicit solution for points Λ of the form $\Lambda = (\gamma, \ldots, \gamma)$, where $\gamma > 0$ is an arbitrary given number.

In fact, in this case,

$$V(\xi_\tau, \Lambda_\gamma) = \lim_{\Lambda \to \Lambda_\gamma} \det M(\xi_\tau, \Lambda) / \prod_{i<j}(\lambda_i - \lambda_j)^8.$$

In order to calculate this limit, use the expansion of the exponential into the Taylor series and elementary properties of the determinant. In Melas (1978) it was proved that this limit is equal to

$$\left(\frac{1}{m}\right)^m e^{-\gamma \sum_{i=2}^m x_i} \prod_{i<j}(x_j - x_i)^2. \tag{2.11}$$

It is easy to check that the value (2.11) coincides with the value of the determinant of the information matrix for linear (by parameters) regression model

$$E(Y|x) = e^{-\gamma x} \sum_{i=1}^m \beta_i x^{i-1},$$

where $\gamma > 0$ is a given number and $\beta_1, \ldots, \beta_m$ are the parameters to be estimated.

As it is known (see Karlin and Studden (1966, Chap. X)), (2.11) has the unique extremal point

$$\tau^* = (x_2^*, \ldots, x_m^*) = \frac{1}{\gamma}(\gamma_1, \ldots, \gamma_{m-1}),$$

where $\gamma_1, \ldots, \gamma_{m-1}$ are the roots of the Laugerre's polynomial of degree $m - 1$ with the associated parameter 1. Thus, we know the unique solution of the equation system (2.10) under $\Lambda = \Lambda_\gamma$. For the case of an arbitrary Λ, it can be proved (see Melas (1978)) that the equation system (2.10) has a unique solution. Denote this solution by $\tau^* = \tau^*(\Lambda)$. With arbitrary k, the unique LDMS design is

$$\xi_* = \xi * (\Lambda) = \xi_{\tau^*(\Lambda)}.$$

Considering the determinant of the matrix F, it is easy to check that for any scalar $h \neq 0$,

$$\varphi(\tau, h\Lambda) = h\varphi\left(\frac{\tau}{h}, \Lambda\right).$$

Therefore, $\tau^*(h\Lambda) = \tau^*(\Lambda)/h$ and we can restrict our attention to vectors Λ with $\sum_{i=1}^{k} \lambda_i = k$. It allows one to reduce the number of parameters.

Let us introduce the new parameters

$$z = (z_1, \ldots, z_{k-1})^T, \ z_i = 1 - \lambda_i, i = 1, \ldots, k - 1.$$

Note that, for $k = 2$, the number of new parameters is equal to 1. Note also that with $\sum_{i=1}^{k} \lambda_i = k$, there exists the one-to-one correspondence between the set of new parameters and the set of vectors Λ:

$$\lambda_i = 1 - z_i, \ i = 1, \ldots, k - 1; \ \lambda_k = k - \sum_{i=1}^{k-1} \lambda_i = 1 + \sum_{i=1}^{k-1} z_i.$$

Denote

$$\bar{\varphi}(\tau, z) = \varphi(\tau, \Lambda(z)),$$

$$g_i(\tau, z) = \frac{\partial}{\partial \tau_i} \bar{\varphi}(\tau, z), \ i = 1, \ldots, m - 1, \tag{2.12}$$

$$g(\tau, z) = (g_1(\tau, z), \ldots, g_{m-1}(\tau, z))^T.$$

Now the equation system (2.10) can be written as the vector equation

$$g(\tau, z) = 0. \tag{2.13}$$

This equation determines the vector function

$$z \to \bar{\tau}^*(z) = \tau^*(\Lambda(z))$$

implicitly, which allows to apply the Implicit Function Theorem (see Section 1.8).

We will now present an extended formulation of this theorem for the vector function $g(\tau, \Lambda)$ of a general form (not necessary connected with the design problem considering here).

Let $g(\tau, z)$, $\tau \in R^{m-1}$, $z \in R^{k-1}$, be an arbitrary vector function $g = (g_1, \ldots, g_{m-1})^T$ with the following properties:

(i) $g(\tau, z)$ is a real analytic vector function in the point $(\tau_{(0)}, z_{(0)})$ (this means that the component of this vector function can be expanded into a convergent multivariate Taylor series in the point).

(ii) $g(\tau_{(0)}, z_{(0)}) = 0$.

(iii) The Jacobi matrix

$$J_{(0)} = \left(\frac{\partial g_i(\tau, z)}{\partial \tau_j} \right)^{m-1}_{i,j=1} \Big|_{\tau = \tau_{(0)}, z = z_{(0)}}$$

is invertible.

In order to formulate the theorem, let us introduce the following notations. Let $Q(u)$ be an arbitrary (scalar or vector) function of one variable that is infinitely many times differentiable in a point $u_{(0)}$. Denote

$$Q_{(0)} = Q(u_{(0)}),$$

$$Q_{(3)} = \frac{1}{s!} \frac{d^s}{du^s} Q(u) \Big|_{u=u_{(0)}} , \quad s = 1, 2, \ldots$$

If the function $Q(u)$ is real analytic in a vicinity of the point $u = u_{(0)}$, then

$$Q(u) = Q_{(0)} + \sum_{s=1}^{\infty} Q_{(s)} (u - u_{(0)})^s$$

in this vicinity.

In the multidimensional case $u = (u_1, \ldots, u_{k-1})$, it is necessary to interpret s as the multi-index $s = (s_1, \ldots, s_{k-1})$ and denote

$$Q_{(s)} = \frac{1}{s_1!} \cdots \frac{1}{s_{k-1}!} \frac{\partial^{s-1}}{\partial u_1^{s_1}} \cdots \frac{\partial^{s_{k-1}}}{\partial u_{k-1}^{s_k}} Q(u) \Big|_{u=u_{(0)}} .$$

Theorem 2.2.1 *Let a vector function $g(\tau, z)$, $\tau \in R^{k-1}$, $z \in R^{k-1}$, possess the properties (i)–(iii). Then in a vicinity (say U) of the point $z_{(0)}$, there exists a vector function $\tilde{\tau} = \tilde{\tau}(z)$ such that the following hold:*

(I) $g(\tilde{\tau}(z), z) = 0$, $z \in U$.

(II) $\tilde{\tau}(z_{(0)}) = \tau_{(0)}$ and $\tilde{\tau}(z)$ is a real analytic vector function in U.

(III) The coefficients $\hat{\tau}_{(s)}$ of the expansion $\tilde{\tau}((z)$ into the Taylor series

$$\tilde{\tau}(z) = \sum_{s_1=0}^{\infty} \cdots \sum_{s_{k-1}=0}^{\infty} \tilde{\tau}_{(s)} (z_1 - z_{1(0)})^{s_1} \ldots (z_{k-1} - z_{k-1(0)})^{s_{k-1}}$$

can be calculated by recurrent formula that in the case $k = 2$ has the form

$$\tilde{\tau}_{(s+1)} = -J_{(0)}^{-1} g_{(s+1)} (\tilde{\tau}_{<s>}(z), z), s = 0, 1, \ldots,$$

where

$$\tilde{\tau}_{<s>}(z) = \tilde{\tau}_{(0)} + \sum_{j=1}^{s} \tilde{\tau}_{(j)} (z - z_{(0)})^j.$$

Note that assertions (I) and (II) are simply a reformulation of Theorem 1.8.1. Assertion (III) was established in Dette, Melas and Pepelyshev (2004b) and will be proved for the case of arbitrary k in Section 2.6.

Let us now apply this theorem to the function $g(\tau, z)$ given by relations (2.12). As is well known, the exponentials are real analytic at R^1 since

$$e^{-\lambda t} = 1 - \lambda t + \frac{(-\lambda t)^2}{2!} + \ldots + \frac{(-\lambda t)^n}{n!} + \ldots$$

and the series is convergent for any λ and t.

Additionally, multiplications and sums of real analytic functions are real analytic and, therefore,

$$\det(\psi_l(x_j))_{l,j=1}^m$$

is a real analytic function in $\Lambda = (\lambda_1, \ldots, \lambda_k)^T$ and $(x_1, \ldots, x_m)$ in R^{k+m}.

Note that the function $\varphi(\tau, \Lambda)$ and the vector function $g(\tau, z)$ are real analytic in a vicinity of the points $(\tau_{(0)}, \Lambda_{(0)})$, and $(\tau_{(0)}, z_{(0)})$, respectively, where $\Lambda_{(0)} = (1, \ldots, 1)$, $\tau_{(0)} = \tau^*(\Lambda_{(0)}) = (\gamma_1, \ldots, \gamma_{n-1})$, and $Z_{(0)} = (0, \ldots, 0)$,

In fact,

$$V(\xi_\tau, \Lambda) = \frac{\det M(\xi_\tau, \Lambda)}{\prod_{i<j}(\lambda_i - \lambda_j)^8}$$

and it can be verified [see Melas (1978)] that this function, codefined with preserving the continuity in points Λ such that $\lambda_i = \lambda_j$ for some $i \neq j$, is real analytic for arbitrary $\tau \in R^{m-1}$ and arbitrary $\Lambda \in R^k$.

Additionally, the function

$$\varphi(\tau, \lambda) = (V(\xi_\tau, \Lambda))^{1/m}$$

is real analytic as a rational degree of the real analytic function. It follows from here by the standard arguments that the function $\bar\varphi(\tau, z)$ and the vector function $g(\tau, z)$ are also real analytic for arbitrary $\tau \in R^{m-1}$ and $z \in R^{k-1}$.

Let us now calculate the matrix

$$J_{(0)} = \left(\frac{\partial g_i(\tau, z)}{\partial \tau_j}\right)_{i,j=1}^{m-1} \Bigg|_{\tau=\tau_{(0)}, z=z_{(0)}}$$

$$= \left(\frac{\partial^2}{\partial \tau_i \partial \tau_j} \bar\varphi(\tau, z)\right)_{i,j=1}^{m-1} \Bigg|_{\tau=\tau_{(0)}, z=z_{(0)}}.$$

Due to (2.11) and the definition of $\bar\varphi(\tau, z)$ given in (2.12), we have

$$m(\varphi(\tau, z_{(0)})^m = e^{-2\sum_{i=1}^{m-1} \tau_i} \left(\prod_{i=1}^{m-1} \tau_i^2\right) \prod_{i<j}^{m-1} (\tau_i - \tau_j)^2.$$

A direct calculation shows that

$$\frac{\partial \bar{\varphi}(\tau, Z_{(0)})}{\partial \tau_j} = \left[-1 + \frac{1}{\tau_j} + \sum_{s \neq j} \frac{1}{\tau_j - \tau_s} \right] \bar{\varphi}(\tau, z_{(0)}), \ i = 1, \ldots, m-1,$$

and the derivatives are equal to zero with $\tau = \tau_{(0)}$ by the definition of the point $\tau_{(0)}$.

Therefore,

$$(J_{(0)})_{ij} = \frac{\partial^2 \bar{\varphi}(\tau, z_{(0)})}{\partial \tau_i \partial \tau_j}\bigg|_{\tau = \tau_{(0)}} = \frac{\bar{\varphi}(\tau_{(0)}, z_{(0)})}{(\gamma_j - \gamma_i)^2} \ (i \neq j),$$

$$(J_{(0)})_{ij} = \frac{\partial^2 \bar{\varphi}(\tau, z_{(0)})}{\partial^2 \tau_j}\bigg|_{\tau = \tau_{(0)}} = -\left(\frac{1}{\gamma^2} + \sum_{s \neq j} \frac{1}{(\gamma_j - \gamma_s)^2} \right) \bar{\varphi}(\tau_{(0)}, z_{(0)}),$$

$i, j = 1, \ldots, m-1$.

Thus, for the matrix $J = J_{(0)}$, we have

$$(J)_{ij} > 0, \ i \neq j, \ J_{ij} < 0, \ i, j = 1, \ldots, m-1,$$

$$\sum_{j=1}^{m-1} (J)_{ij} = -\frac{\bar{\varphi}(\tau_{(0)}, z_{(0)})}{\gamma_j^2} < 0, \ i = 1, \ldots, m-1.$$

Due to the Hadamard criterion (see, e.g., Gantmacher (1998)), for an $(m-1) \times (m-1)$ matrix A

$$\det A \neq 0 \ \text{if} \ (A)_{ii} > \sum_{i \neq j} |A_{ij}|, \ i = 1, 2, \ldots, m-1.$$

The matrix $(-J_{(0)})$ satisfies these conditions and, therefore, $\det J_{(0)} \neq 0$.

Thus, we proved that the function $g(\tau, z)$ determined by equalities (2.11) satisfies the conditions of Theorem 2.2.1 with $z_{(0)} = (0, \ldots, 0)$. $\tau_{(0)} = (\gamma_1, \ldots, \gamma_{m-1})$.

Consider now the case $k = 2$. In this case, the regression function is

$$\eta(x, \theta) = a_1 e^{-\lambda_1 x} + a_2 e^{-\lambda_2 x}, \ a_1, a_2 \neq 0, \lambda_1 \neq \lambda_2,$$

where $\lambda_1 > 0$ and $\lambda_2 > 0$. As will be shown in Chapter 6, LDMS designs are in this case LD among all (approximate) designs. Support points of these designs, as was already shown, do not depend on a_1 and a_2 and if λ_1 and λ_2 are multiplied by the same number $h > 0$, then the points should be divided by this number. Therefore, it will do to consider Λ such that $\lambda_1 + \lambda_2 = 2$ and to study the dependence of the support points of LDMS design on the parameter

$$z = z_1 = 1 - \lambda_1 = (\lambda_2 - \lambda_1)/2.$$

Let $z_{(0)} = 0$ and $\tau_{(0)} = (\gamma_1, \gamma_2, \gamma_3) = (0.467, 1.652, 3.879)$.
Note that the function $\varphi(\tau, z)$ is even,

$$\varphi(\tau, z) = \varphi(\tau, -z).$$

By this reason $\tau^*(z) = \tau^*(-z)$ and all odd coefficients $\tau^*_{(2j+1)}$, $j = 0, 1, \ldots$ are equal to zero. Therefore

$$\tau^*(z) = \tau_{(0)} + \sum_{t=1}^{\infty} \tau^*_{(2k)} z^{2t}. \tag{2.14}$$

The coefficients can be calculated by recurrent formulas of Theorem 2.2.1. These calculations can be easily realized with the help of the software package Maple. Some details of the implementation of the package are given in the Appendix of the present book.

First even coefficients calculated in this way are presented in Table 2.1.

Table 2.1: Coefficients $\tau_{<2t>}$, $t = 0, 1, \ldots, 6$

0	1	2	3	4	5	6
0.46791	0.02919	0.00305	0.00056	0.00022	0.00008	−0.00005
1.65270	0.36419	0.21113	0.15971	0.13371	0.11650	0.10252
3.87938	2.00661	1.86581	1.92887	2.04481	2.16523	2.26335

The method allows one to calculate as many coefficients as we like. Since the coefficients are already obtained, one can construct the corresponding designs simply by several first coefficients in the expansion (2.14).

However, we have a few problems here. The first problem concerns the radius of convergency of the series (2.14). Note that $0 \leq |z| \leq 1$ since

$$z = (\lambda_1 - \lambda_2)/2 \text{ and } (\lambda_1 + \lambda_2)/2 = 1.$$

Numerical studies show that the series are convergent for any $|z| < 1$. However, a strong theoretical proof of this fact is not obtained up to now.

The next problem consists of the determination of how many coefficients should be used in order to calculate support points of LDMS designs with an appropriate precision.

Denote $\tau(z, s) = \tau_{(0)} + \sum_{t=1}^{s} \tau_{2t} z^{2t}$ and

$$I_{(s)} = I_{(s)}(z) = \left(\frac{\det M(\xi_{\tau(z,s)}, z)}{\det M(\xi^*(z), z)} \right)^{1/m}, \quad s = 0, 1, \ldots,$$

where $\xi^*(z) = \xi_{\tau^*(z)}$ is a LDMS design.

The value $I_{(s)}(z)$ is the efficiency of the design $\xi_{\tau(z,s)}$ constructed by s first even coefficients with respect to the LDMS for a given z. This value can be evaluated with the help of Kiefer's inequality (see Section 1.6)

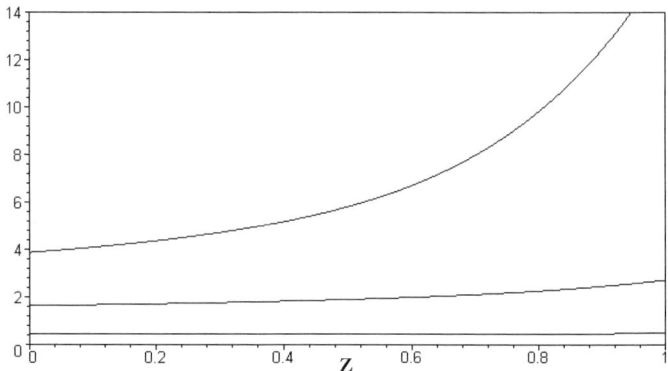

Figure 2.1: The dependence of support points of the LDMS designs on z for the exponential model with $k = 2$

without calculating the LDMS design. Some numerical results are given in Table 2.2. They represent an evaluation of $I_{(s)}(z)$, obtained with the help of Kiefer's inequality. Note that with $0 < z < 0.5$, $I_{(0)} = 1.00$ and there is no reason to calculate more coefficients.

Table 2.2: The efficiency of designs $\xi_{\tau_{<t>}}(z)$

$z \backslash t$	0	1	2	3	4	5	6	7	8	9
0.50	0.98	1.00	1.00	1.00	1.00	1.00	1.00	1.00	1.00	1.00
0.70	0.90	0.98	1.00	1.00	1.00	1.00	1.00	1.00	1.00	1.00
0.80	0.80	0.93	0.97	0.99	1.00	1.00	1.00	1.00	1.00	1.00
0.85	0.72	0.88	0.94	0.97	0.98	0.99	1.00	1.00	1.00	1.00
0.90	0.61	0.79	0.87	0.92	0.95	0.97	0.98	0.99	0.99	1.00
0.95	0.45	0.61	0.71	0.78	0.83	0.87	0.90	0.93	0.94	0.96
0.97	0.35	0.49	0.58	0.65	0.71	0.76	0.80	0.84	0.86	0.89

From Table 2.2 we can conclude that with $|z| \leq 0.7$, it will do to use only one or two nonzero coefficients. However, for $z = 0.9$, we need 20 coefficients in order to obtain the efficiency greater than 0.995. Table 2.2 also shows that with $|z| \leq 0.9$, the expansions allow one to construct locally optimal designs with a very high precision. For $|z| > 0.9$, we can use a similar expansion with $z_{(0)} = 0, 9$ as the initial point. The dependence of support points of the LD designs on z is presented in Figure 2.1. Note that we used 10 nonzero Taylor coefficients in order to construct this figure.

The next important question is: How efficient are LD designs with respect to equidistant designs usually implemented in practice?

Table 2.3: The efficiency of LD designs in respect to the best equidistant design

λ_1	1.1	1.3	1.5	1.7	1.9	1.95
λ_2	0.9	0.7	0.5	0.3	0.1	0.05
I	2.13	2.06	1.92	1.70	1.80	2.20

Denote by

$$\xi_{N,T} = \begin{pmatrix} 0 & T/(N-1) & \cdots & T(N-2)/(N-1) & T \\ \frac{1}{N} & \frac{1}{N} & \cdots & \frac{1}{N} & \frac{1}{N} \end{pmatrix}$$

the design located in N equidistant points at the interval $[0, T]$. For large N, the quality of this design is not very sensitive to the value of N, but it depends on T.

Consider the efficiency of LD designs constructed above with respect to equidistant design with an optimal choice of T; that is, we will take the value of T in such a way that the minimal efficiency of the equidistant designs for $z \in [0.1, 0.9]$ is the maximal one.

Our numerical results are given in Table 2.3. In this table, $T = 10$, $N = 20$;

$$I = \left(\frac{\det M(\xi_{\tau^*(\Lambda)}, \Lambda)}{\det M(\xi_{N,T}, \Lambda)} \right)^{1/m}, \ m = 4.$$

We see from Table 2.3 that in the most cases the efficiency of the LD design with respect to the equidistant design is more than 2 or close to 2. This means that the number of experiments in accordance with a LD design needed in order to achieve a given accuracy is approximately twice less than the same number for the best equidistant design if $\Lambda^{(0)} = \Lambda^*$. However, since Λ^* is unknown, these results describe the efficiency of LD designs only in an asymptotical sense. The influence of the choice of $\Lambda^{(0)}$ on the quality of LD designs can be studied numerically. However, in the following subsection we will show that the application of the functional approach can be used for such a study and allows one to compare LD designs with the maximin efficient ones.

2.2.3 Maximin efficient designs

Assume that it is known that $\Lambda^* \in \Omega$, where Ω is a given bounded and closed set in $R_+^k = \{\Lambda = (\lambda_1, \dots \lambda_k); \lambda_i > 0, i = 1, \dots, k\}$. Then a natural criterion of the efficiency of a given design is the value

$$\min_{\Lambda \in \Omega} \left(\frac{V(\xi, \Lambda)}{V(\xi(\Lambda), \Lambda)} \right)^{1/m}, \tag{2.15}$$

where $\xi(\Lambda)$ is a LD design, and for Λ such that $\lambda_i \neq \lambda_j$ $(i \neq j)$, the value $V(\xi, \Lambda)/V(\xi(\Lambda), \Lambda)$ is equal to $\det M(\xi, \Lambda)/\det M(\xi(\Lambda), \Lambda)$ [see the end of Section 2.2.1 for a discussion on this matter]. The value (2.15) will be called the minimal efficiency and the designs that maximize this value will be called maximin efficient D-optimal designs or, briefly, MME designs.

We will study the MME designs for the exponential model (2.1) and the set $\Omega = \Omega(\delta)$,

$$\Omega(\delta) = \Omega(\delta, c) = \{\Lambda = (\lambda_1, \ldots, \lambda_k) : (1-\delta)x_i \leq \lambda_i \leq (1+\delta)c_i, \ i = 1, \ldots, k\},$$

where $\delta \in (0, 1)$, $c = (c_1, \ldots, c_k)$, $c_i > 0$, and $i = 1, \ldots, k$.

Let us restrict our attention to designs with the minimal support. In the following, it will be shown (see Theorems 2.2.2 and 2.2.3 and numerical results) that MME designs have the minimal support for sufficiently small δ and arbitrary c.

We have already proved that

$$\det M(\xi_\Delta, \Lambda) < \det M(\xi, \Lambda),$$

where

$$\xi_\Delta = \begin{pmatrix} \Delta & x_2 + \Delta & \cdots & x_m + \Delta \\ \frac{1}{m} & \frac{1}{m} & \cdots & \frac{1}{m} \end{pmatrix}, \ \Delta > 0,$$

$$\xi = \begin{pmatrix} 0 & x_2 & \cdots & x_m \\ \frac{1}{m} & \frac{1}{m} & \cdots & \frac{1}{m} \end{pmatrix}.$$

Therefore, MME designs with a minimal support have the from

$$\xi_\tau = \begin{pmatrix} 0 & x_2 & \cdots & x_m \\ \frac{1}{m} & \frac{1}{m} & \cdots & \frac{1}{m} \end{pmatrix}, \ \tau = (\tau_1, \ldots, \tau_{m-1}) = (x_2, \ldots, x_m). \quad (2.16)$$

Let us introduce the function

$$\hat{\varphi}(\tau, \Lambda) = \left(\frac{V(\xi_\tau, \Lambda)}{V(\xi_{\tau^*(\Lambda)}, \Lambda)} \right)^{1/m},$$

where $\xi_{\tau^*(\Lambda)}$ is a LDMS design.

Theoretical studies (see Theorems 2.2.2 and 2.2.3) show that for sufficiently small $\delta > 0$,

$$\min_{\Lambda \in \Omega(\delta, c)} \hat{\varphi}(\tau, \Lambda) = \min\{\hat{\varphi}(\tau, (1-\delta)c), \hat{\varphi}(\tau, (1+\delta)c\}$$

$$= \min_{0 \leq \alpha \leq 1} \alpha \hat{\varphi}(\tau, (1-\delta)c) + (1-\alpha)\hat{\varphi}(\tau, (1+\delta)c).$$

Based on this, let us introduce the following class of designs. Let us say that a design is a maximin efficient design with a minimal structure or, briefly, MMEMS design, if this design is of the form (2.16), where $\tau = \hat{\tau}$ and $\hat{\tau}$ maximizes the value

$$\min_{0 \leq \alpha \leq 1} \alpha \hat{\varphi}(\tau, (1-\delta)c) + (1-\alpha)\hat{\varphi}(\tau, (1+\delta)c)$$

at the set of all vectors τ with positive coordinates.

In the case when intervals of possible values are the same for all parameters λ_i, $i = 1, \ldots, k$ (i.e., $c_1 = c_2 = \cdots = c_k$) the MMEMS designs can be found explicitly.

In order to describe these designs, let us denote

$$u = (\tau, \alpha) = (\tau_1, \ldots, \tau_{m-1}, \alpha),$$

$$\Phi(u, \delta) = \alpha \hat{\varphi}(\tau, (1 - \delta)c) + (1 - \alpha) \hat{\varphi}(\tau, (1 + \delta)c),$$

$$\hat{\xi} = \xi_{\hat{\tau}} \text{ - MMEMS design.}$$

Let $\gamma_1, \ldots, \gamma_{m-1}$ be, as above, the roots of Laugerre's polynomial of degree $m - 1$ with the associated parameter 1,

$$h = h(\delta) = 2\delta / \ln\left(\tfrac{1+\delta}{1-\delta}\right),$$

$$I(\delta) = [h(\delta)e^{(1-h(\delta))}]^{m(m-1)/2},$$

$$H = \left(\det M(\xi_{\tau^*(c)}, c)\right)^{1/m}.$$

Remember that $\tau^*(\gamma c) = \tau^*(c)/\gamma$ for any $\gamma > 0$. Also, it follows from here that

$$\hat{\varphi}(\tau, (1 - \delta)c) = (\det M(\xi_\tau, (1 - \delta)c))^{1/m} / (H(1 - \delta)),$$

$$\hat{\varphi}(\tau, (1 + \delta)c) = (\det M(\xi_\tau, (1 + \delta)c))^{1/m} / (H(1 + \delta)).$$

This simplifies theoretical and numerical studies of the MMEMS designs.

An explicit solution of the problem in the case $c_1 = c_2 = \cdots = c_k$ is given by the following theorem.

Theorem 2.2.2 *Consider model (2.1) and the set $\Omega = \Omega(\delta, c)$ of the form (2.4), where $c_1 = \cdots = c_k$. In this case the following hold:*

(I) *There exists a unique MMEMS design for any fixed $c_1 > 0$ and $\delta < 1$. This design is*

$$\hat{\xi} = \xi_{\hat{\tau}}, \ \hat{\tau} = (\hat{\tau}_1, \ldots, \hat{\tau}_{m-1}),$$

$$\hat{\tau}_i = \gamma_i / (c_1 h(\delta)), \ i = 1, \ldots, m - 1,$$

and

$$\Phi(\hat{u}, \delta) = I(\delta).$$

(II) *This design is a locally D-optimal design for $\Lambda = c/h(\delta)$.*

(III) *For any sufficiently small positive δ, this design is MME among all (approximate) designs and its minimal efficiency is equal to $I(\delta)$.*

Proof. Note that for $\Lambda = (c_1, \ldots, c_k)$, $c_1 = \cdots = c_k$ the value of $V(\xi, \Lambda)$ coincides with the value of determinant of the information matrix for the linear (by parameters) regression function

$$e^{-c_1 x}(\beta_1 + \beta_2 + \cdots + \beta_m x^{m-1}),$$

where $\beta = (\beta_1, \ldots, \beta_m)$ is the vector of estimating parameters and c_1 is a given number (as we already mentioned a detailed proof can be found in Melas (1978)). For this reason assertions (I) and (II) follows immediately from the results of Dette, Haines and Imhof (2003). Assertion (III) is a special case of Theorem 2.2.3(II).   ∎

Note that the set of δ values for which assertion (III) holds can be found numerically. In particular, we found in such a way that assertion (III) is true for $k = 1$ with $\delta \leq 0.54$, for $k = 2$ it holds with $\delta \leq 0.22$, and for $k = 3$ it holds with $\delta \leq 0.18$. Thus, under realistic values of δ in the case $c_1 = \cdots = c_k$, MMEMS designs described in Theorem 2.2.2 are in fact MME designs among all (approximate) designs. It is also worth mentioning that in all of the cases, mentioned above, the minimal efficiency proves to be grater than 0.9, which can be easily checked by the explicit formula for $I(\delta)$.

In the case of arbitrary values $c_1, \ldots, c_k$, it seems does not possible to find MMEMS designs explicitly. However, the dependence of such designs on δ with a given c can be investigated with the help of constructing Taylor series in a way very similar to that was already applied to LDMS designs.

As is well known, the function of minimum is continuous. Also, we have already shown that the value $V(\xi_\tau, \Lambda)$ tends to zero with $\tau_{m-1} \to \infty$. Therefore, the function

$$\min_{0 \leq \alpha \leq 1} \Phi(u, \delta), \; u = (\tau, \alpha) \tag{2.17}$$

is bounded with $\tau \in R_+^{m-1}$ and there exists an MMEMS design (i.e., the design that maximizes (2.17) by $\tau \in R_+^{m-1}$).

Consider the equation system

$$\frac{\partial}{\partial u_i} \Phi(u, \delta) = 0, \; i = 1, \ldots, m. \tag{2.18}$$

Let $\hat{J}(\delta)$ be the Jacobi matrix of this system,

$$\hat{J}(\delta) = \left(\frac{\partial^2}{\partial u_i \partial u_j} \Phi(u, \delta) \right)_{i,j=1}^{m} \bigg|_{u=u(\delta)},$$

where $u(\delta)$ is a solution of (2.18); the existence of this solution is provided by the following theorem.

Theorem 2.2.3 *Consider the regression model (2.1) for the set* $\Omega = \Omega(c, \delta)$ *defined in (2.4); the following assertions take place:*

(I) *There exists a unique MMEMS design. Moreover, there exists a unique solution of the equation system (2.18), first* $m - 1$ *components of this solution generate the vector* $\hat{\tau}$, *and the matrix* $J(\delta)$ *is invertible. The solution is a real analytic function of* δ.

(II) *If in a vicinity of* $\Lambda = c$ *the unique LDSM design is locally D-optimal among all (approximate) designs, then the MMEMS design is MME among all (approximate) designs for sufficiently small positive* δ.

A proof of this theorem will be given in Section 2.6.3.

Note that as in the case of Theorem 2.2.2(III), the set of δ values for which assertion (II) is valid can be found numerically. For example, with $k = 2$ and $c = (1, 5)$, the MMEMS designs prove to be MME among all designs for all $\delta \leq 0.27$.

Theorem 2.2.3 justifies studying MMEMS and MME designs along the following steps:

1. Find numerically the MMEMS design for some value $\delta = \delta_0$ (in our calculations, we took $\delta_0 = 0.5$).

2. With the help of the recurrent formulas, construct the Taylor expansions for functions $\hat{\alpha}(\delta)$ and $\hat{\tau}_1(\delta), \ldots, \hat{\tau}_{m-1}(\delta)$.

3. Check whether the designs constructed are MME designs among all approximate designs for different values of δ by the equivalence theorem from Dette, Haines and Imhof (2003).

Let us illustrate the approach by examples.

With $k = 1$, the MMEMS designs are given by Theorem 2.2.2:

$$\hat{\xi} = \xi_{\hat{\tau}} = \left(\begin{array}{cc} 0 & \hat{\tau}_1 \\ 1/2 & 1/2 \end{array} \right),$$

where $\hat{\tau}_1 = 1/(c_1 h(\delta))$. A numerical calculation shows that this design is MME among all approximate designs if $\delta \leq 0.54$.

Let $k = 2$ and the set Ω be

$$\Omega = \Omega(z) = \{(\lambda_1, \lambda_2); c_i(1 - \delta) \leq \lambda_i \leq c_i(1 + \delta), i = 1, 2\}.$$

Without loss of generality, assume that $1 = c_1 \leq c_2$. Set $c_2 = 5$ (for other cases we obtain similar results).

Taylor coefficients for the functions $\hat{x}_i(\delta)$, $i = 2, 3, 4$ and $\hat{\alpha}(\delta)$ in a vicinity of $\delta = \delta_0 = 0.5$ are given in Table 2.4. Note that the series are convergent for $\delta \in [0, 1)$, and with $\delta < 0.8$, we need only three first coefficients to calculate MMEMS with a good precision. The values of the functions received by usage of the first 11 coefficients are depicted at Figure2.2. Note that the minimum efficiency is always achieved at the two points $(1 - \delta)c$ and $(1 + \delta)c$. The behavior of the function $\varphi(\hat{\tau}, \Lambda)$ with $\hat{\tau} = \hat{\tau}(0.5)$, $\Lambda \in \Omega(0.5)$ is shown in Figure 2.2. The dependence of the minimal efficiency on δ is presented in Figure 2.2.

Table 2.4: Coefficients in the Taylor expansion for the functions $\hat{x}_2(\delta)$, $\hat{x}_3(\delta)$, $\hat{x}_4(\delta)$, and $\alpha(\delta)$ by degrees of $(\delta - 0.5)$.

j	$\hat{x}_2$	$\hat{x}_3$	$\hat{x}_4$	α	j	$\hat{x}_2$	$\hat{x}_3$	$\hat{x}_4$	α
0	0.17	0.69	2.06	0.44	6	0.49	5.33	13.56	-0.54
1	0.05	0.34	1.16	-0.14	7	0.88	9.55	23.80	-0.94
2	0.09	0.67	2.13	-0.09	8	1.60	17.26	42.59	-1.66
3	0.11	0.99	2.87	-0.14	9	2.95	31.35	77.19	-2.97
4	0.17	1.73	4.76	-0.19	10	5.50	57.22	141.40	-5.37
5	0.28	2.99	7.85	-0.32	11	10.32	104.91	261.17	-9.74

The verification by the equivalence theorem mentioned above shows that the MMEMS designs are MME among all approximate designs with $\delta \leq 0.28$. Additionally, our calculations (not presented here) show that MMEMS designs have the minimal efficiency at 40–50% more than the best equidistant designs (such designs are often used in practice).

However, for $\delta > 0.28$, it is possible to construct even more efficient designs. For example, with $\delta = 0.5$ we constructed numerically a design that is MME among all approximate designs. This design has six support points and is approximately equal to

$$\begin{pmatrix} 0 & 0.140 & 0.440 & 1.048 & 1.75 & 3.25 \\ 0.24 & 0.18 & 0.19 & 0.16 & 0.13 & 0.10 \end{pmatrix}.$$

The minimal efficiency of this design is equal to 0.8431, whereas such efficiency for the MMEMS design is 0.7045. Note that for the LD design at the central point $\Lambda = (1, 5)$, this value is 0.6150, and for the best equidistant design, it is 0.5904.

For model (2.1) with three exponentials, we obtained similar results. However, the critical value of δ, for which the MMEMS designs remains MME among all designs, is smaller than that for the two exponential models.

2.3   Description of the Model

In this section we will introduce assumptions on the regression functions providing the application of the functional approach. The corresponding class of nonlinear regression models includes, in particular, the exponential models, considered in Section 2.2, as well as rational models and the three

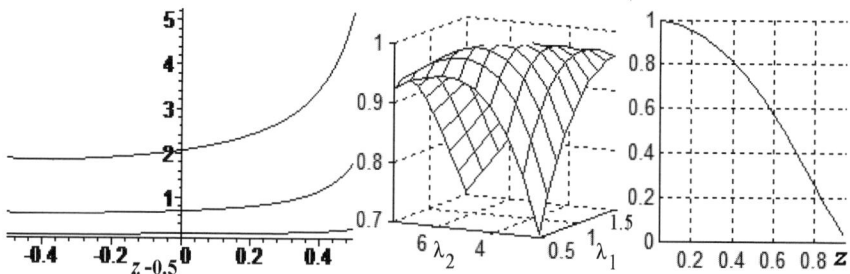

Figure 2.2: Functions $\hat{x}_2(\delta)$, $\hat{x}_3(\delta)$, and $\hat{x}_4(\delta)$; $z = \delta$ (top) and the minimal of efficiency of MMEMS design (bottom right) with $\delta \in (0, 1)$. The efficiency of MMEMS design with $\delta = 0.5$ over $\Omega(0.5)$ (bottom left).

parameters logistic model. One more example is the Monod model to be studied in Chapter 8. For this class of models we introduce the basic equation determining the support points of locally D-optimal designs as implicit functions of values of the model parameters.

2.3.1 Assumptions and notation

Let us consider the general nonlinear regression model

$$y_j = \eta(x_j, \Theta) + \varepsilon_j, \ j = 1, \dots, N, \tag{2.19}$$

where $y_1, \dots, y_N \in R^1$ are experimental results, $\Theta = (\theta_1, \dots, \theta_m)^T$ is the vector of unknown parameters, $\eta(x, \Theta)$ is a function of known form continuously differentiable along the parameters, $x_j \in \mathfrak{X}$, $\mathfrak{X}$ is a given set, and $\varepsilon_1, \dots, \varepsilon_N$ are independent and identically distributed random values with zero expectation and a finite (unknown) variance $\sigma^2 > 0$.

Let us introduce the following notation (it was already given in Sections 1.7 and 2.2 but will be represented here for the sake of convenience of the reader):

$$f_i(x, \Theta) = \tfrac{\partial}{\partial \theta_i} \eta(x, \Theta), \ i = 1, \dots, m,$$

$$f(x, \Theta) = (f_1(x, \Theta), \dots, f_m(x, \Theta))^T,$$

$$M(\xi, \Theta) = \int f(x, \Theta) f^T(x, \Theta) \xi(dx),$$

the information matrix,

$$\xi = \left(\begin{array}{ccc} x_1 & \cdots & x_n \\ \omega_1 & \cdots & \omega_n \end{array} \right), \ x_i \neq x_j \ (i \neq j), \ x_i \in \mathfrak{X}, \ \omega_i > 0, \ \sum \omega_i = 1,$$

approximate experimental design.

Denote by Θ^* the proper vector of the parameters. Designs maximizing the determinant of the information matrix for a fixed vector Θ will be called the LD designs. Usually, such designs depend only on a part of parameters (see Section 2.2). Without loss of generality, assume that these parameters are $\theta_{k+1}, \ldots, \theta_m$ and call them *nonlinear parameters*. Denote $\Theta_1 = (\theta_1, \ldots, \theta_k)^T$ and $\Theta_2 = (\theta_{k+1}, \ldots, \theta_m)^T$. Let us fix Θ_1 and consider the matrix $M(\xi, \Theta_2) = M(\xi, \Theta)$.

2.3.2 The basic equation

In many practical problems, $\mathcal{X} = [a, b]$, and we will restrict our attention by this case.

The triple (n_1, n_2, n_3), where $n_1(n_3)$ is the number of support points of design at the left (right) bound, $n_1, n_3 = 0$ or 1, and $n_2 = n - n_1 - n_2$ will be called *a type of design*.

Let us consider designs LD among designs with the minimal support (i.e., with $n = m$). We call them LDMS designs. They often prove to be LD among all approximate designs. As was shown in Section 2.2, for such designs $\omega_1 = \cdots, = \omega_m = 1/m$.

Let $\Theta_2 \in \Omega$, where Ω is a given open set of possible values of Θ_2^*.

Assume that LDMS designs under $\Theta_2 \in \Omega$ have a fixed type (n_1, n_2, n_3), $n_1 + n_2 + n_3 = m$. Consider the case $n_1 = 1$ and $n_3 = 0$ (for all other cases, we can proceed in a very similar way). In this case, we will define the vector τ and the design ξ_τ as follows

$$\tau = (x_2, \ldots, x_m) = (\tau_1, \ldots, \tau_{m-1}),$$

$$\xi_\tau = \begin{pmatrix} x_1 & x_2 & \cdots & x_m \\ 1/m & 1/m & \cdots & 1/m \end{pmatrix}, \ x_1 = a.$$

Assume that the set Ω contains r linearly independent vectors and there are no $r + 1$ linearly independent vectors belonging to Ω. For example, for

$$\Omega = \{(\theta_{k+1}, \ldots, \theta_m)^T : \theta_i > 0, \ \sum_{i=k+1}^{m} \theta_i = m - k\}$$

$r = m - k - 1$.

Let Q be a given real analytic vector function on Ω such that

$$\Theta_2 \to z = Q(\Theta_2) \in R^r$$

is a one-to-one correspondence and, therefore, the inverse function $Q^{-1}(z)$ at the set $Z = Q(\Omega)$ is well defined. As an example, we can point out the vector function

$$z_i = 1 - \theta_{k+i}, \ i = 1, \ldots, r; \ r = m - k - 1, \tag{2.20}$$

introduced in Section 2.2.

Denote $\Theta^T(z) = \left(\Theta_1^T, (Q^{-1}(z))^T\right)$. Let $\mathcal{N}$ be the set of all vectors $z \in Z = Q(\Omega)$ such that

$$\det M(\xi_\tau, \Theta(r)) = 0$$

for any $\tau \in [a, b]^{m-1}$. For the case of exponential models described in Section 2.2, we have $\mathcal{N} = Q(\bar{\Omega})$, where $\bar{\Omega}$ is the set of all vectors $\Theta_2 \in \Omega$ such that two or more coordinates coincide with each other and Q is given by (2.20).

Let us introduce the following definition.

Definition 2.3.1 A vector function

$$\tau^*(z) : Z \to V,$$

where

$$V = \{\tau = (\tau_1, \ldots, \tau_{m-1}) : a < \tau_1 < \cdots < \tau_{m-1} < b\}$$

will be called the optimal design function if for any $z \in Z \setminus \mathcal{N}$, the design $\xi_{\tau*(z)}$ is a LDMS design for $\Theta^T = (\Theta_1^T, \Theta_2^T(z))$, $\Theta_2(z) = Q^{-1}(z)$ and for any sequence $z_{(1)}, z_{(2)}, \ldots$ such that $z_{(i)} \in Z \setminus \mathcal{N}$, $z_i \to \bar{z} \in \mathcal{N}$, $i \to \infty$,

$$\lim_{i \to \infty} \tau^*(z_i) = \tau^*(\hat{z}).$$

This definition is given for the case $n_1 = 1$ and $n_3 = 0$. The modification for other design types seems to be obvious.

Let us define the function

$$\varphi(\tau, z) = [\det M(\xi_\tau, \Theta(z))]^{1/m}; \tag{2.21}$$

the degree $1/m$ is introduced in order to secure a local convexity in a vicinity of the extreme points.

Due to the above assumption for any fixed $z \in Z \setminus \mathcal{N}$, the maximal value of the function $\varphi(\tau, z)$ by $\tau \in [a, b]^{m-1}$ is achieved in V. Therefore, a necessary condition for ξ_τ to be an LDMS design consists of vanishing of the derivatives

$$\frac{\partial}{\partial \tau_i} \varphi(\tau, z) = 0, \ i = 1, \ldots, m - 1. \tag{2.22}$$

Set

$$g_i = g_i(\tau, z) = \frac{\partial}{\partial \tau_i} \varphi(\tau, z), \ i = 1, \ldots, m - 1,$$

$$g = (g_1, \ldots, g_{m-1})^T.$$

The equation system (2.22) can be now written in the form

$$g(\tau, z) = 0. \tag{2.23}$$

This equation will be called *the basic equation of the functional approach*. It allows one to reduce the LDMS designs problem to the analysis of implicit functions. Such an analysis will be performed in Section 2.4. Now we will describe a class of regression functions for which this equation has a unique solution.

2.3.3   The uniqueness and the analytical properties

Let Z, N, and Q be as described above. Let us introduce the following assumptions:

A1. The functions
$$f_i(x, \Theta(z)), \; i = 1, \ldots, m,$$
are real analytic by the variables $\{x_1, z_1, \ldots, z_r\}$ at $(a, b) \times Z$.

A2. For $\Theta_2 \in \Omega$ all LDMS designs have the same type (n_1, n_2, n_3). For certainty, we will consider the case $n_1 = 1$ and $n_3 = 0$. Denote $H(\tau) = \prod_{1 \leq i \leq j \leq m} (x_i - x_j)^2$, $\tau = (x_2, \ldots, x_m)$, $x_1 = a$.

A3. There exists an algebraic polynomial $\Psi(z)$ such that
$$\inf_{z \in Z \backslash N} \; \inf_{\tau \in V} \; \frac{\varphi^m(\tau, z)}{\Psi(z) H(\tau)} > 0,$$
$$\sup_{z \in Z \backslash N} \; \sup_{\tau \in V} \; \frac{\varphi^m(\tau, z)}{\Psi(z) H(\tau)} < \infty.$$

Note that if the closure of Z does not intersect N, we can take $\Psi(z) \equiv 1$. In this case, the assumption A3 means simply that the functions
$$f_1(x, \Theta), \ldots, f_m(x, \Theta)$$
generate an extended Chebyshev system of order m on $[a, b]$ (see Section 1.9 for the definition) for all $\Theta_z = (\Theta_1, \Theta_2)$, $\Theta_z \in \Omega$.

Note also that the exponential regression functions introduced in Section 2.2 possess this property and all other assumptions were justified in that section.

Let us codefine the function
$$\bar{\varphi}(\tau, z) = \frac{\varphi(\tau, z)}{(\Psi(z))^{1/m}} = \left[\frac{\det M(\xi_{tau}, \Theta(z))}{\Psi(z)} \right]^{1/m}$$

by continuity with $z \in N$. This is possible due to assumption A3.

A4. There exists a vector $z_{(0)} \in Z$ such that the equation system
$$\frac{\partial}{\partial \tau_i} \bar{\varphi}(\tau, z_{(0)}) = 0, \; i = 1, \ldots, m - 1,$$
has a unique solution with $\tau \in V$.

In Section 2.2, we have shown that this assumption holds for the exponential models with $z_{(0)} = (0, \ldots, 0)$.

Now, the basic theorem of the functional approach can be formulated in the following way.

Theorem 2.3.1 *Let assumptions A1–A4 be fulfilled. Then the following hold:*

(I) *There exists a unique optimal design function $\tau^*(z) : Z \to V$. It is a real analytic vector function in Z.*

(II) *Taylor coefficients of this vector function can be calculated by recurrent formulas given in Section 2.4.*

A proof of this theorem will be given in Section 2.6.

2.4 The Study of the Basic Equation

In this section we will study (2.23) for a vector function $g(\tau, z)$ of a general form not necessarily connected with studying optimal experimental designs. We will obtain results stronger than that of Theorem 2.2.1(I, II), namely we will prove that under certain conditions, the function $\tau(z)$ determined implicitly by this equation is unique.

2.4.1 Properties of implicit functions

Assume that m and r are arbitrary natural numbers, and $m \geq r$ and $m \geq z$. Let

$$\hat{V} = \{\tau = (\tau_1, \ldots, \tau_{m-1})^T : a \leq \tau_1 \leq \cdots \leq \tau_{m-1} \leq b\},$$

$$V = \{\tau = (\tau_1, \ldots, \tau_{m-1})^T : a < \tau_1 < \cdots < \tau_{m-1} < b\},$$

and Z be an open one-connected set in R^r.

Let $\varphi(\tau, z)$, $\tau \in \hat{V}$, $z \in Z$, be a function of a general form real analytic in $V \times Z$, and $\varphi(\tau, z) \geq 0$.

Consider the case when $\varphi(\tau, z) = 0$ for some points $z \in Z$. Let $\mathcal{N}$ be the set of all such points. Assume that there exists an algebraic polynomial $\Psi(z)$ such that $\Psi(z) = 0$ for $z \in \mathcal{N}$ and the function

$$\bar{\varphi}(\tau, z) = \varphi(\tau, z)/\Psi(z)$$

can be codefined in points $z \in \mathcal{N}$ by continuity.

Let $\bar{\varphi}(\tau, z)$ be the function codefined in the points $z \in \mathcal{N}$ in this way. Assume that $\bar{\varphi}(\tau, z) > 0$, $\tau \in V$, $z \in Z$, and

$$\inf_{\tau \in V} \frac{(\bar{\varphi}(\tau, z))^m}{H(\tau)} > 0,$$

$$\sup_{\tau \in V} \frac{(\bar{\varphi}(\tau, z))^m}{H(\tau)} < \infty,$$

for any $z \in Z$, where

$$H(\tau) = \prod_{i=1}^{m-1} (\tau_i - a)^2 \prod_{1 \leq i \leq j \leq m-1} (\tau_i - \tau_j)^2.$$

Let us denote

$$g(\tau, z) = (g_1(\tau, z), \ldots, g_{m-1}(\tau, z)),$$

$$g_i(\tau, z) = \frac{\partial}{\partial \tau_i} \varphi(\tau, z),$$

$$G(\tau, z) = \left(\frac{\partial^2}{\partial \tau_i \partial \tau_j} \varphi(\tau, z) \right)_{i,j=1}^{m-1},$$

$$\bar{g}(\tau, z) = g(\tau, z)/\Psi(z),$$

$$\bar{G}(\tau, z) = G(\tau, z)/\Psi(z) = \left(\frac{\partial^2}{\partial \tau_i \partial \tau_j} \bar{\varphi}(\tau, z) \right)_{i,j=1}^{m-1}.$$

Consider the equations

$$g(\tau, z) = 0,$$

$$\bar{g}(\tau, z) = 0,$$

(2.24)

$z \in Z$, $\tau \in V$. For $z \in Z \setminus \mathcal{N}$, these equations are equivalent to each other. Let us introduce the following assumptions:

(a) There exists a point $z_{(0)} \in Z$ such that (2.24) has a unique solution belonging to V.

(b) For any point z and any solution $z = \tau(z)$ of (2.24),

$$\det \bar{G}(\tau, z) \Big|_{\tau = \tau(z)} \neq 0.$$

Theorem 2.4.1 *Let the assumptions formulated above be satisfied. Then there exists a unique vector function $\tau^*(z) : Z \to V$ such that*

$$\bar{g}(\tau^*(z), z) = 0.$$

This vector function is real analytic for $z \in Z$ and satisfies the equation

$$G(\tau^*(z), z)\tau'_{z_i}(z) = (g(\tau, z))'_{z_i} \Big|_{\tau = \tau^*(z)}, \quad i = 1, \ldots, m - 1.$$

Proof. Due to assumptions (a) and (b) and the Implicit Function Theorem (Theorem 1.8.1), there exists a vicinity of the point $z_{(0)}$ such that there exists a unique vector function, say $\bar{\tau}(z)$, satisfying (2.24). This vector function is real analytic. Let U be a union of all such vicinities. Then $\bar{\tau}(z)$ can be extended to U in a unique way and this extended function is real analytic in U.

Suppose that $U \neq Z$. Denote the closure of U by $\bar{U}$. Since $U \neq Z$, there exists a point $\bar{z} \in \bar{U} \setminus U$, $\bar{z} \in Z$. Then there exists a sequence $z_{(1)}, z_{(2)}, \ldots,$ such that $z_{(i)} \in U$ and $\lim_{i \to \infty} z_{(i)} = \bar{z}$. Denote by $\bar{\tau}$ the limit

$$\lim_{i \to \infty} \bar{\tau}(z_{(i)}).$$

Then we have

$$\bar{g}(\bar{\tau}, \bar{z}) = 0.$$

Suppose that $\bar{\tau} \in V$. Then, due to assumption (b),

$$\det \bar{G}(\bar{\tau}, \bar{z}) \neq 0$$

and there exists a vicinity of point $\bar{z}$, say W, and vector function $\tau_{(1)}(z)$ such that $\tau_{(1)}(\bar{z}) = \bar{\tau}$ and this vector function is real analytic in this vicinity. Moreover, for sufficiently large i, $z_{(i)}$ belongs to this vicinity. It follows from here that $\tau_{(1)}(z)$ and $\bar{\tau}(z)$ coincide in $W \cap Z \neq \emptyset$. Therefore, $\bar{\tau}_{(1)}(z)$ is a real analytical extension of $\bar{\tau}(z)$ to W and $W \bar{\subset} U$. This is impossible by our supposition and we obtained a contradiction.

Now, let $\bar{\tau} \in \hat{V} \setminus V$. Denote $\tau_{(i)} = \tau(z_{(i)})$, $i = 1, 2, \ldots$. Then

$$\lim_{j \to \infty} \frac{\partial}{\partial \tau_i} \left[\frac{(\bar{\varphi}(\tau, z_j))^m}{Q(\tau)} \right] \Bigg|_{\tau = \tau_{(j)}}$$

$$= \lim_{j \to \infty} \left\{ \left\{ \frac{\partial}{\partial \tau_i} \left(\bar{\varphi}(\tau, z_{(1)}) \right)^m \right\} \Bigg|_{tau = \tau_{(j)}} \Bigg/ Q(\tau_{(j)}) \right.$$

$$\left. - \left(\bar{\varphi}(\tau_{(j)}, z_{(i)}) \right)^m \frac{\partial Q(\tau)/\partial \tau_i}{Q^2(\tau)} \Bigg|_{\tau = \tau_{(i)}} \right\} = \infty.$$

However, due to our assumption, the function

$$\frac{\bar{\varphi}(\tau, z)}{Q(\tau)}$$

is real analytic in $V \times Z$, and the limit should be finite. The obtained contradiction shows that $U = Z$. In a similar way, it can be proved that for any $z \in Z$, (2.24) has a unique solution. ∎

In order to apply Theorem 2.4.1 to the function $\varphi(\tau, z)$ defined in Section 2.3, we need only to verify property (b). To this end we will introduce a representation for the Jacobi matrix of (2.24).

2.4.2 Jacobian of the basic equation

First, we analyze the Jacobian of the basic equation for functions $\varphi(\tau, z)$ of a general kind that can be represented as the minimum of some convex function.

Let m, r, and t be arbitrary natural numbers, $T \subset \mathbf{R}^{m-1}, Z \subset \mathbf{R}^r$, and $\mathfrak{A} \subset \mathbf{R}^t$ be arbitrary open sets, and $\mathfrak{A}$ be convex.

Consider the function $q(\tau, a, z), \tau \in T, a \in \mathfrak{A}, z \in Z$, that satisfies the following conditions: Function $q(\tau, a, z)$ is twice continuously differentiable along τ and a; function $q(\tau, a, z)$ is strictly convex along a.

Moreover, let function $\varphi(\tau, z)$ have the form

$$\varphi(\tau, z) = \min_{a \in \mathfrak{A}} q(\tau, a, z), \tag{2.25}$$

where the minimum is attained for any $\tau \in T$ and $z \in Z$. Since the function $q(\tau, a, z)$ is strictly convex along a, this minimum is attained on the unique vector $a = \tilde{a} = \tilde{a}(\tau, z)$. Therefore, function $\varphi(\tau, z)$ is twice continuously differentiable along τ.

For any fixed z, let there exist a point $\tilde{\tau} = \tilde{\tau}(z)$ satisfying the equation $\frac{\partial}{\partial \tau} \varphi(\tau, z) = 0$.

Consider the following matrices:

$$E = \left(\frac{\partial^2}{\partial \tau_j \partial \tau_i} q(\tau, a, z) \right)_{i,j=1}^{m-1},$$

$$B = \left(\frac{\partial^2}{\partial \tau_j \partial a_i} q(\tau, a, z) \right)_{i,j=1}^{t, m-1}, \tag{2.26}$$

$$D = \left(\frac{\partial^2}{\partial a_j \partial a_i} q(\tau, a, z) \right)_{i,j=1}^{t}$$

at $\tau = \tilde{\tau}$ and $a = \tilde{a}(\tilde{\tau}, z)$. It follows from the above conditions that matrix D is positive definite and hence the inverse matrix D^{-1} exists.

Theorem 2.4.2 *Under the above conditions, the following formula is valid:*

$$J(\tilde{\tau}(z), z) = E - B^T D^{-1} B.$$

Let us apply this theorem to the function $\varphi(\tau, z)$, defined by (2.21).

Denote the set of all positive definite $m \times m$ matrices $A = (a_{ij})$, such that $a_{mm} = 1$ by $\mathcal{A}$. Assign a number $\nu = \nu(i, j)$ in alphabetical order to each pair of indices $(i, j), i \leq j, i, j = 1, \dots, m$, where $(i, j) \neq (m, m)$. For any vector $a \in \mathbf{R}^t, t = m(m+1)/2 - 1$, define a matrix $A(a)$ that satisfies the following relations:

$$a_{ji} = a_{ij} = a_{\nu(i,j)}, \ a_{mm} = 1, \ i, j = 1, \dots, m, \ i \leq j.$$

Define set $\mathfrak{A}$ as

$$\mathfrak{A} = \{a \in \mathbf{R}^t : A(a) \in \mathcal{A}\}.$$

Evidently, $\mathfrak{A}$ is open and convex in $\mathbf{R}^t$. Introduce the function

$$q(\tau, a, z) = (\det A(a))^{-1/m} \operatorname{tr} (A(a) M(\xi, z)) / m. \tag{2.27}$$

Consider the function $\varphi(\tau, z) = (\det M(\xi, z))^{1/m}$. It is known (Karlin and Studden, 1966, Chap. 10.2) that (2.25) is valid for this function. It can also be checked that the function (2.27) possesses the required properties. Therefore, by Theorem 2.4.2,

$$J(\tilde{\tau}(z), z) = E - B^T D^{-1} B, \qquad (2.28)$$

where $\tilde{\tau}(z) = \tau^*(z)$. Set $\delta(a) = (\det A(a))^{-1/m}$. It is easy to verify by direct differentiation that the following formulas are valid for matrices B and E:

$$E = \mathrm{diag}\{E_{11}, \ldots, E_{m-1m-1}\},$$

$$E_{ii} = \delta(a^*)\frac{\partial^2}{\partial x^2}(f^T(x)A(a^*)f(x))\bigg|_{x=x_{i+1}^*}, \quad i = 1, \ldots, m-1,$$

$$A(a^*) = \mathrm{const}\left(M(\xi_{\tau^*(z)}, z)\right)^{-1}, \qquad (2.29)$$

$$B = (b_{\nu k})_{\nu, k=1}^{t, m-1},$$

$$b_{\nu k} = 2\delta(a^*)\frac{\partial}{\partial x}(f_i(x)f_j(x))\bigg|_{x=x_k^*}, \quad \nu = \nu(i, j).$$

Remark 2.4.1 Note that the matrix $J = J(\tau^*(z), z)$ is negative definite and hence nonsingular provided at least one of the following conditions is satisfied:

    (1) All diagonal elements of matrix E are negative;
    (2) Matrix B is of full rank.

    Indeed, matrix $B^T D^{-1} B$ has the form SS^T; hence, it is nonnegative definite in the general case and positive definite if matrix B has full rank. Since $J = E - B^T D^{-1} B$, J is negative definite if either of conditions (1) and (2) is valid.

    This remark will be applied in Section 2.6.2 in order to prove that the matrix J is invertible under assumptions A1–A4.

2.4.3 On the representation of implicit functions

It is well known that derivatives of implicit functions can be calculated with the help of indefinite coefficients techniques, as introduced by Euler. In this subsection we offer recurrent formulas convenient for the implementation in software packages such as Maple and Mathcad. These formulas are a generalization for the multidimensional case of formulas introduced in Dette, Melas and Pepelyshev (2004b).

    Let us assume that $s = (s_1, \ldots, s_r)$, where $s_i \geq 0$, $i = 1, \ldots, r$, are integers. For an arbitrary (scalar, vector, or matrix) function $\mathcal{F}$, denote

$$(\mathcal{F}(z))_{(s)} = \frac{1}{s_1! \cdots s_r!}\frac{\partial^{s_1}}{\partial z_1^{s_1}} \cdots \frac{\partial^{s_r}}{\partial z_r^{s_r}}\mathcal{F}(z)\big|_{z=z_{(0)}},$$

where $z_{(0)}$ is a given point.

Introduce also the notation

$$S_t = \left\{ s = (s_1, \ldots, s_r); \; s_i \geq 0, \; \sum_{i=1}^{r} s_i = t \right\},$$

$t = 0, 1, \ldots$, and

$$(z - z_{(0)})^s = (z_1 - z_{1(0)})^{s_1} \ldots (z_r - z_{r(0)})^{s_r}.$$

Let the function $\psi(z)$ be of the form

$$\psi(z) = (z - z_{(0)})^l \bar{\psi}(z),$$

where $l = (l_1, \ldots, l_r), l_i \geq 0$, $i = 1, \ldots, r$, are integers, and $\bar{\psi}(z)$ is a homogeneous polynomial of degree $p \geq 0$,

$$\bar{\psi}(z) = \sum_{s \in S_p} a_{(s)} (z - z_{(0)})^s,$$

such that $a_{(p,0,\ldots,0)} \neq 0$.

Let

$$I_t = U_{j=0}^t S_j,$$

$$\tau_{<I_t>}(z) = \sum_{s \in I_t} \tau_{(s)} (z - z_{(0)})^s, \quad \tau_{(s)} = (\tau(z))_{(s)},$$

$$J_{(l)} = (J(\tau_{(0)}, z))_{(l)}.$$

First, let $p = 0$. Note that under condition (a), the matrices $J_{(s)}$, $s_i \leq l_i$, $i = 1, \ldots, r$, $s \neq l$, are zero matrices and $\det J_{(l)} \neq 0$.

Theorem 2.4.3 *Under conditions (a) and (b) for the function $\tau(z)$, defined in Theorem 2.4.1, the following formulas hold:*

$$(\tau(z))_{(s)} = -J_{(l)}^{-1} g(\tau_{<I>}(z), z)_{(s+l)}, \tag{2.30}$$

where $I = I_{t-1}$, $s \in S_t$, $t = 1, 2, \ldots, K - 1$.

If condition (c) is also fulfilled, then these formulas hold for $t = 1, 2, \ldots$.

Thus, if $\tau_{(0)}$ is known, coefficients $\{\tau_{(s)}\}$ can be calculated in the following way. At the step t ($t = 1, 2, \ldots$), calculate all coefficients with indices from S_t by (2.30). This calculation can be easily performed by a computer with the help of packages such as Maple or Mathcad.

Consider now the case $p > 0$. Define the set

$$\hat{S}_t = \left\{ s = (s_1, \ldots, s_r); \; s_i \geq 0, i = 1, \ldots, r, s_1 + 2 \sum_{i=2}^{r} s_i = t \right\}.$$

Let

$$\hat{I}_t = U^t_{j=0} \hat{S}_j, \quad u = (p, 0, \ldots, 0),$$

$$J_{(l+u)} = \left(J(\tau_{(0)}, z) \right)_{(l+u)}.$$

It can be verified that, under condition (a), $\det J_{(l+u)} \neq 0$.

Theorem 2.4.4 *With $p > 0$, Theorem 2.4.3 remains true with (2.30) replaced by*

$$(\tau(z))_{(s)} = -J^{-1}_{(l+u)} g(\tau_{<I>}(z), z)_{(s+l+u)},$$

where $s \in \hat{S}_t$, $I = \hat{I}_{t-1}$, and $t = 1, 2, \ldots$.

Note that u can be replaced by any vector of the form $(0, \ldots, p, 0, \ldots, 0)$.

2.4.4 The monotony property

Let us obtain another representation for the Jacobi matrix. It is based on the known formula for the derivative of the matrix determinant. This representation is to help us to derive the monotony of coordinates of the optimal design function for some forms of regression.

At first, let $x \in \mathfrak{X}$ and $z \in Z$, where Z is some bounded set in $\mathbf{R}^r$ and $f_i(x, z), i = 1, \ldots, m$, are arbitrary twice differentiable with respect to x functions.

Let the function $\varphi(\tau, z)$ be defined by (2.21). Let assumptions A1–A4 be satisfied. We will use the formula of differentiating the matrix determinant (see, e.g., Fedorov (1972)).

$$\frac{\partial}{\partial \alpha} \det M(\alpha) = \det M(\alpha) \left(\operatorname{tr} M^{-1}(\alpha) \frac{\partial}{\partial \alpha} M(\alpha) \right),$$

as well as the formula

$$\frac{\partial}{\partial \alpha} M^{-1}(\alpha) = -M^{-1}(\alpha) \frac{\partial M(\alpha)}{\partial \alpha} M^{-1}(\alpha)$$

and the explicit form of matrix $M(\xi, z)$:

$$M(\xi, z) = \sum_{i=1}^{m} f(x_i) f^T(x_i)/m,$$

where $f(x) = f(x, z)$ and

$$\xi = \begin{pmatrix} x_1 & \cdots & x_{m-1} & x_m \\ \frac{1}{m} & \cdots & \frac{1}{m} & \frac{1}{m} \end{pmatrix}.$$

Let us calculate the derivatives of the function

$$\varphi(\tau, z) = (\det M(\xi_\tau, z))^{1/m},$$

$$\xi_\tau = \begin{pmatrix} \tau_1 & \cdots & \tau_{m-1} & b \\ \frac{1}{m} & \cdots & \frac{1}{m} & \frac{1}{m} \end{pmatrix}.$$

We obtain

$$\frac{\partial \varphi(\tau, z)}{\partial \tau_i} = \frac{1}{m^2} \varphi(\tau, z) \operatorname{tr} M^{-1}(\xi, z)(f(\tau_i) f^T(\tau_i))'$$

$$= \frac{2}{m^2} \varphi(\tau, z) f^T(\tau_i) M^{-1}(\xi, z) f'(\tau_i),$$

$i = 1, \ldots, m-1$. Let z be fixed and τ be such that

$$g(\tau, z) = \left(\frac{\partial}{\partial \tau_i} \varphi(\tau, z) \right)_{i=1}^{m-1} = 0. \tag{2.31}$$

Moreover, let τ be a local maximum of the function $\varphi(\tau, z)$. Set $F = (f_j(x_i))_{i,j=1}^m$. Then the relation $M = FF^T/m$ is valid,

$$f^T(\tau_i) M^{-1}(\xi, z) f(\tau_j) = m f^T(\tau_i)(F^{-1})^T F^{-1} f(\tau_j)$$

$$= m \, e_{i+1}^T e_{j+1}^T = \begin{cases} 0 & i \neq j \\ m & i = j \end{cases}.$$

Let us consider the matrix

$$G = \left(\frac{\partial^2}{\partial \tau_i \partial \tau_j} \varphi(\tau, z) \right)_{i,j=1}^{m-1}, \quad \tau = \tau(z),$$

where $\tau(z)$ is the unique solution of (2.31).

Using these relations and the formula of the inverse matrix differentiation, derive

$$(G)_{ij} = \frac{\partial^2}{\partial \tau_i \partial \tau_j} \varphi(\tau, z)$$

$$= -\frac{4}{m^3} \varphi(\tau, z) \left(f^T(\tau_i) M^{-1}(\xi, z) f'(\tau_j) \right) \left(f^T(\tau_j) M^{-1}(\xi, z) f'(\tau_i) \right)$$

$$= -\frac{4}{m} \varphi(\tau, z) \left(e_i^T F^{-1} f'(\tau_j) \right) \left(e_j^T F^{-1} f'(\tau_i) \right)$$

for $i \neq j$, $i, j = 1, \ldots, m-1$. For calculating the diagonal elements of the matrix, let us also use the following relation:

$$f^T(\tau_i) M^{-1}(\xi, z) f'(\tau_i) = \frac{m^2}{2\varphi(\tau, z)} \frac{\partial \varphi(\tau, z)}{\partial \tau_i} = 0, \; i = 1, \ldots, m-1.$$

The direct differentiation gives the following result

$$(G)_{ii} = \frac{\partial^2}{\partial \tau_i \partial \tau_i} \varphi(\tau, z)$$

$$= \frac{2}{m^2} \varphi(\tau, z) f^T(\tau_i) M^{-1}(\xi, z) f''(\tau_i)$$

$$= \frac{2}{m} \varphi(\tau, z) e_i^T F^{-1} f''(\tau_i), \; i = 1, \ldots, m-1.$$

Now, assume the that functions $f_i(x) = f_i(x, z), i = 1, \ldots, m$, form an *ET*-system (see Section 1.9 for the definition) of the first order under any fixed $z \in Z$.

Since the matrix F^{-1} is formed by the cofactors of the elements of matrix F, divided by its determinant, then, for $j > i$, $i > 2$, we have

$$e_i^T F^{-1} f'(\tau_j)$$

$$= \frac{\det \left(f(x_1) \vdots f(x_2) \vdots \ldots \vdots f(x_{i-1}) \vdots f'(x_j) \vdots f(x_{i+1}) \vdots \ldots \vdots f(x_m) \right)}{\det F}$$

(with the evident changes for $i \leq 2$). Inserting a column $f'(x_j)$ between a line $f(x_j)$ and the following one, derive

$$e_i^T F^{-1} f'(\tau_j) = (-1)^{j-i} \det \tilde{F} / \det F,$$

$$\tilde{F} = \left(f(x_1) \vdots f(x_2) \vdots \ldots \vdots f(x_{i-1}) \vdots f(x_{i+1}) \vdots \ldots \vdots f(x_j) \vdots f'(x_j) \vdots \ldots \vdots f(x_m) \right).$$

By definition of the *ET*-system of the first order, $\det \tilde{F} > 0$. Thus,

$$\text{sign} \left[e_i^T F^{-1} f'(\tau_j) \right] = (-1)^{j-i}.$$

Similarly, for $i < j$, we have

$$\text{sign} \left[e_{j+1}^T F^{-1} f'(\tau_i) \right] = (-1)^{i-j}.$$

Therefore, for $i \neq j$

$$\text{sign} \, (G)_{ij} = (-1)(-1)^{j-i-1}(-1)^{i-j} = 1.$$

It will be proved in Section 2.6.2 that the matrix G is negative definite. Let us use the following statement (see Szegö, 1959): If matrix A is positive definite and each of its off-diagonal elements is negative, then all of the elements of the matrix A^{-1} are positive. Since the matrix G is negative definite and its off-diagonal elements are positive, then the matrix $A = -G$ possesses the required properties. Applying the above statement, we have that all of the elements of the matrix G^{-1} are negative.

Thus, we have derived the following result.

Lemma 2.4.1 *If functions $f_i(x, z), i = 1, \ldots, m, x \in \mathfrak{X}, z \in Z$, are twice continuously differentiable on $\mathfrak{X}$ and form an ET-system of the first order for any fixed $z \in Z$, the matrix G is invertible and all of the elements of matrix G^{-1} are negative.*

Let the conditions of Lemma 2.4.1 be satisfied. By Theorem 2.3.1, the optimal design function $\tau(z) : Z \to V$ is uniquely determined. Let L_j stand for the vector

$$\frac{\partial}{\partial z_j} g(\tau, z) = \left(\frac{\partial^2}{\partial \tau_i \partial z_j} \varphi(\tau, z) \right)_{i=1}^{s-u}, \quad j = 1, 2, \ldots, r.$$

By the Implicit Function Theorem, we have

$$\tau'_{z_j} = -G^{-1}L_j. \tag{2.32}$$

Thus, if all of the elements of vector L_j are positive, then

$$(\tau_i(z))'_{z_j} < 0, \ i = 1, \ldots, m - 1;$$

that is, all of the coordinates of function $\tau(z)$ monotonously decrease with respect to z_j.

Let us introduce a class of regression functions for which all the unfixed points of a locally D-optimal design monotonously depend on each parameter. We will show further that this class contains the exponential models considered in Section 2.2, as well as some rational models.

Consider a real function $K(x, y)$, defined for $(x, y) \in \mathfrak{X} \times \mathfrak{X}_1$, where $\mathfrak{X}$ and $\mathfrak{X}_1$ are intervals. Let function $K(x, y)$ be an extended strictly positive kernel of the m-th order (ESP(m)) along both variables. The corresponding definition can be found in Karlin and Studden (1966, Chap. I).

Consider the regression function

$$\eta(x, \Theta) = \sum_{i=1}^{k} \theta_i K(x, \theta_{i+k}), \ \theta_i \neq 0, i = 1, \ldots, k, \ m = 2k.$$

Let the functions $f_i(x, \Theta) = \frac{\partial}{\partial \theta_i} \eta(x, \Theta)$, $i = 1, \ldots, m$, for $\Theta_2 = (\theta_{k+1}, \ldots, \theta_m)^T = (z_1, \ldots, z_k)^T \in Z \subset \mathfrak{X}_1^k$ be real analytic.

By the definition of ESP(m) functions $f_i(x, z), i = 1, \ldots, m$ (for fixed Θ_1) at any fixed $z \in Z$ form an ET-system. Therefore, the optimal design function $\tau(z)$ is uniquely determined at $z \in Z$.

Assume that for some point $z_{(0)}$ for $z = z_{(0)}, \tau = \tau(z_{(0)})$ the following inequality is valid:

$$\frac{\partial^2 \varphi(\tau, z)}{\partial \tau_i \partial z_j} > 0, \ i = 1, \ldots, s - u, \ j = 1, \ldots, k. \tag{2.33}$$

Theorem 2.4.5 *Under the above conditions, all of the components of the vector function $\tau(z)$ decrease with respect to each of $z_1, \ldots, z_k$ in a strictly monotonous way.*

Proof. By Lemma 2.4.1 and formula (2.32), it is sufficient to prove that

$$\frac{\partial^2 \varphi(\tau, z)}{\partial \tau_i \partial z_j} > 0, \ i = 1, \ldots, m - 1, \ j = 1, \ldots, k.$$

for any $z \in Z$.

Let $\bar{z}_1 < \bar{z}_2 < \cdots < \bar{z}_k$. Set $v = z_j$ and consider the function $\partial \varphi(\tau, z)/\partial \tau_\nu$ $(\nu = 1, \ldots, m-1)$ as a function of v under fixed $z_i, i = 1, \ldots, k$,

$i \neq j$, $x_1, \ldots, x_{m-1}$. Denote this function by $h(v)$. Note that the function $h(v)$ has second-order zeros at points z_i, $i \neq j, i = 1, \ldots, k$, since the determinants corresponding to $h(v)$ and $h'(v)$ have common lines. Moreover, $h(\bar{z}_j) = 0$. Since $\varphi^m(\tau, z)$ equals

$$\text{const} \det (K(x_j, z_1), K'_z(x_j, z_1), \ldots,$$

$$K(x_j, z_k), K'_z(x_j, z_k))^m_{j=1}, \tag{2.34}$$

then the function $h(v)$ has no more than $2k - 1$ zeros (counting with their multiplicities). Therefore, $h'(\bar{z}_j) \neq 0$. The case that some $\bar{z}_i$ coincide with one another can be processed in a similar way (here determinant (2.34) is to be modified as is stated in the definition of the ESP kernel). Thus, the functions

$$\frac{\partial^2}{\partial \tau_i \partial z_j} \varphi(\tau, z)$$

do not vanish. Since, by assumption, condition (2.33) is valid for $z = z_{(0)}$, it is valid for any $z \in Z$. ∎

Consider two examples.

Example 2.4.1 Algebraic sum of simplest fractions.

Let

$$K(x, y) = \frac{1}{x + y}, \quad \mathfrak{X} \subset [0, \infty), \ \mathfrak{X}_1 \subset [0, \infty).$$

It is known that such a function is an ESP kernel of any order (Karlin and Studden 1966, Chap. I). The corresponding regression function takes the form

$$\eta(x, \Theta) = \sum_{i=1}^{k} \frac{\theta_i}{x + \theta_{i+k}}, \ x \in [0, \infty),$$

$\theta_{i+k} > 0$, $\theta_i \neq 0$, $i = 1, \ldots, k$. For corresponding basis functions $f_i(x, \Theta), i = 1, \ldots, k$, condition (2.33) can be verified directly. It can be demonstrated also that condition A4 is satisfied. These models will be thoroughly investigated in Chapter 5.

Example 2.4.2 Algebraic sum of exponential functions.

Let

$$K(x, y) = e^{xy}, \quad \mathfrak{X} \subset (-\infty, \infty), \ \mathfrak{X}_1 \subset (-\infty, \infty).$$

This function is an ESP kernel of any order [Karlin and Studden (1966, Chap. I)]. The corresponding regression function takes the form

$$\eta(x, \Theta) = \sum_{i=1}^{k} \theta_i e^{-\theta_{i+k} x},$$

$\theta_i \neq 0, i = 1, \ldots, k$. ,k

2.5   Three-Parameter Logistic Distribution

Consider the function

$$\eta(t, \alpha, \beta, \gamma) = \frac{\alpha e^{\gamma t + \beta}}{1 + e^{\gamma t + \beta}}.$$

It is called a three-parameter logistic distribution. By the substitution $x = e^t$, $\theta_1 = \alpha$ and $\theta_2 = \gamma$, $\theta_3 = e^{-\beta}$ this function is reduced to

$$\eta(x, \Theta) = \frac{\theta_1 x^{\theta_2}}{\theta_3 + x^{\theta_2}}, \tag{2.35}$$

which is called the Hill equation in microbiological studies (see Bezeau and Endrenyi (1986)).

We will construct locally D-optimal designs for model (2.35) using the functional approach described above.

Assume that $x \in [a, b]$, $a \geq 0$, $\theta_1 \neq 0$, $\theta_3 > 0$. By a direct calculation, we obtain

$$\det M(\xi, \Theta) = \theta_1^4 \theta_3^2 \det \bar{M}(\zeta_\xi, \theta_3),$$

where

$$\xi = \begin{pmatrix} x_1 & x_2 & x_3 \\ 1/3 & 1/3 & 1/3 \end{pmatrix}, \quad \zeta_\xi = \begin{pmatrix} t_1 & t_2 & t_3 \\ 1/3 & 1/3 & 1/3 \end{pmatrix},$$
$$t_i = x_i^{\theta_2}, \ i = 1, 2, 3,$$

$$\bar{M}(\zeta_\xi, \theta_3) = \sum_{i=1}^{3} f(t_i, \theta_3) f^T(t_i, \theta_3)/3,$$

$$f(t, \theta) = \left(\frac{t}{\theta + t}, \frac{t}{(\theta + t)^2}, \frac{t \ln t}{(\theta + t)^2} \right)^T.$$

Set

$$z = 1/\theta_3, \ r = 1, \ \Omega = [0, \infty), \ \psi(z) = z^6, \ \mathcal{N} = \{0\}. \tag{2.36}$$

Assumption A1 follows here from the properties of elementary functions, A2 and A3 follows from the results of Dette, Melas, and Wong (2004b). It was also proved there that a locally D-optimal design has the type $(0, 2, 1)$ and is unique. It can be also proved that A4 holds for the considered model.

Thus, due to Theorem 2.3.1, it follows that support points of locally D-optimal designs are real analytic functions of z with $z \in [0, 1)$.

Let us consider the case $[a, b] = [0, 1]$, $\theta_2 = 1$. For arbitrary $0 \leq a < b, \theta_2$ optimal designs can be calculated by a scale transformation. With $\theta_3 \to \infty$ and $z = \frac{1}{\theta_3} \to 0$, we obtain

$$\frac{\det^2(f_i(x_j, \theta_3))}{z^6} \to \det^2 \begin{pmatrix} x_1^2 & x_2^2 & 1 \\ x_1 & x_2 & 1 \\ x_1 \ln x_1 & x_2 \ln x_2 & 0 \end{pmatrix} := Q(x_1, x_2)$$

and

$$(x_1^*(z), x_2^*(z)) \to \arg \max_{0 < x_1 < x_2 < 1} Q(x_1, x_2).$$

Thus, it is easy to calculate numerically that $x_1^*(0) = 0.15370$ and $x_2^*(0) = 0.61680$.

By the recurrent formulas (2.30) given in Section 2.4, we calculated the Taylor coefficients with $z_{(0)} = 0$. The first coefficients are represented in Table 2.5.

Table 2.5: Coefficients of the Taylor expansions for x_1 and x_2 in a vicinity of point $z = 0$

	0	1	2	3	4	5	6
x_1	0.15370	−0.09435	0.06747	−0.05117	0.04089	−0.03371	0.02845
x_2	0.61680	−0.20012	0.08251	−0.03885	0.02085	−0.01212	0.00754

Let $\xi_{<n>}(z)$ be the design constructed by using n first coefficients and let $\bar{z}_n$ be the maximal z such that

$$\max_{x \in [0,1]} |d(x, \xi_{<n>}(z)) - 3| \leq 10^{-5},$$

$$d(x, \xi) = f^T(x) M^{-1}(\xi, z) f(x), \tag{2.37}$$

where

$$f(x) = \frac{\partial \eta(x, \Theta)}{\partial \theta_i}, \quad M(\xi, z) := M(\xi, \Theta(z)), \quad \Theta(z) = (1, 1, 1/z)^T.$$

Note that due to the Kiefer–Wolfowitz equivalence theorem (see Section 1.5), a design satisfying condition (2.37) will be very close to a locally D-optimal design. Numerical calculations show that $\bar{z}_{10} \approx 0.705$ and $\bar{z}_{20} \approx 0.865$.

In a similar way we constructed expansions of the vector function $\tau^*(z) = (x_1^*(z), x_2^*(z))^T$ in a vicinity of point $z_{(0)} = 1$ by degrees of $(z - 1)$ and $(1/z - 1)$. The corresponding coefficients are presented in Tables 2.6 and 2.7, respectively. It proves that for the first expansion with 20 coefficients, the inequality (2.37) holds with $0 < z \leq 2.7$. For the second expansion with the same number of the coefficients, it holds for $0.6 \leq z \leq 13.8$.

The behavior of the design points for $0 \leq z \leq 10$ is presented in Figure 2.3. We used the first expansion for $z \leq 1$ and the second for $1 \leq z \leq 10$ to construct Figure 2.3.

Note also that the efficiency of the limiting design (at the point $z_{(0)} = 0$) measured by the quantity

$$I(\xi, z) = \left(\frac{\det M(\xi, z)}{\det M(\xi_{\tau(z)}, z)} \right)^{1/3}, \quad \xi = \xi_{\tau_{(0)}} := \xi(0),$$

proves to be very high with $z \leq 1$ ($\theta_3 \geq 1$). This efficiency is presented in Table 2.8.

Table 2.6: Coefficients of the Taylor expansions for x_1 and x_2 in a vicinity of point $z = 1$ by degrees of $(z - 1)$

	0	1	2	3	4	5	6
x_1	0.09723	-0.03401	0.01308	-0.00530	0.00222	-0.00095	0.00041
x_2	0.47233	-0.10533	0.02743	-0.00791	0.00245	-0.00080	0.00027

Table 2.7: Coefficients of the Taylor expansions for x_1 and x_2 in a vicinity of point $z = 1$ by degrees of $(1/z - 1)$

	0	1	2	3	4	5	6
x_1	0.09723	0.03401	-0.02093	0.01314	-0.00844	0.00555	-0.00375
x_2	0.47233	0.10533	-0.07790	0.05838	-0.04431	0.03404	-0.02647

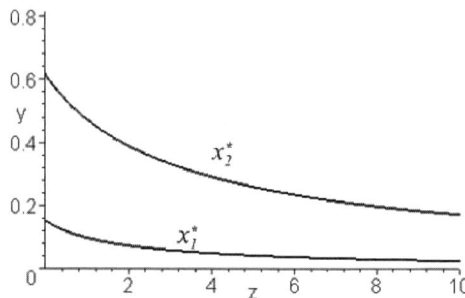

Figure 2.3: The dependence of the support points x_1 and x_2 on z

Table 2.8: Efficiency of designs $\xi(0)$ and $\xi(1)$ and the points of locally D-optimal designs

z	0.2	0.4	0.6	0.8	1.0
x_1	0.13690	0.12387	0.11333	0.10460	0.09723
x_2	0.57956	0.54751	0.51943	0.49456	0.47233
$\left(\dfrac{\det M(\xi(0),z)}{\det M(\xi_z,z)}\right)^{1/3}$	0.99343	0.97771	0.95681	0.93310	0.90801
$\left(\dfrac{\det M(\xi(1),z)}{\det M(\xi_z,z)}\right)^{1/3}$	0.94919	0.97468	0.98995	0.99774	1

At the same time, the minimal efficiency of the design $\xi(1) = \xi_{\tau^*(1)}$ with $0 < z \leq 1$ is even more than that of $\xi(0) = \xi_{\tau^*(0)} = \xi_{\tau_{(0)}}$; see Table 2.8. Moreover, numerical calculations show that the design $\xi(z^*) = \xi_{\tau^*(z^*)}$ with $z^* = 0.5$ has a maximum of the minimal efficiency at the interval $(0, 1]$ among locally D-optimal designs at points $z = 0.1, \ldots, 0.9, 1$. Its minimal efficiency is equal to 0.981.

Note that a maximin efficient D-optimal design that is the design maximizing the minimum by $z \in [0.1, 1]$ of the efficiency among all (approximate) designs, was constructed numerically in Dette, Melas, and Wong (2004b). This design is very close to $\xi(0.5)$ and has the minimal efficiency 0.982.

A similar calculation was performed for the interval $[1, 10]$ for z. It showed that the design $\xi(4)$, the best design among $\xi(1), \xi(2),\ldots,\xi(10)$, has minimal efficiency 0.8407. The maximin efficient design calculated in Dette, Melas and Wong (2004b) has four support points with unequal weights and its minimal efficiency equals 0.885. However, for example, design $\xi(1)$, the locally optimal design for z=1, has the minimal efficiency 0.5430 on $[1, 10]$. This design is rather bad! It requires almost twice as many observations as $\xi(4)$ to achieve the same accuracy of the estimates of the parameters if the true value of z equals 10.

Thus, we see that the approach allows very efficient calculation of locally D-optimal designs and gives an opportunity to study their efficiency.

We conclude also that locally D-optimal designs could be very efficient if the initial values are chosen in an optimal way inside given intervals of possible values.

2.6 Appendix: Proofs

We begin with the proofs for the theorems of Section 2.4.

2.6.1 Proof of Theorems 2.4.2, 2.4.3, and 2.4.4

Proof of Theorem 2.4.2. Due to the necessary condition for an extremum point, we have

$$\frac{\partial}{\partial a} q(\tau, a, z) = 0$$

with an arbitrary fixed $z \in Z$ and with $\tau = \tilde{\tau} = \tilde{\tau}(z)$ and $a = \tilde{a} = \tilde{a}(z, \tilde{\tau}(z))$.

Consider this vector equality at fixed z and arbitrary a and τ as an equation system that implicitly defines a function $a(\tau)$. The Jacobian of this system at the points $(\tilde{\tau}, \tilde{a})$ equals $\det D \neq 0$. Therefore, by the Implicit Function Theorem, in a vicinity of $\tilde{\tau}$ there exists a unique continuous vector function $a(\tau)$ such that $a(\tilde{\tau}) = \tilde{a}$. This function is continuously differentiable and

$$\left. \frac{\partial a(\tau)}{\partial \tau} \right|_{\tau = \tilde{\tau}} = -D^{-1}B.$$

An immediate calculation now gives

$$\left(\left. \frac{\partial^2}{\partial \tau_j \partial \tau_i} q(\tau, a(\tau), z) \right|_{\tau = \tilde{\tau}} \right)_{i,j=1}^{m-1} = E - B^T D^{-1} B.$$

For any fixed $z \in Z$, we have

$$\varphi(\tau, z) = \min_{a \in \mathfrak{A}} q(\tau, a, z) = q(\tau, a(\tau), z),$$

with τ from a vicinity of $\tilde{\tau} = \tilde{\tau}(z)$.

Differentiating this equality twice by τ, we obtain

$$J(\tilde{\tau}(z), z) = J(\tilde{\tau}, z) = E - B^T D^{-1} B.$$

∎

Proof of Theorem 2.4.3. Let $\tau(z)$ be an arbitrary $K - 1$ times continuously differentiable vector function in a vicinity of a point $z_{(0)}$, $z_{(0)} \in \mathbf{R}^r$, $\tau(z) = (\tau_1(z), \ldots, \tau_{m-1}(z))$. Consider the following auxiliary result.

Lemma 2.6.1 *Under condition (b) and with $p = 0$ and $l = 0$ the following equalities are valid:*

$$\frac{\partial^t}{\partial z_1^{s_1} \cdots \partial z_k^{s_k}} \left[g\left(\tau_{<I>}(z), z\right) - g\left(\tau(z), z\right) \right] |_{z=z_{(0)}} = 0,$$

for $k \geq 1$, $s \in S_t$, where $I = I_t$, $t = 1, 2, \ldots, K - 1$.

Proof of Lemma 2.6.1. At first, consider $k = 1$. Since

$$\frac{\partial}{\partial z} g\left(\tau(z), z\right) = \frac{\partial}{\partial \tau} g(\tau, z)|_{\tau=\tau(z)} \times \tau'(z) + \frac{\partial}{\partial z} g(\tau, z)|_{\tau=\tau(z)},$$

we obtain for $t = 1, \ldots, K - 1$:

$$\frac{\partial^t}{\partial z^t} g(\tau(z), z)|_{z=z_{(0)}}$$

$$= t! J_{(0)} \tau_{(t)} + \frac{\partial^t}{\partial z^t} g(\tau_{(0)}, z_{(0)}) + \cdots \qquad (2.38)$$

$$+ \sum_{i_1, \ldots, i_t=1}^{m} \frac{\partial^t}{\partial \tau_{i_1} \cdots \partial \tau_{i_t}} g(\tau_{(0)}, z_{(0)}) \tau_{i_1}(1) \cdots \tau_{i_t}(1) i_1! \cdots i_t!,$$

where the right-hand side depends only on $\tau_{(0)}, \ldots, \tau_{(t)}$ and does not depend on $\tau_{(t+1)}, \ldots$. Therefore,

$$\frac{\partial^t}{\partial z^t} g\left(\tau(z), z\right)|_{z=z_{(0)}} = \frac{\partial^t}{\partial z^t} g\left(\tau_{(t)}(z), z\right)|_{z=z_{(0)}}.$$

In the case $k > 1$, the proof is similar. ∎

Return to the proof of Theorem 2.4.3. Let $k = 1$ and $l = 0$. Note that on the right-hand side of (2.38), only the first term depends on $\tau_{(t)}$, as the other ones depend only on $\tau_{(s)}$, $s \leq t - 1$. Since $g(\tau^*(z), z) \equiv 0$ in a vicinity of $z_{(0)}$,

$$-\frac{\partial^t}{\partial z^t} g\left(\tau^*_{<t-1>}(z), z\right)|_{z=z_0} = t! J_{(0)} \tau^*_{(t)}.$$

For $k > 1$, $l \neq 0$, the proof is similar. ∎

Proof of Theorem 2.4.4. At first, consider $l = 0$. Note that

$$\left(g(\tau_{<I>}(z), z)\right)_{(s+u)} = \sum_{w+v=s+u} a_{(w)} \tilde{g}(\tau_{<I>}(z), z)_{(v)} \tag{2.39}$$

for any collection of indexes I,

$$\tau_{<I>}(z) = \sum_{s \in I} \tau_{(s)} (z - z_{(0)})^s.$$

For $w = u$, vector s is the only vector v such that $w + v = s + u$. Let $s \in \hat{S}_n$, $I = \hat{I}_n$. Note that for $w \neq u$, any vector v such that $w + v = s + u$ belongs to set $\hat{S}_t$, $t \leq n - 1$, from which it follows that the right-hand side of (2.39) has the form

$$a_{(u)} \tilde{g} \left(\tau^*_{<\hat{I}_n>}(z), z\right)_{(s)}.$$

It can be verified by direct calculation that $J_{(u)} = a_{(u)} \tilde{J}_{(0)}$. Therefore, Theorem 2.4.4 is valid at $l = 0$. For arbitrary l, its validity can be verified by direct calculation. ∎

2.6.2 Proof of Theorem 2.3.1

Consider a vector function $\tilde{\tau}(z) = (\tilde{\tau}_1(z), \ldots, \tilde{\tau}_{m-1}(z))^T$, $\tilde{\tau}(z) : Z \to R^{m-1}$ such that $\xi_{\tilde{\tau}}$ with $\tilde{\tau} = \tilde{\tau}(z)$ is a saturated locally D-optimal design at the point $\Theta^{0^T} = \left(\Theta_1^{0^T}, (q^{-1}(z))^T\right)$. This function should satisfy equation (2.10) and due to the Implicit Function Theorem (Gunning and Rossi, 1965) we need only to prove that the Jacobi matrix, J, is invertible. For this, it will do to prove that matrix B is of full rank. Suppose, oppositely, that it is not the case. Then there exists a vector $d \in R^{m-1}$, $d \neq 0$, such that $d^T B = 0$ and therefore

$$\sum_{s=2}^{m} \left[f_i(x_s^*) f_j'(x_s^*) + f_i'(x_s^*) f_j(x_s^*)\right] d_s = 0, \tag{2.40}$$

$i, j = 1, \ldots, m$, $(i, j) \neq (m, m)$, $x_s^* = \tilde{\tau}_{s-1}(z)$, $f_i(x) := f_i(x, z)$, $i = 1, \ldots, m$, $s = 1, \ldots, m - 1$.

Note that (2.40) holds also for $(i, j) = (m, m)$. In fact, since

$$\xi_{\tilde{\tau}} = \begin{pmatrix} x_1^* & \cdots & x_{m-1}^* & b \\ 1/m & \cdots & 1/m & 1/m \end{pmatrix}$$

is a saturated locally D-optimal design, we have

$$\frac{\partial}{\partial x_s} \det M(\xi_{\bar{\tau}}, z) = \sum_{i,j=1}^{m} (f_i(x_s^*) f_j(x_s^*))' d_{ij} = 0, \qquad (2.41)$$

where $d_{ij} = (M^{-1}(\xi_{\bar{\tau}}, z))_{i,j}$, $s = 1, \ldots, m-1$.

Multiplying (2.40) by d_s and summing the results, we obtain

$$\sum_{i,j=1}^{m} \left(\sum_{s=2}^{m} (f_i(x_s^*) f_j(x_s^*))' d_s \right) d_{ij} = 0.$$

Substituting (2.40) in the above equation, we obtain

$$\left(\sum_{s=2}^{m} (f_m^2(x_s^*))' d_s \right) d_{mm} = 0.$$

Since $(M(\xi_{\bar{\tau}}, z))^{-1}$ is a positive definite matrix,

$$d_{mm} = e_m^T (M(\xi_{\bar{\tau}}, z))^{-1} e_m \neq 0, \; e_, = (0, \ldots, 0, 1)^T,$$

and, thus, (2.40) holds for $(i, j) = (m, m)$.

Define a vector ν by the equality

$$\nu^T f(x) = \det \begin{pmatrix} f_1(x_1^*) & \cdots & f_m(x_1^*) \\ \cdots\cdots\cdots \\ f_1(x_{m-1}^*) & \cdots & f_m(x_{m-1}^*) \\ f_1(x) & \cdots & f_m(x) \end{pmatrix}.$$

Certainly, $\nu^T f(x_i^*) = 0$, $i = 1, \ldots, m-1$, and we obtain from (2.40) that

$$\sum_{s=2}^{m} \nu^T f'(x_s^*) f_j(x_s^*) d_s = 0, \; j = 1, \ldots, m.$$

Due to assumption A1, we have $q^T f'(x_s^*) \neq 0$, $s = 1, \ldots, m-1$. Therefore,

$$L_{(t)} \alpha = 0, \; t = 1, \ldots, m, \qquad (2.42)$$

where $\alpha = \left(d_s \nu^T f'(x_s^*) \right)_{s=1}^{m-1}$; $L_{(t)}$ is obtained from the matrix $(f_i(x_j^*))_{i,j=1}^{m,m-1}$ by rejecting the t-th line. It follows from (2.45) that $\det L_{(t)} = 0$, $t = 1, \ldots, m$, and it implies $\det (f_i(x_j^*))_{i,j=1}^{m} = 0$. However, the last equality is impossible.   ∎

Note that if $f_1(x), \ldots, f_{m-1}(x)$ generate a Chebyshev system on [a,b], then we need not use (2.41) and the points x^*i, $i = 2, \ldots, m$ need not to be support points of a locally D-optimal design in order for the matrix B be of full rank. This remark will be needed in the following for the consideration of MMEMS designs.

2.6.3   Proof of Theorem 2.2.3

Let us begin with the proof of part (I). A direct calculation shows that the matrix $J = J(\delta)$ is of the form

$$J = \begin{pmatrix} A & l \\ l^T & 0 \end{pmatrix},$$

where

$$A = \left(\frac{\partial^2}{\partial \tau_i \partial \tau_j} \Phi(u, \delta) \right)_{i,j=1}^{m-1} \bigg|_{u=\hat{u}(\delta)},$$

$$l = (l_1, \ldots, l_{m-1})^T,$$

$$l_i = \frac{R_1(\hat{\tau}_i)}{1 - \delta} - \frac{R_2(\hat{\tau}_i)}{1 + \delta}, \; i = 1, \ldots, m - 1,$$

$$R_s(\hat{\tau}_i) = \frac{\partial}{\partial \tau_i} \left(\det M(\xi_{\hat{\tau}}, \Lambda_{(s)}) \right)^{1/m}$$

$$= \left(\det M(\xi_{\hat{\tau}}, \Lambda_{(s)}) \right)^{1/m}$$

$$f^T(\hat{\tau}, \Lambda_{(s)}) M^{-1}(\xi_{\hat{\tau}}, \Lambda_{(s)}) f(\hat{\tau}, \Lambda_{(s)}),$$

$$s = 1, 2, \; \Lambda_{(1)} = (1 - \delta)c, \; \Lambda_{(2)} = (1 + \delta)c.$$

Let us prove that $l \neq (0, \ldots, 0)^T$. Suppose, oppositely, that $l = (0, \ldots, 0)^T$. With $u = \hat{u}$, from the definition of $\hat{u}$, we have

$$\frac{\partial}{\partial \alpha} \Phi(u, \delta) = 0, \; \frac{\partial}{\partial \tau_i} \Phi(u, \delta) = 0, \; i = 1, \ldots, m - 1$$

for $u = \hat{u}$. From the first equality we obtain

$$\frac{\det M(\xi_{\hat{\tau}}, \Lambda_{(1)})}{1 - \delta} = \frac{det M(\xi_{\hat{\tau}}, \Lambda_{(2)})}{1 + \delta}.$$

Due to other $m - 1$ equalities, we have

$$\frac{1}{1 + \delta} R_2(\hat{\tau}_i) + \alpha \left\{ \frac{R_1(\hat{\tau}_i)}{1 - \delta} - \frac{R_2(\hat{\tau}_i)}{1 + \delta} \right\} = 0, \; i = 1, \ldots, m - 1.$$

Now, it follows from the supposition $l = (0, \ldots, 0)^T$ that

$$\frac{\partial}{\partial \tau_i} \varphi(\tau, \Lambda_{(s)}) = 0, \; i = 1, \ldots, m, \; s = 1, 2$$

with $\tau = \hat{\tau}$.

A direct calculation shows that

$$\varphi(\tau, \Lambda_{(2)}) = \varphi \left(\frac{\tau}{1 + \delta} (1 - \delta), \Lambda_{(1)} \right) \frac{1 + \delta}{1 + \delta}.$$

Therefore,

$$\frac{\partial}{\partial \tau_i} \varphi(\hat{\tau}, \Lambda_{(1)}) = 0, \ i = 1, \dots, m-1,$$

$$\frac{\partial}{\partial \tau_i} \varphi(h\hat{\tau}, \Lambda_{(1)}) = 0, \ h = \frac{1+\delta}{1-\delta}, \ i = 1, \dots, m-1. \tag{2.43}$$

However, in Melas (1978) it was proved that the equation system (2.43) has a unique solution in the set V. The contradiction obtained proves that $l \neq (0, \dots, 0)^T$.

Let us now study the matrix A. Similar to the proof of Theorem 2.4.2, it can be proved that the matrix A has the form

$$A = E - \alpha B_{(1)}^T \mathcal{D}_{(1)}^{-1} B_{(1)} - (1-\alpha) B_{(2)}^T \mathcal{D}_{(2)}^{-1} B_{(2)},$$

where E is a diagonal matrix, $\mathcal{D}_{(1)}$ and $\mathcal{D}_{(2)}$ are positive definite, and $\alpha = \hat{\alpha}$.

Repeating the arguments from the proof of Theorem 2.3.1, obtain that the matrices $B_{(1)}$ and $B_{(2)})$ have full rank and $(E)_{ii} \leq 0, \ i = 1, \dots, m-1$. Therefore the matrix A is negative definite and invertible.

Now, we have

$$\det J = -l^T A l \neq 0.$$

Now, assertion (I) of Theorem 2.2.3 follows from Theorem 2.4.1.

Let us prove part (II). From the general equivalence theorem for max-imin efficient designs (see Dette, Haines and Imhof (2003) or Müller and Pazman (1998)) it follows that the MMEMS design $\xi_{\hat{\tau}}$ is MME design among all approximate designs if and only if the two following conditions are satisfied:

$$\hat{\alpha} f^T(x, \Lambda_{(1)}) M^{-1}(\xi_{\hat{\tau}}, \Lambda_{(1)}) f(x, \Lambda_{(1)})$$

$$+ (1-\hat{\alpha}) F^T(x, \Lambda_{(2)}) M^{-1}(\xi_{\hat{\tau}}, \Lambda_{(2)}) f(x, \Lambda_{(2)}) \leq m \tag{2.44}$$

with $x \geq 0$, where $\Lambda_{(1)} = (1-\delta)c$, $\Lambda_{(2)} = (1+\delta)c$, and $\hat{\tau} = \hat{\tau}(\delta)$, and

$$\min_{\Lambda \in \Omega(\delta)} (\tau, \Lambda) = \min_{0 \leq \alpha \leq 1} \alpha Q(\tau, \Lambda_{(1)}) + (1-\alpha) Q(\tau, \Lambda_{(2)}), \tag{2.45}$$

where

$$Q(\tau, \Lambda) = \varphi(\tau, \Lambda)/\varphi(\tau^*(\Lambda), \Lambda).$$

In a vicinity of $\Lambda = c$ let LDMS designs be locally D-optimal among all approximate designs. Then, due to the standard continuity arguments, inequality (2.44) holds for sufficiently small δ.

In order to prove (2.45), we will need the following auxiliary result.

Lemma 2.6.2 *Consider a general function* $\varphi : V \times \Omega \to R$, *where* $V \subset R^s$ *and* $\Omega \subset R^k$ *are open sets.*

Suppose that the following assumptions are satisfied:

(a1) The function φ is positive and twice continuously differentiable.

(a2) For any $\Lambda \in \Omega$, the equation

$$g(\tau, \Lambda) = 0,$$

where $g(\tau, \Lambda) = \left(\frac{\partial}{\partial \tau_1} \varphi(\tau, \Lambda), \ldots, \frac{\partial}{\partial \tau_3} \varphi(\tau, \Lambda)\right)^T$ possess a unique solution $\tau^* = \tau^*(\Lambda)$.

(a3) For all $\Lambda \in \Omega$, the matrix

$$K = B^T J(\Lambda) B,$$

where

$$J(\Lambda) = \left(\frac{\partial^2}{\partial \tau_i \partial \tau_j} \varphi(\tau, \Lambda)\right)_{i,j=1}^{s} \Big|_{tau=\tau^*(\Lambda)},$$

$$B = \left(\frac{\partial \tau_i(\Lambda)}{\partial \lambda_j}\right)_{i,j=1}^{sk}, \quad \lambda_j = (\Lambda)_j,$$

consists of negative elements.
Let $\Omega = \Omega(\delta)$, where

$$\Omega(\delta) = \left\{\Lambda \big| \Lambda = (\lambda_1, \ldots, \lambda_k), (1 - \delta)c_i \le \lambda_i \le (1 + \delta)c_i, \, i = 1, \ldots, k\right\},$$

$0 < \delta < 1$, *and*

$$Q(\tau, \Lambda) = \varphi(\tau, \Lambda)/\varphi(\tau^*(\Lambda), \Lambda).$$

Then for τ sufficiently close to $\tau^*(\Lambda)$

$$\min_{\Lambda \in \Omega(\delta)} Q(\tau, \Lambda) = \min_{0 \le \alpha \le 1} \alpha(Q(\tau, \Lambda_{(1)}) + (1 - \alpha)Q(\tau, \Lambda_{(2)})),$$

where $\Lambda_{(1)} = (1 - \delta)c$, $\Lambda_{(2)} = (1 + \delta)c$.

Proof of the lemma. The proof is similar to that of Proposition A1 in Dette, Melas and Pepelyshev (2003). A direct calculation shows that

$$\frac{\partial}{\partial \lambda_i} \varphi(\tau^*(\Lambda), \Lambda) = \frac{\partial}{\partial \lambda_i} \varphi(\tau, \Lambda) \Big|_{\tau=\tau^*(\Lambda)}$$
$$+ \sum_{j=1}^{s} \frac{\partial}{\partial \tau_j} \varphi(\tau, \Lambda) \Big|_{\tau=\tau^*(\Lambda)} \frac{\partial}{\partial \lambda_i}(\tau_j^*(\Lambda)) \quad , \tag{2.46}$$

$i = 1, 2, \ldots, k$.
    Due to assumption (a2), we obtain that

$$\frac{\partial^2}{\partial \lambda_i \partial \lambda_j} \varphi(\tau^*(\Lambda), \Lambda) = \frac{\partial^2}{\partial \lambda_i \partial \lambda_j} \varphi(\tau, \lambda) \Big|_{tau=\tau^*(\Lambda)}$$
$$+ \sum_{u,v=1}^{s} \frac{\partial^2}{\partial \tau_u \partial \tau_v} \varphi(\tau, \Lambda) \Big|_{rau=tau^*(\Lambda} \frac{\partial}{\partial \lambda_i} \tau_u^*(\Lambda) \frac{\partial}{\partial \lambda_j} \tau_v^*(\Lambda), \tag{2.47}$$

$i, j = 1, \ldots, k.$

Now, it is easy to calculate

$$\frac{\partial^2}{\partial \lambda_i \partial \lambda_j} Q(\tau, \Lambda) = Q_{1ij}(\tau, \Lambda) + Q_{2ij}(\tau, \Lambda),$$

where $Q_{1ij}(\tau, \Lambda)$ is such that $Q_{1ij}(\tau^*(\Lambda), \Lambda) = 0,$

$$Q_{2ij}(\tau, \Lambda) = \left[\left(\frac{\partial^2}{\partial \lambda_i \partial \lambda_j} \varphi(\tau, \Lambda) \right) H(\Lambda) - \varphi(\tau, \Lambda) \frac{\partial^2}{\partial \lambda_i \partial \lambda_j} H(\Lambda) \right] \Big/ H^2(\Lambda),$$

$$H(\lambda) = \varphi(\tau^*(\Lambda), \Lambda),$$

$i, j = 1, 2, \ldots, k.$

From (2.47) and the above formulas it follows that the matrix

$$\left(\frac{\partial^2}{\partial \lambda_1 \partial \lambda_j} Q(\tau, \Lambda) \right)_{i,j=1}^{k} \Big|_{\tau = \tau^*(\Lambda)}$$

is equal to K.

Since all elements of this matrix are negative by assumption (a3), the minimum of $Q(\tau, \Lambda)$ by $\Lambda \in \Omega(\delta)$ is achieved at the set $\{\Lambda_{(1)}, \Lambda_{(2)}\}$ for sufficiently small δ and with τ sufficiently close to $\tau^*(\Lambda)$. This is equivalent to the assertion of the lemma.   ∎

Now, let

$$\varphi(\tau, \Lambda) = \left(\frac{\det M(\xi_\tau, \Lambda)}{\prod_{i<j} (\lambda_i - \lambda_j)^8} \right)^{1/m}. \tag{2.48}$$

We can assume that the function in the points Λ with $\lambda_i = \lambda_j$ for some $i \neq j$ is codetermined with preserving the continuity (it can be done due to the discussion in Section 2.2).

Condition (a1) is evidently satisfied for this function and conditions (a2) and (a3) are proved in Melas (1978).

Thus, due to Lemma 2.6.2 for the function $\varphi(\tau, \Lambda)$ determined by (2.48) condition (2.45) takes place for sufficiently small δ. This completes the proof of part (II) of Theorem 2.2.3.

Chapter 3

Polynomial Models

Here, we study e_k-optimal designs (i.e., designs to be optimal for individual coefficients) and E-optimal designs in polynomial regression models.

On the basis of the functional approach, a full analytical solution is obtained for e_k-optimal designs on arbitrary intervals and E-optimal designs on symmetrical segments of arbitrary length.

3.1 Introduction

The present chapter is devoted to constructing and studying optimal designs for polynomial regression models on arbitrary intervals on the basis of the functional approach. Here, we will demonstrate some peculiarities of the approach for criteria different from D-criterion.

We will study e_k-optimal designs (i.e., designs optimal for estimating individual coefficients) as well as E-optimal designs. The results seem to be rather completed.

Polynomial models are the classical tool for approximating dependencies of different kinds to be experimentally studied. These models possess a great universality and applicability. They remain actual in spite of various alternatives, elaborated in the last years, such as rational models, spline functions, wavelet models, and exponential and other nonlinear (by parameters) regression models.

Basic results for optimal designs for the polynomial models can be found in Karlin and Studden (1966), Fedorov (1972), Pukelsheim (1993). A great number of articles are also devoted to such optimal designs (see, e.g., Studden (1980a), Pukeisheim and Torsney (1991), Dette (1993), Heiligers (1994), Huang, Chang, and Wong (1995), Lopez-Fedalgo and Rodriquez-Diez (2004) among many others). However, if we consider the three popular criteria: D-, c-, and E-optimality we discover that only for D-criterion, the problem is completely solved (under the condition that the observation errors are uncorrelated and their variances are either the same or described by one of the standard weight functions).

The presence of a simple and complete solution in this case caused by the invariance of the problem with respect to shift and scale transformations (the support points of the D-optimal designs can be received by the same shift and scale transformation). On the standard design interval $[-1, 1]$ the problem of D-optimal designs is reduced to that of maximization of the corresponding Wandermonde determinants. The last problem was solved in view of other purposes in the 19th century by Stilties (see, Karlin and Studden (1966, Chap. X)).

The problem of c-optimal designs, (i.e., the designs minimizing the variance of a given linear combination of parameters) was studied mainly for the case of estimating individual coefficients (Studden, 1968; Sahm, 1998; Dette, Melas, and Pepelyshev, 2004b). A full solution of the problem for this case was obtained in the last of these papers with the help of the functional approach. This solution will be presented in the next section.

The substantial difference of this case from the case of the D-criterion consists of the following. The e_k-optimal designs nontrivially depend on a parameter (equal to the ratio of the origin of the design interval to its length) and can be of different types. The application of the functional approach allows one to study this dependence.

E-Optimal designs have a similar peculiarity, but they are even more difficult for the study. An E-optimal design is defined as a design maximizing the minimal eigenvalue of the information matrix. Also, the form of E-optimal designs depends substantially on the multiplicity of this value (see Melas, 1995).

The first to consider E-optimal designs for the general polynomial model was Kovrigin (see Kovrigin (1980); unfortunately only an abstract of his work was published in English). In that paper E-optimal designs were constructed for the standard design interval $[-1, 1]$. This result in a slightly more general form (truncated E-optimal designs were also considered) was rediscovered in Pukelsheim and Studden (1993). In Heiligers (1991), E-optimal designs were constructed for rather small symmetrical intervals and segments lying on the nonnegative or nonpositive semiaxis. A generalization of these results for heteroscedastic models can be found in Dette (1993) and Heiligers (1994). In all of the cases, mentioned above, support points of E-optimal design are extremal points of the polynomial least deviated from zero on the corresponding interval. These points can be received from extremal points of the Chebyshev polynomial of the first kind by the shift and scale transformations. Such designs naturally can be called Chebyshev designs. However, Chebyshev designs are the solution of the problem only if its information matrix have multiplicity one or is a limit of such matrices (Melas, 1995). Even, in the case of rather large symmetrical intervals Chebyshev designs can not be E-optimal (Heiligers, 1991). Methods of constructing non-Chebyshev E-optimal designs were elaborated in Melas and

Krylova (1998) and Melas (2000) on the basis of the functional approach. Sections 3.3 and 3.4 are devoted to the elaboration of these methods.

3.2 Designs for Individual Coefficients

Individual coefficients in the polynomial model $\eta(t, \theta) = \sum_{i=0}^{m} \theta_i t^i$ are proportional to derivatives of the regression function at the point $t = 0$: $\theta_0 = \eta(0, \theta)$, $\theta_1 = \eta'(0, \theta)$, and $\theta_2 = \frac{1}{2}\eta''(0, \theta), \ldots, \theta_{m-1} = \eta^{(m-1)}(0, \theta)/(m-1)!$. At least the first and second derivative have the natural interpretation as velocity and acceleration. By this reason, estimating of individual coefficients can be of certain practical value. A more detailed discussion on this question can be found in Sahm (1998). The theoretical interest in the problem was caused, particularly, by the fact that the information matrix of optimal design can be singular.

The main peculiarity of the application of the functional approach in this case is the necessity of a joint study of the direct and the dual problems. Both problems are incorporated in the minimax representation of the criterion. In the present section we will introduce and study an equation system that determines solutions of both problems as implicit functions of the value $(a + b)/(b - a)$, where a and b are the bounds of the design interval. It will be shown that the function can be expanded into a Taylor series using general recurrent formulas introduced in Section 2.4. Also, it will be shown through examples that these expansions allow one to calculate optimal designs with a high precision*.

3.2.1 Statement of the problem

Consider the homoscedastic polynomial model

$$E[y(t)] = \beta^T f(t) = \sum_{i=0}^{d} \beta_i t^i,$$

$$\text{Var}[y|t] = \sigma^2 > 0,$$

where the variable t belongs to the compact design interval $[a, b]$ ($-\infty < a < b < \infty$), $\beta = (\beta_1, \ldots, \beta_n)^T$ is the vector of unknown parameters, and $f(t) = (1, t, \ldots, t^d)^T$ – the vector of basic regression functions; results of different observations assumed uncorrelated.

As in the previous chapter, we will call (approximate) *experimental design* a discrete probability measure

$$\xi = \begin{pmatrix} t_1 & \cdots & t_n \\ \omega_1 & \cdots & \omega_n \end{pmatrix},$$

*In this section materials (theorems, tables, and figures) are taken from Dette H., Melas, V.B., Pepelyshev, A.N. (2004b). Optimal designs for estimating individual coefficients in polynomial regression — a functional approach. *J. Statist. Plan. Inference*, **118**, 201–219. ©2002 Elsevier B.V. with permission of Elsevier Publisher.

where $t_i \neq t_j$ $(i \neq j)$ are observation points and $\omega_i > 0$, $\sum_{i=1}^{m} \omega_i = 1$ are weight coefficients. As in Chapters 1 and 2, by $M(\xi)$ we will denote the information matrix

$$M(\xi) = \sum_{i=1}^{n} f(t_i) f^T(t_i) \omega_i.$$

As the criterion of optimality we will use the value

$$\Phi_k(\xi) = \begin{cases} e_k^T M^-(\xi) e_k, & e_k \in \text{range } M(\xi) \\ \infty, & \text{otherwise,} \end{cases}$$

$k = 1, \ldots, m$, where A^- denotes a generalized inverse for A and $e_1 = (1, 0, \ldots, 0)^T$, $\ldots$, $e_m = (0, 0, \ldots, 0, 1)^T$.

A design will be called e_k-optimal if it minimizes $\Phi_k(\xi)$ in the class of all approximate designs. Note that $\Phi_k(\xi)$ equals the variance of the best linear unbiased estimator of β_k.

3.2.2 e_k-Optimal designs

In this section, we briefly review the known results about e_k-optimal designs that form the basis for our analytic approach in the following section. Because the cases $k = 0$ (estimation of the intercept) and $k = d$ (estimation of the highest coefficient) are well known (see, e.g., Sahm (1998) or Studden (1980a)), we restrict ourselves to the case $1 \leq k \leq d - 1$. Sahm (1998) introduced the sets

$$A_i = (-\nu_{d-k+1-i}, \nu_{i+1}), \quad i = 0, \ldots, d-k,$$
$$B_{1,i} = -B_{2,i} = [\nu_i, \rho_i], \quad i = 1, \ldots, d-k, \tag{3.1}$$
$$C_i = (\rho_i, -\rho_{d-k+1-i}), \quad i = 1, \ldots, d-k,$$

where $\nu_{d-k+1} = \infty$ and $\nu_1 < \nu_2 < \cdots < \nu_{d-k}$ are the roots of the k-th derivative of the polynomial

$$(x+1)U_{d-1}(x) \tag{3.2}$$

and $U_j(x) = \sin((j+1)\arccos x)/\sin(\arccos x)$ is the j-th Chebyshev polynomial of the second kind. The points ρ_i are obtained from these roots via the transformation

$$\rho_i = \nu_i + (1 + \nu_i) \frac{1 - \cos(\pi/d)}{1 + \cos(\pi/d)}.$$

Note that the union of these sets defines a partition of the real axis and Sahm (1998) proved that the location of the parameter $s^* = \frac{b+a}{b-a}$ determines the structure of the optimal design as follows. If

$$s^* \in \bigcup_{i=0}^{d-k} A_i,$$

the e_k-optimal design is supported at $d+1$ points, including the boundary points a and b. If

$$s^* \in \bigcup_{i=1}^{d-k} B_{1,i},$$

the optimal design for estimating the parameter β_k is supported at d points, including the boundary point a, and the case

$$s^* \in \bigcup_{i=1}^{d-k} B_{2,i}$$

is essentially obtained by symmetry arguments interchanging the role of a and b. In these cases, the e_k-optimal design can be described explicitly in terms of transformed Chebyshev points $t_j = \cos(\pi j/d)$ and we refer to Sahm (1998) for more details. In the remaining case,

$$s^* \in \bigcup_{i=1}^{d-k} C_i, \tag{3.3}$$

the situation is substantially more difficult. Here, the design is supported at d points including both boundary points of the design space but an explicit representation of the weights and support points is not available. Sahm (1998) characterized the solution for this case by a constrained optimization problem, which is difficult to use for the numerical construction of the optimal design. Additionally, he proved the existence of points

$$\mu_i \in C_i, \qquad i = 1, \ldots, d-k, \tag{3.4}$$

for which the solution of the design problem can be found explicitly. The points μ_i are the zeros of the k-th derivative of the polynomial

$$(x^2 - 1)U_{d-2}(x), \tag{3.5}$$

and for $s^* = \mu_i$, the e_k-optimal design is obtained as the optimal design for estimating β_k in a polynomial regression of degree $d-1$, where the case $s^* \in \cup_{i=0}^{d-k-1} A_i$ is applicable (see Section 3.2.3 for more details). Further, we will propose an analytic approach that allows the (numerical) determination in all cases specified by (3.3) and, therefore, closes the final gap in the solution of the e_k-optimal design problem on arbitrary intervals.

3.2.3 Analytical properties of e_k-optimal designs

Throughout this section we restrict ourselves to the (unsolved) case (3.3). In this case, as it was mentioned above, the optimal design is of the form

$$\xi_k^* = \begin{pmatrix} a, t_2^*, \ldots, t_{d-1}^*, b \\ \omega_1^*, \omega_2^*, \ldots, \omega_{d-1}^*, \omega_d^* \end{pmatrix}. \tag{3.6}$$

If a is fixed and we vary b such that (2.3) is satisfied, the weights and support points in (3.1) are functions of the right boundary point b, (i.e., $t_j^* = t_j^*(b)$, $j = 2, \ldots, d - 1$, $w_j^* = w_j^*(b)$,and $j = 1, \ldots, d$). We collect the information given by the e_k-optimal design ξ_k^* in the vector

$$\tau^* = \tau^*(b) = (t_2^*(b), \ldots, t_{d-1}^*(b), w_2^*(b), \ldots, w_d^*(b)) \qquad (3.7)$$

and note that this function is well defined due to the uniqueness of the e_k-optimal design ξ_k^* in (3.1) for $1 \leq k \leq d$ (see Sahm (1998)). Note that formally the optimality criterion (1.3) could be considered as a function of nontrivial weights and support points

$$\tau = (t_2, \ldots, t_{d-1}, w_2, \ldots, w_d), \qquad (3.8)$$

where the points t_i and w_i correspond to the support points and weights of a design of the form (3.1), and the optimal design is implicitly determined as a solution of the equations

$$\frac{\partial}{\partial \tau} \Phi_k = 0.$$

However, a direct differentiation of the optimality criterion with respect to support points and weights seems to be intractable due to the singularity of the corresponding information matrix of the d-point design. In order to circumvent this problem, we will relate the design problem to a dual extremal problem for polynomials. This duality is used to derive a necessary and sufficient condition for the parameters of the design and the coefficients of the extremal polynomial by differentiating an appropriate function. We begin with a slightly different formulation of the equivalence theorem for e_k-optimal designs than is usually stated in the literature (for a proof of the following lemma, see, e.g., Pukelsheim (1993)).

Lemma 3.2.1 *Let $f_k(t) = (1, t, \ldots, t^{k-1}, t^{k+1}, \ldots, t^d)^T$ denote the vector obtained from $f(t) = (1, t, \ldots, t^d)^T$ by omitting the monomial t^k. A design ξ_k^* is e_k-optimal on the interval $[a, b]$ if and only if there exist a positive number h_k and a vector $q^* \in \mathbb{R}^d$ such that the polynomial $\rho_k(t) = t^k - f_k^T(t)q^*$ satisfies the following conditions:*

(1) $h_k \rho_k^2(t) \leq 1 \quad \forall\, t \in [a, b]$,

(2) supp $(\xi_k^*) \subset \{t \in [a, b] \mid h_k \rho_k^2(t) = 1\}$,

(3) $\int \rho_k(t) f_k(t) d\xi_k^*(t) = 0 \in \mathbb{R}^d$.

Moreover, in this case, $h_k = \Phi_k(\xi_k^*)$. The polynomial $\rho_k(t)$ is called an extremal polynomial.

Throughout this section let

$$T = \left\{ (t_2, \ldots, t_{d-1}, \omega_2, \ldots, \omega_d)^T \mid a < t_1 < \cdots < t_{d-1} < b; \right.$$
$$\left. \omega_i > 0; \sum_{j=2}^{d} \omega_j < 1 \right\} \tag{3.9}$$

define for any $\tau \in T$ the design ξ_τ by

$$\xi_\tau = \begin{pmatrix} a & t_1 & \ldots & t_{d-1} & b \\ \omega_1 & \omega 2 & \ldots & \omega_{d-1} & \omega_d \end{pmatrix}, \tag{3.10}$$

where $\omega_1 = 1 - \sum_{j=2}^{d} \omega_j$, and a vector $d_q \in \mathbb{R}^{d+1}$ for $q \in \mathbb{R}^d$ by the equation

$$f^T(t) d_q = t^k - f_k^T(t) q \quad (q \in \mathbb{R}^d). \tag{3.11}$$

For any $\tau \in T$, we will consider the function

$$\varphi(q, \tau, b) = d_q^T M(\xi_\tau) d_q. \tag{3.12}$$

It is well known (see Karlin and Studden (1966) or Pukelsheim (1993)) that the optimal design ξ_{τ^*} satisfies a saddle point characterization; that is,

$$\varphi^{-1}(q^*, \tau^*, b) = \min_{\tau \in T} \max_{q \in \mathbb{R}^d} \varphi^{-1}(q, \tau, b)$$
$$= \max_{q \in \mathbb{R}^d} \min_{\tau \in T} \varphi^{-1}(q, \tau, b). \tag{3.13}$$

Here, the vector q^* is the optimal solution of the extremal problem

$$\inf_{q \in \mathbb{R}^d} \sup_{x \in [a,b]} |x^k - f_k^T(x) q|, \tag{3.14}$$

and the saddle point is unique, because the extremal polynomial $\rho_k(x) = x^k - f_k^T(x) q^*$ must attain its extremal values at the support points of the e_k-optimal design ξ_k^*, which is unique whenever $1 \leq k \leq d$. Note that formally the minimum has to be taken over the set of all vectors $\tau \in T$ such that e_k is estimable by the design ξ_τ, i.e., $e_k \in \text{Range}(M(\xi_\tau))$. However, it is straightforward to see that in the case $e_k \notin \text{Range}(M(\xi_\tau))$, we have

$$\max_{q \in \mathbb{R}^d} \varphi^{-1}(q, \tau, b) = \infty$$

(see also Studden (1968)). Consequently, the optimization over the slightly larger set T in (3.9) will yield a solution τ^*, q^* such that e_k is estimable by the design ξ_{τ^*}, even if this restriction is not incorporated in the definition of the set T. The following result is an immediate consequence of this discussion and its proof is therefore omitted.

Lemma 3.2.2 *The design ξ_{τ^*} is e_k-optimal and the vector q^* corresponds to the solution of the generalized Chebyshev problem (3.14) if and only if the point $(q^*, \tau^*) \in \mathbb{R}^d \times T$ is the unique solution of the system*

$$\frac{\partial}{\partial \tau} \varphi(q, \tau, b) = 0,$$
$$\frac{\partial}{\partial q} \varphi(q, \tau, b) = 0 \tag{3.15}$$

in the set of all pairs $(q, \tau) \in \mathbb{R}^d \times T$ such that e_k is estimable by the design ξ_τ and such that

$$|d_q^T f(t)|^2 \;=\; |t^k - q^T f_k(t)|^2 \;\leq\; d_q^T M(\xi_\tau) d_q \qquad \text{for all } t \in [a, b] .$$

Here, $\frac{\partial}{\partial \tau} \varphi$ and $\frac{\partial}{\partial q} \varphi$ denote the gradient of φ with respect to $\tau \in T$ and $q \in \mathbb{R}^d$, respectively.

Lemma 3.2.2 determines a vector differential equation, which implicitly determines τ^*, q^* as a vector-valued function of the boundary point b such that (3.3) is satisfied (where the left boundary of the design space has been fixed). In the following discussion, we will show that the Jacobian matrix of equation (3.15) is nonsingular, which allows the application of the Implicit Function Theorem to study the functions $\tau^*(b)$ and $q^*(b)$ as analytic functions of the right boundary b such that (2.3) is satisfied. To this end, define

$$\Theta = (\Theta_1, \ldots \Theta_{3(d-1)}) = (q^T, \tau^T),$$
$$\Theta^*(b) = (q^{*T}(b), \tau^{*T}(b)) \tag{3.16}$$

as the vector containing the parameters of the e_k-optimal design and the coefficients of the solution of the corresponding Chebyshev problem and

$$\bar{\varphi}(\Theta, b) = \varphi(q, \tau, b);$$

then the basic equation (3.15) can be rewritten as

$$\frac{\partial}{\partial \Theta} \bar{\varphi}(\Theta, b) = 0 \in \mathbb{R}^{3(d-1)}. \tag{3.17}$$

Theorem 3.2.1 *For any fixed $a \in \mathbb{R}$, define a function $s : (a, \infty) \to \mathbb{R}$ by*

$$s(b) = \frac{(a + b)}{(a - b)}$$

and $B_i = s^{-1}(C_i)$; then the components of the function

$$\Theta^* : \begin{cases} \bigcup_{i=1}^{d-k} B_i & \to \quad \mathbb{R}^{3(d-1)} \\ b & \to \quad \Theta^*(b) \end{cases}$$

are real analytic functions. Moreover, the vector function Θ^* is a solution of the system

$$G(\Theta(b), b) \cdot \Theta'(b) = Q(\Theta(b), b) \tag{3.18}$$

with initial conditions

$$\Theta(b_0) = \Theta^*(b_0), \tag{3.19}$$

where b_0 is any arbitrary point such that (3.3) is satisfied for $s_0 = s(b_0)$ and the functions G and Q are defined by

$$G(\Theta, b) = \left(\frac{\partial^2}{\partial \Theta_i \partial \Theta_j} \bar{\varphi}(\Theta, b) \right)_{i,j=1}^{3(d-1)}, \tag{3.20}$$

$$Q(\Theta, b) = \left(\frac{\partial^2}{\partial b \partial \Theta_i} \bar{\varphi}(\Theta, b) \right)_{i=1}^{3(d-1)}. \tag{3.21}$$

Proof. We will prove that the Jacobi matrix

$$J(b) = G(\Theta^*(b), b) \in \mathbb{R}^{3(d-1) \times 3(d-1)} \tag{3.22}$$

is nonsingular. The assertion of Theorem 3.2.1 then follows by a straightforward application of the Implicit Function Theorem (see, e. g., Gunning and Rossi (1965)). For this Jacobi matrix, we obtain the representation

$$J = J(b) = 2 \begin{pmatrix} D & B_1^T & B_2^T \\ B_1 & E & 0 \\ B_2 & 0 & 0 \end{pmatrix}_-, \tag{3.23}$$

where A_- denotes the $3(d-1) \times 3(d-1)$ matrix obtained from $A \in \mathbb{R}^{(3d-2) \times (3d-2)}$ by deleting the $(k+1)$-st row and $(k+1)$-st column. The matrices $D, B_1, B_2,$ and E in (3.23) are defined as follows ($t_1^* = a, t_d^* = b$):

$$D = M(\xi_{\tau^*}) \in \mathbb{R}^{d+1 \times (d+1)},$$

$$B_1^T = \left(w_2^* f'(t_2^*) \cdot d_{q^*}^T f(t_2^*), \ldots, w_{d-1}^* f'(t_{d-1}^*) \cdot d_{q^*}^T f(t_{d-1}^*) \right)$$

$$= c \left(w_2^* f(t_2^*), -w_3^* f(t_3^*), \ldots, (-1)^{d-1} w_{d-1}^* f'(t_{d-1}^*) \right),$$

$$B_2^T = \left(d_{q^*}^T f(t_2^*) \cdot \{ f(t_2^*) - f(t_1^*) \}, \ldots, d_{q^*}^T f(t_d^*) \cdot \{ f(t_d^*) - f(t_1^*) \} \right)$$

$$= c \left(f(t_2^*) - f(t_1^*), (-1) \{ f(t_3^*) - f(t_1^*) \}, \ldots, (-1)^d \{ f(t_d^*) - f(t_1^*) \} \right),$$

$$E = \mathrm{diag} \left(w_2^* d_{q^*}^T f(t_2^*) \cdot d_{q^*}^T f''(t_2^*), \ldots, w_{d-1}^* d_q^* f(t_{d-1}^*) \cdot d_q^* f''(t_{d-1}^*) \right)$$

$$= \mathrm{diag} \left(w_2^* \rho''(t_2^*) \rho(t_2^*), w_3^* \rho''(t_3^*) \rho(t_3^*), \ldots, w_{d-1}^* \rho''(t_{d-1}^*) \rho(t_{d-1}^*) \right), \tag{3.24}$$

where $c = \Delta \Phi^{-1/2}(\xi_{\tau^*})$, $\Delta \in \{-1, 1\}$ is a fixed constant, and the polynomial ρ is defined by $\rho(t) = d_{q^*}^T f(t)$ (all other entries in the matrix J are 0). The Jacobi matrix J in (3.23) is essentially obtained by direct differentiation and the properties of the extremal polynomial $\rho(t) = d_{q^*}^T f(t)$ presented in Lemma 3.2.1. For example, consider the calculation of the matrix B_1^T and let $I_- \in \mathbb{R}^{d+1 \times d}$ denote the identity matrix with the deleted $(k+1)$-st column. We obtain, by straightforward calculation,

$$\frac{\partial^2 \varphi}{\partial t \partial q} = \frac{\partial}{\partial t} 2 I_-^T M(\xi_\tau) \, dq$$

$$= 2 I_-^T \left(w_{j+1} d_q^T f'(t_{j+1}) \cdot f(t_{j+1}) + w_{j+1} d_q^T f(t_{j+1}) \cdot f'(t_{j+1}) \right)_{j=1}^{d-2}.$$

Now, for $q = q^*$, we have $\rho(t_j^*) = d_{q^*}^T f(t_j^*) = \Delta(-1)^j \Phi^{-1/2}(\xi_{\tau^*})$ ($j = 2, \ldots, d$) for some $\Delta \in \{-1, 1\}$. This follows from Lemma 3.2.1, which shows that ρ is equioscillating and implies $\rho'(t_j^*) = d_{q^*}^T f'(t_j^*) = 0$. Consequently, we obtain

$$B_1^T = \frac{2 \cdot \Delta}{\Phi_k^{1/2}(\xi_{\tau^*})} \left((-1)^{j+1} w_{j+1}^* f'(t_{j+1}^*) \right)_{j=1}^{d-2},$$

which proves the representation of the block B_1^T in (3.23). The other cases are treated similarly and left to the reader.

On the basis of the representation (3.23), the proof of the nonsingularity of the Jacobi matrix $J(b)$ is straightforward. Note that the matrix D_- is non-negative definite, because it is obtained from the non-negative definite matrix $M(\xi_{\tau^*})$ by deleting the $(k+1)$-st row and column. Similarly, the matrix E defined in (3.24) is negative definite, which follows because it essentially contains the second derivatives $\rho''(t_i)$ ($i = 2, \ldots, d-1$) of the extremal polynomial $\rho(t) = t^k - f_k^T(t) q^*$ specified in Lemma 3.2.1. To be precise, we note that the results of Sahm (1998) show that for the case $b = \mu_i$, this polynomial is of degree $d - 1$ whereas in the case $b \in C_i \setminus \{\mu_i\}$, the polynomial is of degree d with one extremum outside the interval $[a, b]$. A careful counting of the multiplicities of the zeros of the polynomial $\Phi(\xi_{\tau^*}) \rho^2(t) - 1$ shows

$$\rho''(t_i) = d_{q^*}^T f''(t_i) \neq 0, \quad i = 2, \ldots, d-1. \tag{3.25}$$

Moreover, by the oscillating property of the extremal polynomial, the second derivative must alternate in sign, yielding $\rho''(t_i) \rho(t_i) < 0$ ($i = 2, \ldots, d-1$), and the definition of the matrix E in (3.24) shows that this matrix has negative diagonal elements.

From these auxiliary results, it follows that the matrix

$$D_- - \tilde{B}_1^T E^{-1} \tilde{B}_1$$

is positive definite, where $\tilde{B}_1^T$ denotes the matrix obtained from B_1^T by deleting the $(k+1)$-st row. Similarly, let $\tilde{B}_2^T$ be obtained from B_2^T by deleting the $(k+1)$-st row, then it follows by the Frobenius formula (see, e.g., Fedorov (1972)) and the representation (3.23) that $\det J(b)$ is equal to

$$\det \begin{pmatrix} D_- & \tilde{B}_1^T & \tilde{B}_2^T \\ \tilde{B}_1 & E & 0 \\ \tilde{B}_2 & 0 & 0 \end{pmatrix}$$

$$= -\det \begin{pmatrix} D_- & \tilde{B}_1^T \\ \tilde{B}_1 & E \end{pmatrix} \cdot \det\left\{ (\tilde{B}_2 \mid 0) \begin{pmatrix} D_- & \tilde{B}_1^T \\ \tilde{B}_1 & E \end{pmatrix}^{-1} \begin{pmatrix} \tilde{B}_2^T \\ 0 \end{pmatrix} \right\}$$

$$= -\det E \cdot \det(D_- - \tilde{B}_1^T E^{-1} \tilde{B}_1) \cdot \det\{ \tilde{B}_2 (D_- - \tilde{B}_1^T E^{-1} \tilde{B}_1)^{-1} \tilde{B}_2^T \}.$$
$$(3.26)$$

Now, the matrix $\tilde{B}_2^T$ is of rank $d-1$ (because of the Chebyshev property of the polynomials $1, x, \ldots, x^d$) and the matrix $D_- - \tilde{B}_1^T E^{-1} \tilde{B}_1$ is positive definite by the preceding discussion. Consequently, all determinants in (3.26) are different from zero, which proves the nonsingularity of the Jacobi matrix $J(b)$. ∎

Theorem 3.2.2 *Let $b_0 \in B_i = s^{-1}(C_i)$ for some $i = 1, \ldots, d-k$, and Θ^* be the function defined in Theorem 3.2.1; then the coefficients in the Taylor expansion*

$$\Theta^*(b) = \Theta^*(b_0) + \sum_{j=1}^{\infty} \Theta^*(j, b_0)(b - b_0)^j$$

in a neighborhood of the point b_0 can be obtained recursively by the formulas

$$\Theta^*(n+1, b_0) = -\frac{1}{(n+1)!} J^{-1}(b_0) \left(\frac{d}{db} \right)^{n+1} g(\Theta^*_{(n)}(b), b) \Big|_{b=b_0}, \qquad (3.27)$$

where $n = 0, 1, 2, \ldots$, the polynomial $\Theta^*_{(s)}$ of degree s^* is defined by

$$\Theta^*_{(s)}(b) = \Theta^*(b_0) + \sum_{j=1}^{s} \Theta^*(j, b_0)(b - b_0)^j,$$

and the function g is given by

$$g(\tilde{\Theta}, b) = \frac{\partial}{\partial \Theta} \bar{\varphi}(\Theta, b) \big|_{\Theta = \tilde{\Theta}}. \qquad (3.28)$$

This theorem is a particular case of Theorem 2.4.3.

In general, Theorems 3.2.1 and 3.2.2 show that for any b_0 such that (3.3) is satisfied, the functions

$$t_j^* : b \to t_j^*(b), \quad j = 2, \ldots, d-1,$$
$$w_j^* : b \to w_j^*(b), \quad j = 2, \ldots, d,$$
$$q_j^* : b \to q_j^*(b), \quad j = 1, \ldots, d$$

(here q_j^* denotes the j-th component of the vector of coefficients q^* of the extremal polynomial) can be expanded into a Taylor series in a neighborhood of the point b_0. The coefficients of these expansions can be directly computed from the recurrence formulas (3.27) and, therefore, the remaining case in the optimal design problem for estimating the individual coefficients in a polynomial regression on an arbitrary interval can be easily solved numerically if we are able to find a point b_0 such that (3.2) is satisfied and for which the solution of the e_k-optimal design problem is known. But such a point has been identified by Sahm (1998), who showed that there exist $d - k$ points

$$\mu_i = s(b_i) = \frac{a + b_i}{a - b_i} \in C_i, \quad i = 1, \ldots, d - k,$$

such that the optimal design for estimating the parameter β_k is supported at the d Chebyshev points

$$t_j^*(b_i) = \frac{b_i - a}{2} \left\{ \cos\left(\frac{j - 1}{d - 1} \pi \right) - \mu_i \right\}, \quad j = 1, \ldots, d, \tag{3.29}$$

with weights

$$w_j^*(b_i) = 2\gamma_{j-1} \frac{\sum_{\ell=0}^{k}(d - \ell - 1)\gamma_\ell \cos(\frac{(j-1)\ell}{d-1}\pi) C_{k-\ell}^{(d-k-1)}(\mu_i)}{C_k^{(d-k-1)}(\mu_i)}, \tag{3.30}$$

$j = 1, \ldots, d$, where $\gamma_0 = \gamma_{d-1} = 1/2(d-1), \gamma_j = 1/(d-1), \quad j = 1, \ldots, d-2$, and $C_n^{(\lambda)}(x)$ denotes the n-th ultraspherical polynomial (see, e.g., Szegö (1959)). Moreover, the points μ_i (or, equivalently, b_i) are determined as the zeros of the polynomial in (3.5). For this reason, we are able to find in each interval $s^{-1}(C_i)$ a Taylor expansion for the weights and support points of the e_k-optimal design, which is based on the location $b_i = s^{-1}(\mu_i)$ ($i = 1, \ldots, d - k$). This technique provides a numerical solution for the open design problem and will be illustrated in the following subsection. We finally remark that the Taylor expansion is not necessarily valid on the whole interval B_i. From a theoretical point of view, it is possible that further expansions have to be performed in order to cover the whole range of B_i. However, in all of our numerical examples only one expansion was sufficient (although we cannot prove this in general).

3.2.4   A numerical example

Consider the case $d = 4$. We are interested in the estimation of the coefficient of β_1, β_2, and β_3 in the case that cannot be treated by the results of Sahm (1998). We concentrate on the case $a = -1$ and vary the parameter b, which corresponds to the situation considered in Section 3.2.3. The general case can be reduced to this case by an appropriate scaling of the symmetry parameter $s^* = s(b) = (a + b)/(a - b)$.

(a) If $k = 3$, we have one critical interval for the symmetry parameter s^* given by

$$C_1 = (\rho_1, -\rho_1) = (-0.1213, 0.1213),$$

and $\mu_1 = 0$, which corresponds in the b scale to the interval

$$B_1 = s^{-1}(C_1) = (0.7836, 1.2761) \tag{3.31}$$

and $b_1 = s^{-1}(0) = 1$. The first six coefficients in the Taylor expansion for the coefficients of the extremal polynomial, interior support points and weights are listed in Table 3.1 and are calculated by the procedure described at the end of Section 3.2.3 using the recursive relation (3.27). For example, if $b = 1.2$, we obtain for the e_3-optimal design on the interval $[-1, 1.2]$,

$$\xi_3^* = \begin{pmatrix} -1 & -0.595 & 0.395 & 1.2 \\ 0.239 & 0.412 & 0.261 & 0.088 \end{pmatrix}$$

and the extremal polynomial is given by

$$\rho_3(t) = t^3 - 0.654t^4 + 0.685t^2 - 0.808t - 0.134.$$

Similarly, the optimal design for estimating the coefficient of x^3 on the interval $[-1, 0.9]$ is given by

$$\xi_3^* = \begin{pmatrix} -1 & -0.406 & 0.506 & 0.9 \\ 0.121 & 0.290 & 0.379 & 0.210 \end{pmatrix}$$

and the extremal polynomial is

$$\rho_3(t) = t^3 + 0.426t^4 - 0.333t^2 - 0.650t + 0.052.$$

The corresponding weights and support points are displayed in Figures 3.1 and 3.2 for a parameter b running continuously through the interval $(-1, \infty)$. Note that this covers all cases in (3.1) and that the structure of the design changes four times.

In order to investigate the efficiency of the determined designs, we use a lower bound for the efficiency derived in Dette (1996). For the situation of e_k-optimality, this bound is given by

$$\begin{aligned} \text{eff}(\xi_3^*) &= \frac{(e_k^T M^-(\xi_3^*)e_k)^{-1}}{\sup_\xi (e_k^T M^-(\xi)e_k)^{-1}} \\ &\geq \frac{1}{\sup_{t \in [-1,b]} \sqrt{h_3^*|t^3 - f_3^T(t)q^*|}} \end{aligned} \tag{3.32}$$

with $h_3^* = e_3^T M^-(\xi_3^*)e_3$. We obtain for the cases $b = 1.2$ and $b = 0.9$ the lower bounds 0.975 and 0.997, respectively, which demonstrates that the designs determined numerically by the Taylor expansion are nearly the optimal ones.

Table 3.1: First six coefficients of the Taylor expansions of the coefficients of the extremal polynomial $t^3 + q_4^* t^4 + q_3^* t^2 + q_2^* t + q_1^*$ the interior support points t_2^* and t_3^* and weights w_2^*, w_3^*, and w_4^* of the e_3-optimal design in a polynomial regression of degree 4 on the interval $[-1, b]$, where $b \in (0.8836, 1.2761)$. The center of the expansion is $b_1 = 1$.

	0	1	2	3	4	5
q_1^*	0.0000	−0.6250	−0.6250	5.4688	0.0000	−87.2500
q_2^*	−0.7500	−0.7500	2.5625	0.0000	−8.0000	8.0000
q_3^*	0.0000	3.5000	0.0000	−18.0000	18.0000	314.5000
q_4^*	0.0000	−4.0000	4.0000	13.0000	−30.0000	−217.2500
t_2^*	−0.5000	−0.7500	2.0000	−1.0000	−15.5000	23.7500
t_3^*	0.5000	−0.2500	−2.0000	1.0000	15.5000	−23.7500
w_2^*	0.3333	0.4444	−0.0741	−2.2099	1.4053	26.1485
w_3^*	0.3333	−0.4444	0.3704	1.9136	−5.0021	−18.6588
w_4^*	0.1667	−0.4444	0.0741	2.2099	−1.4053	−26.1485

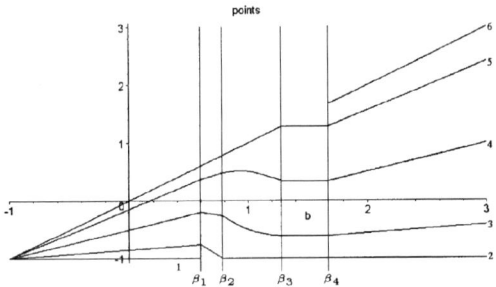

Figure 3.1: Support points of e_3-optimal designs for the cubic regression model

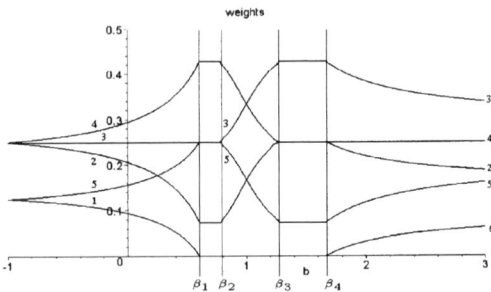

Figure 3.2: Weight coefficients of e_3-optimal designs for the cubic regression model

(b) If $k = 2$, we have two critical intervals for the symmetry parameter s^* given by

$$C_1 = (-0.5687, -0.1213), \quad C_2 = (0.1213, 0.5687),$$

and the specific points (where the solution is known) are $\mu_1 = -0.4564$ and $\mu_2 = 0.4564$. This corresponds to the intervals

$$B_1 = s^{-1}(C_1) = (1.9677, 3.6374),$$

$$B_2 = s^{-1}(C_2) = (0.2749, 0.5082)$$

(3.33)

and the points $b_1 = s^{-1}(\mu_1) = 2.6794$ and $b_2 = s^{-1}(\mu_2) = 0.3732$ for the parameter b, which can be used for the Taylor expansion in the respective intervals. The corresponding support points and weights are depicted in Figures 3.3 and 3.4, where the parameter b runs continuously through the interval $(-1, \infty)$. Note that this covers all cases in (3.1) and that the structure of the design changes eight times.

For example, if $b = 2.2$, the optimal design for estimating the parameter β_2 on the interval $[-1, 2.2]$ is

$$\xi_2^* = \begin{pmatrix} -1 & 0.109 & 1.582 & 2.2 \\ 0.217 & 0.415 & 0.283 & 0.085 \end{pmatrix}$$

and the extremal polynomial is given by

$$p_2(t) = -0.498 - 0.217t + t^2 + 0.061t^3 - 0.194t^4 .$$

The e_2-optimal design on the interval $[-1, 0.45]$ is given by

$$\xi_2^* = \begin{pmatrix} -1 & -0.716 & -0.044 & 0.45 \\ 0.083 & 0.280 & 0.417 & 0.220 \end{pmatrix}$$

and the extremal polynomial is given by

$$p_2(t) = -0.103 + 0.090t + t^2 + 0.010t^3 - 0.902t^4$$

(note that for $b = 0.45$, we have $s^* = 11/29$, which corresponds to the case $b \in B_2$). The corresponding lower bounds in (3.32) for the efficiencies are given by 0.9906 in the case $b = 2.2$ and 0.9745 for $b = 0.45$.

For the sake of completeness, the first coefficients of the corresponding expansions are listed in Tables 3.2 and 3.3.

(c) If $k = 3$, the critical intervals for the symmetry parameter are given by

$$C_1 = (-0.8504, -0.6925),$$

$$C_2 = (-0.2060, 0.2060),$$

$$C_3 = (0.6925, 0.8504)$$

and the specific points (for which the solution is known) are $\mu_1 = 0.7906$, $\mu_2 = 0$, and $\mu_3 = 0.7906$, respectively. This gives in the b scale

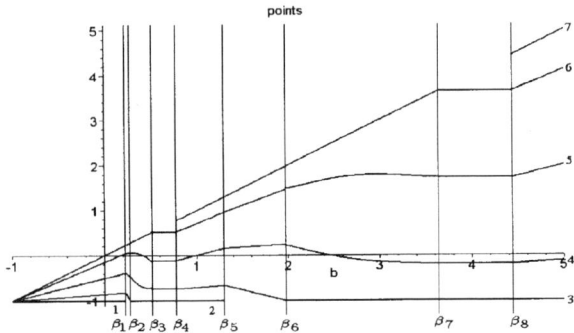

Figure 3.3: Support points of e_2-optimal designs for the cubic regression model

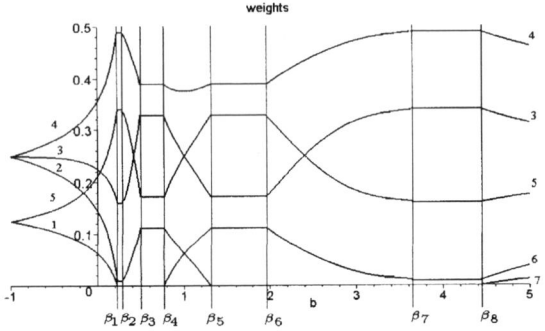

Figure 3.4: Weight coefficients of e_2-optimal designs for the cubic regression model

Table 3.2: First six coefficients of the Taylor expansions of the coefficients of the extremal polynomial $t^2 + q_4^* t^4 + q_3^* t^3 + q_2^* t + q_1^*$ and the interior support points t_2^* and t_3^* and weights w_2^*, w_3^*, and w_4^* of the e_2-optimal design in a polynomial regression of degree 4 on the interval $[-1, b]$, where $b \in (1.9677, 3.6374)$. The center of the expansion is $b_1 \approx 2.6794$.

	0	1	2	3	4	5
q_1^*	−0.6111	−0.0837	0.2284	−0.2006	−0.0144	0.1035
q_2^*	0.1679	0.6556	−0.3245	−0.1527	−0.0306	0.5245
q_3^*	−0.3970	−0.6202	0.4635	−0.0246	−0.0262	−0.3029
q_4^*	0.0000	0.2550	−0.2492	0.0956	−0.0232	0.0922
t_2^*	−0.0801	−0.2936	0.2332	0.0121	−0.1175	−0.0640
t_3^*	1.7596	0.2064	−0.4092	−0.0212	0.3126	0.0944
w_2^*	0.4550	0.0679	−0.0338	−0.0097	−0.0004	0.0283
w_3^*	0.2116	−0.1159	0.0804	0.0117	−0.0308	−0.0320
w_4^*	0.0450	−0.0679	0.0338	0.0097	0.0004	−0.0283

Table 3.3: First six coefficients of the Taylor expansions of the coefficients of the extremal polynomial $t^2 + q_4^* t^4 + q_3^* t^3 + q_2^* t + q_1^*$ and the interior support points t_2^* and t_3^* and weights w_2^*, w_3^*, and w_4^* of the e_2-optimal design in a polynomial regression of degree 4 on the interval $[-1, b]$, where $b \in (0.2749, 0.5082)$. The center of the expansion is $b_2 \approx 0.3732$.

	0	1	2	3	4	5
q_1^*	-0.0851	-0.3725	1.2528	10.3417	-33.0375	-172.2301
q_2^*	-0.0627	1.5888	6.2421	-37.8069	188.1016	2915.9334
q_3^*	1.0636	-14.7798	7.5615	212.6852	-1336.5488	-10635.3851
q_4^*	0.0000	-13.1439	13.4439	168.2561	-1132.4771	-7551.2227
t_2^*	-0.6567	-1.2064	7.8709	-24.0211	-237.7108	2949.0192
t_3^*	0.0299	-0.7064	-4.4867	13.6928	75.3346	-1269.9645
w_2^*	0.2116	0.8321	1.9158	-20.5706	26.1568	1026.1537
w_3^*	0.4550	-0.4878	-0.4334	9.4172	-57.9234	-266.3371
w_4^*	0.2884	-0.8321	-1.9158	20.5706	-26.1568	-1026.1537

the intervals

$$B_1 = s^{-1}(C_1) = (5.5041, 12.369),$$

$$B_2 = s^{-1}(C_2) = (0.6583, 1.5190),$$

$$B_3 = s^{-1}(B_3) = (0.0808, 0.1817)$$

and $b_1 = s^{-1}(\mu_1) = 8.5511$, $b_2 = s^{-1}(0) = 1$ and $b_3 = s^{-1}(\mu_3) = 0.1169$, respectively. The corresponding tables for the coefficients in the Taylor expansions and the figures for the support points and weights as functions of the parameter b can be found in the technical report of Dette, Melas, and Pepelyshev (2000). The interpretation of these graphs and tables is exactly the same as in the previous examples and therefore omitted.

3.3 E-Optimal Designs: Preliminary Results

As it was pointed out in Chapter 1 E-optimal designs minimize the variance of the least favorable linear combination of the parameters and by this reason can be of a practical interest. The main task of the present and the next section is to derive a solution of the problem for polynomial models of arbitrary degree on arbitrary symmetrical segments. Some results for nonsymmetrical segments are also will be presented. Moreover we will begin with a dual theorem for general linear model to provide a tool for the study.

3.3.1 Statement of the problem and a dual theorem

Consider the standard homoscedastic linear regression model

$$E(y|t) = \theta^T f(t), \tag{3.34}$$

where $f(t) = (f_1(t), \ldots, f_m(t))^T$ is the vector of known functions that are linear independent and continuous on a compact topological space $\mathfrak{X}$ and $\theta = (\theta_1, \ldots, \theta_m)^T$ is the vector of parameters to be evaluated.

Let ξ be a discrete probability measure on $\mathfrak{X}$ (approximate experimental design) given by the table

$$\xi = \begin{pmatrix} t_1 & \cdots & t_n \\ \omega_1 & \cdots & \omega_n \end{pmatrix},$$

where $t_i \neq t_j$ $(i \neq j)$, $\omega_i > 0$, $\sum_{i=1}^m \omega_i = 1$, and

$$M(\xi) = \int f(t) f^T(t) \xi(dt).$$

A design is called E-optimal if it maximizes

$$\lambda_{\min}(M(\xi)),$$

where $\lambda_{\min}$ denotes the minimal eigenvalue of M in the class of all approximate designs.

Our main problem is constructing and studying E-optimal designs for the case $\mathfrak{X} = [-r, r]$, $f_i(t) = t^{i-1}$, $i = a, \ldots, m$. For this purpose, we will need a duality theorem obtained independently in Pukelsheim (1980) (see also Pukelsheim (1993)) and Melas (1982). In the last paper a more direct approach was implemented and we will use it throughout the section.

Since the set of information matrices is compact (e. g., see Karlin and Studden (1966, Chap. X)), an E-optimal design for the standard model of linear regression exists.

Let ξ_α be an E-optimal design. Let $\mathcal{P}_\alpha$ denote a linear subset of R^m, spanned by the eigenvectors corresponding to the minimal eigenvalue of $M(\xi_\alpha)$. Let $\mathcal{P} = \cap \mathcal{P}_\alpha$, where the intersection is realized over all E-optimal designs.

Let $\mathcal{A}$ be the class of non-negative definite matrices A, such that $\operatorname{tr} A = 1$. The results of Melas (1982) can be stated in the following way.

Theorem 3.3.1 (Duality and equivalence theorem) *For the model described above, the E-optimality of a design ξ^* is equivalent to the existence of a matrix $A^* \in \mathcal{A}$ such that*

$$\max_{t \in \mathfrak{X}} f^T(t) A^* f(t) \leq \lambda_{\min}(M(\xi^*)).$$

Moreover,

$$\min_{A \in \mathcal{A}} \max_{t \in \mathfrak{X}} f^T(t) A f(t) = \max_{\xi} \lambda_{\min}(M(\xi)),$$

where the minimum on the right-hand side is taken over all experimental designs, and equality

$$f^T(t)A^* f(t) = \lambda_{\min}(M(\xi^*))$$

is valid for all support points $t \in \text{supp}\, \xi^*$ of any E-optimal design ξ^*. The polynomial

$$\rho(t) = f^T(t)A^* f(t)$$

will be called an extremal polynomial.

Theorem 3.3.2 *Any matrix A^* from Theorem 3.3.1 is of the form*

$$A^* = \sum_{i=1}^{\nu} \alpha_i p_{(i)} p_{(i)}^T,$$

where $\nu = \dim \mathcal{P}$, $\alpha_i \geq o$, $\sum \alpha_i = 1$, and $\{p_{(i)}\}$ is some orthonormed basis in $\mathcal{P}$. This basis will be called an extremal basis.

Note that the matrix A^* is a solution of the dual problem,

$$\max_{t \in \mathfrak{X}} f^T(t) A f(t) \to \min_{A \in \mathcal{A}}.$$

3.3.2   The number of support points

Now, we will consider polynomial models on arbitrary design intervals; that is, we will assume $\mathfrak{X} = [a, b]$, $f_i(t) = t^{i-1}$, $i = 1, \dots, m$.

The case $m = 1$ is trivial since the regression function is a constant and does not depend on the choice of design.

For the case $m = 2$,

$$\theta^T f(t) = \theta_1 + \theta_2 t; \tag{3.35}$$

a full solution is given by the following theorem.

Theorem 3.3.3 *Consider the linear model (3.35) on the arbitrary design interval $[a, b]$, $-\infty < a < b < \infty$.*

I. If $ab \geq -1$, then an E-optimal design is unique; this design is

$$\xi^* = \begin{pmatrix} a & b \\ \omega & 1 - \omega \end{pmatrix},$$

where $\omega = (2 + b^2 + ab)/[4 + (a + b)^2]$, and

$$\lambda_{\min}(M(\xi^*)) = r^2/(r^2 + v^2),$$

where $r = (b - a)/2$ and $v = (b + a)/2$.

II. If $ab < -1$, then any design of the form

$$\xi_{u_1,u_2} = \left(\begin{array}{cc} u_1 & u_2 \\ \dfrac{u_2}{u_2-u_1} & -\dfrac{u_1}{u_2-u_1} \end{array} \right),$$

where $0 > u_1 \geq a$, $0 < u_2 \leq b$, $|u_1 u_2| \geq 1$ is E-optimal, and

$$\lambda_{\min}\left(M(\xi_{u_1,u_2})\right) = 1.$$

Proof. Let $ab < -1$. Then designs of the form ξ_{u_1,u_2} exist, and for them,

$$M(\xi_{u_1,u_2}) = \left(\begin{array}{cc} 1 & 0 \\ 0 & |u_1 u_2| \end{array} \right), \quad \lambda_{\min}(M(\xi_{u_1,u_2})) = 1$$

since $|u_1 u_2| \geq 1$. Note that for an arbitrary design ξ,

$$\lambda_{\min}(M(\xi)) \leq e_1^T M(\xi) e_1 = 1.$$

Thus designs ξ_{u_1,u_2} are E-optimal designs.

Now, let $ab \geq -1$. By an immediate calculation, we obtain that

$$\lambda_{\min}(M(\xi^*)) = \left[\frac{1}{r^2} + \frac{v^2}{r^2} \right]^{-1},$$

where $r = (b-a)/2$ $v = (b+a)/2$.

Set

$$q = \left(-\frac{u}{r}, \frac{1}{r} \right)^T, \quad A^* = qq^T / q^T q.$$

Then we have

$$\max_{t\in[a,b]} f^T(t) A^* f(t) = \max_{t\in[a,b]} \left(\frac{t-u}{r} \right)^2 / q^T q = 1 / q^T q = \lambda_{\min}(M(\xi^*)),$$

and by Theorem 3.3.1, the design ξ^* is an E-optimal design. A verification that the E-optimal design is unique can be done with the help of Theorem 3.3.1 and is left to the reader. ∎

Now, let $m > 2$. Denote by ξ^*,

$$\xi^* = \left(\begin{array}{ccc} t_1^* & \cdots & t_n^* \\ \omega_1^* & \cdots & \omega_n^* \end{array} \right),$$

an E-optimal design and set

$$\lambda^* = \lambda_{\min}\left(M(\xi^*)\right).$$

Lemma 3.3.1 *If $m > 2$, then all E-optimal designs have the same support with the interval endpoints included.*

An extremal polynomial is uniquely determined by the expression

$$\rho(t) = \lambda^* + \gamma(t - a)(t - b) \prod_{i=2}^{m-1} (t - t_i^*)^2,$$

where $a < t_2^* < \cdots < t_{m-1}^* < b$ are support points of an E-optimal design and $\gamma > 0$ is a constant. Additionally, $\lambda^* < 1$.

Proof. Let us fix a matrix A^* from Theorem 3.3.1. Let ξ^* be an arbitrary E-optimal design and t^* a support points of ξ^*. Then we have

$$f^T(t)A^*f(t) \le \lambda^*, \ t \in [a, b],$$

$$f^T(t^*)A^*f(t^*) = \lambda^*.$$

Suppose that

$$\rho(t) = f^T(t)A^*f(t) \not\equiv \lambda^*.$$

Note that $\rho(t)$ is a polynomial of degree $\le 2m - 2$. If $t^* = a$ or b, then t^* is a root of the polynomial $\tilde{\rho}(t) = \lambda^* - \rho(t)$. Also, if $t^* \ne a, b$, then t^* is a root of multiplicity ≥ 2 of this polynomial.

If $n < m$, then $\lambda_{\min}(M(\xi)) = 0$. Therefore, an E-optimal design possesses no fewer than m support points. This means that the polynomial $\tilde{\rho}(t)$ has degree $2m - 2$ and the same number of real roots on $[a, b]$ with regard to their multiplicity. It follows from here that $\rho(t)$ has the form pointed out in the lemma and all E-optimal designs have the same support, a and b belong to ξ^*, and $n = m$.

Let us now demonstrate that the polynomial equals λ^* identically cannot be an extremal one. Suppose, oppositely, that

$$\rho(t) = f^T(t)A^*f(t) \equiv \lambda^*. \tag{3.36}$$

According to Theorem 3.3.2, any matrix A^* is of the form $\sum_{i=1}^{\nu} p_{(i)}p_{(i)}^T \alpha_i$, where $\{p_{(i)}\}$ is some orthogonal bases of $\mathcal{P}$, $\alpha_i \ge 0$, and $\sum \alpha_i = 1$. Therefore, equality (3.36) can be represented in the form

$$\sum_{i=1}^{\nu} \left(p_{(i)}^T f(t) \right)^2 \alpha_i \equiv \lambda^*.$$

This equality is valid if and only if $e_1 = (1, 0, \ldots, 0)^T \in \mathcal{P}$ and $\lambda^* = 1$. Let us demonstrate that if $m > 2$, then $\lambda^* < 1$ for any interval $[a, b]$. Indeed,

$$\lambda^* = \lambda_{\min}(M(\xi^*)) \le e_1^T M(\xi^*)e_1 = 1,$$

and since $M(\xi^*)e_1 \ne e_1$ for $m > 2$, $\lambda^* < 1$. ∎

The following result is needed in order to prove the uniqueness of an E-optimal design.

Lemma 3.3.2 *Let* m *and* s *be arbitrary integers. If vectors* $b_{(i)} = (b_1^{(i)}, \ldots, b_s^{(i)})$, $i = 1, \ldots, m$, *are not all null vectors, then there exists a vector* $\beta = (\beta_1, \ldots, \beta_s)$ *such that*

$$\sum_{j=1}^{s} \beta_j b_j^{(i)} \neq 0, \ i = 1, 2, \ldots .m.$$

Proof. It $b = (b_1, \ldots, b_s) \neq 0$, the equation $\sum_{j=1}^{s} \beta_j b_i = 0$ is a hyperplane equation. It is obvious that there exists a vector β that lies outside of m hyperplanes.   ∎

Theorem 3.3.4 *For the model of polynomial regression on an arbitrary compact interval with* $m > 2$, *an E-optimal design is unique. By Lemma 3.3.1, it is concentrated at* m *points, including* a *and* b.

Proof. By Lemma 3.3.1, all E-optimal designs have the same support points $\{a = t_1^*, t_2^*, \ldots, t_{m-1}^*, t_m^* = b\}$. Let us demonstrate that the weight vector $\omega = (\omega_1^*, \ldots, \omega_m^*)^T$ is determined uniquely.

According to Theorems 3.3.1 and 3.3.2,

$$\sum_{i=1}^{\nu} \left(p_{(i)}^T f(x_l^*) \right)^2 \alpha_i = \lambda^*, \ l = 1, \ldots, m,$$

where $\{p_{(i)}\}$ is some orthonormal basis of $\mathcal{P}$, $\alpha_i \geq 0$, and $\sum \alpha_i = 1$. Without loss of generality, assume that $\alpha_1, \ldots, \alpha_{\nu'} > 0$, $\nu' \leq \nu$. Vectors $b_{(j)} = \left(p_{(1)}^T f(x^*)_j, \ldots, p_{(\nu')}^T f(x_j^*) \right)$ are not null vectors for any $j = 1, \ldots, m$. By Lemma 3.3.2, there exists a vector $\beta = (\beta_1, \ldots, \beta_{s'})$ such that $\sum_{j=1}^{s'} \beta_j b_j^{(l)} \neq 0, l = 1, \ldots, m$.

Consider the vector $p = \sum_{j=1}^{\nu'} \beta_j p_{(j)}$. Since this vector is a linear combination of vectors $p_{(j)} \in \mathcal{P}$, we have $p \in \mathcal{P}$ and

$$M(\xi^*)p = \lambda^* p.$$

Denote

$$L = (f_i(x_l^*)d_l)_{i,l=1}^m,$$

where

$$d_l = \sum_{i=1}^{m} f_i(x_l)p_i = \sum_{j=1}^{\nu'} \beta_j b_j^{(l)} \neq 0, \ l = 1, \ldots .m.$$

Note that

$$L\omega^* = \left(\sum_{j=1}^{m} \sum_{l=1}^{m} f_i(x_l^*)f_j(x_l^*)p_j\omega_l^* \right)_{i=1}^m = M(\xi^*)p.$$

We obtain

$$Lw^* = \lambda^* p.$$

The matrix L is invertible due to

$$\det L = d_1 \cdots d_m \det \left(x_i^{*j-1} \right)_{i,j=1}^m \neq 0.$$

Thus, the vector $w^* = \lambda^* L^{-1} p$ is uniquely determined.　　■

As in the previous section, the triple (n_1, n_2, n_3), where $n_1(n_3)$ is the number of support points equal to the left (right) bound of the design interval and $n_2 = n - n_1 - n_3$, will be called *a type of design*.

Theorem 3.3.4 asserts that all E-optimal designs are of one and the same type $(1, m - 2, 1)$.

However, for E-optimal designs, the concept of design type is not sufficient because the method of design constructing depends also on one more value: the multiplicity of the minimal eigenvalue of the information matrix.

3.3.3　Chebyshev designs

Consider now a model, that is a generalization of the polynomial model.

Assume that $\mathfrak{X} = [a, b]$, $-\infty < a < b, \infty$, and the functions $\{f_i(t)\}_{i=1}^M$ generate a Chebyshev system on $[a, b]$. Remember that a system $\{f_i(t)\}_{i=1}^m$ is called a Chebyshev system if for any $t_1, \ldots, t_m$ such that $a \leq t_1 < t_2 < \cdots < t_m \leq b$, the determinant $\det (f_i(t_j))_{i,j=1}^m$ does not vanish. Since Vandermonde's determinants do not vanish, the system $1, t, \ldots, t^{m-1}$ is a Chebyshev one on the arbitrary interval $[a, b]$.

We will need the following results concerning Chebyshev systems (see Section 1.9).

For any Chebyshev system, there exist points $a \leq \bar{t}_1 < \bar{t}_2 < \cdots < \bar{t}_m \leq b$ and a vector $\gamma = (\gamma_1, \ldots, \gamma_m)^T$ such that

$$\gamma^T f(\bar{t}_i) = (-1)^i, \quad i = 1, \ldots, m,$$

$$|\gamma^T f(t)| \leq 1, \quad t \in [a, b].$$

The function $\gamma^T f(t)$ is often called a *Chebyshev polynomial* and points $\bar{t}_1, \ldots, \bar{t}_m$ are called *Chebyshev points*. The Chebyshev polynomial is determined uniquely. As for Chebyshev points, they are determined uniquely under the condition that a constant belongs to the linear space spanned by the functions $f_1(t), \ldots, f_m(t)$.

Let us introduce the following notation:

$$F = \left(f_i(\bar{t}_j)(-1)^j\right)_{i,j=1}^m,$$

$$A_j = (e_i^T F^{-1}\gamma), \quad i = 1, \ldots, m,$$

$$\bar{\omega}_i = |A_i| / \sum_{i=1}^m |A_j|, \quad i = 1, \ldots, m, \tag{3.37}$$

$$\bar{\omega} = (\bar{\omega}_1, \ldots, \bar{\omega}_m)^T, \quad \bar{\lambda} = 1/\gamma^T\gamma.$$

The design

$$\bar{\xi} = \begin{pmatrix} \bar{t}_1 & \cdots & \bar{t}_m \\ \bar{\omega}_1 & \cdots & \bar{\omega}_m \end{pmatrix}$$

concentrated in the Chebyshev points will be called the Chebyshev design (in the E-optimal design problem).

Denote by A_- the matrix A with the last row and last column rejected. Let

$$\lambda_2(A) = \lambda_{\min}(A_-).$$

In many cases, the Chebyshev design is an E-optimal design, as can be seen from the following theorem.

Theorem 3.3.5 *Consider the regression model (3.34), where $\mathfrak{X} = [a, b]$ and the basic functions $f_1(t), \ldots, f_m(t)$ generate a Chebyshev system on $[a, b]$. Then the following assertions hold:*

(a) $\bar{\omega} = F^{-1}\gamma/\gamma^T\gamma$ *and $\bar{\xi}$ is a c-optimal design with $c = \gamma$.*

(b) If $\lambda_2(M(\bar{\xi})) > \bar{\lambda}$, then $\bar{\xi}$ is an E-optimal design and $\dim \mathcal{P} = 1$.

(c) If $\dim \mathcal{P} = 1$ and Chebyshev points are determined uniquely, then $\bar{\xi}$ is the unique E-optimal design and $\lambda_2(M(\bar{\xi})) > \bar{\lambda}$.

Proof. (a) By the definition of $\bar{\omega}_i$, we have $\bar{\omega}_i \geq 0$, $i = 1, \ldots, m$. At first, assume that $\bar{\omega}_i \neq 0$, $i = 1, \ldots, m$.

Let $\bar{\xi}_\omega$ be an arbitrary design with the support in points $\bar{t}_1, \ldots, \bar{t}_m$:

$$\bar{\xi}_\omega = \begin{pmatrix} \bar{t}_1 & \cdots & \bar{t}_m \\ \omega_1 & \cdots & \omega_m \end{pmatrix}, \quad \omega_i \geq 0, \; i = 1, \ldots, m.$$

Denote $H = \mathrm{diag}\{\omega_1, \ldots, \omega_m\}$.

Due to Cauchy's inequality, we have

$$\gamma^T M^{-1}(\bar{\xi}_\omega)\gamma = \gamma^T (F^{-1})^T H^{-1}(F^{-1})\gamma$$

$$= \sum_{i=1}^m (e_i^T F^{-1}\gamma)^2 / \omega_i$$

$$\geq \left(\sum_{i=1}^m |e_i^T F^{-1}\gamma|\right)^2,$$

and the equality holds if and only if $\omega = \bar{\omega}$.

In addition, due to the inequality

$$p^T A p q^T A^{-1} q \geq (p^T q)^2,$$

which is true for arbitrary vectors p and q and a positive definite matrix A (see, e.g., Gantmacher (1998)), and the properties of the Chebyshev polynomial $\gamma^T f(t)$ we have

$$\gamma^T M(\bar{\xi}_\omega) \gamma \gamma^T M^{-1}(\bar{\xi}_\omega) \gamma$$

$$= \left[\sum_{i=1}^{m} \left(\gamma^T f(\bar{t}_i) \right)^2 \omega_i \right] \gamma^T M^{-1}(\bar{\xi}_\omega) \gamma$$

$$= \gamma^T M^{-1}(\bar{\xi}_\omega) \gamma \geq \gamma^T \gamma,$$

and the equality holds if and only if

$$M^{-1}(\bar{\xi}_\omega) \gamma = (\gamma^T \gamma) \gamma.$$

By multiplying both sides of the last equality by $M(\bar{\xi}_\omega)$, we obtain that it is equivalent to the equality

$$M(\bar{\xi}_\omega) \gamma = \gamma(\gamma^T \gamma),$$

and since $f^T(\bar{t}_i) \gamma = (-1)^i$, it follows that

$$M(\bar{\xi}_\omega) \gamma = F\omega = \gamma / (\gamma^T \gamma)$$

Thus, the minimum of $\gamma^T M^{-1}(\bar{\xi}_\omega) \gamma$ by all ω such that $\omega_i > 0$, $i = 1, \ldots, m$, and $\sum_{i=1}^{m} \omega_i = 1$ is achieved if and only if

$$\omega = F^{-1} \gamma / \gamma^T \gamma.$$

Therefore, $\bar{\omega} = F^{-1} \gamma / \gamma^T \gamma$.

The case when $\bar{\omega}_i = 0$ for some i can be considered similarly with replacement of the inverse matrix by the pseudoinverse in the sense of Moor–Penrous.

Note that for the vector γ and the design $\bar{\xi}$, we have

$$|f^T(t)\gamma| \leq 1, \quad t \in [a, b],$$

$$|f^T(\bar{t}_i)\gamma| = 1, \quad i = 1, \ldots, m$$

and

$$F\bar{\omega} = \gamma / \gamma^T \gamma.$$

Due to an obvious extension of Lemma 3.2.1, the design $\bar{\xi}$ is c-optimal for the vector $c = \gamma$.

(b) Denote $\bar{M} = M(\bar{\xi})$. Let $\lambda_2(M(\bar{\xi})) > \bar{\lambda}$ and let $p \in R^{m-1}$ be an arbitrary vector. Then, with $\bar{p}^T = (p^T \vdots 0) \in R^m$, we have

$$\frac{\bar{P}^T \bar{M} \bar{p}}{\bar{p}^T \bar{p}} \geq \lambda_{\min}(\bar{M}_-) = \lambda_2(\bar{M}) > \bar{\lambda} = 1/\gamma^T \gamma.$$

Additionally, as it was proved above,

$$\bar{M}\gamma = F\bar{\omega} = \bar{\lambda}\gamma.$$

Therefore, for arbitrary vector $q \in R^m$, we have a representation $q = \alpha\bar{p} + \beta\gamma$ for some $\alpha, \beta \in R^1$ and

$$\frac{q^T \bar{M} q}{q^T q} = \frac{\alpha^2 \bar{p}^T \bar{M} p + 2\alpha\beta\bar{\lambda}(\bar{p}^T \gamma) + \beta^2}{\alpha^2 \bar{p}^T \bar{p} + 2\alpha\beta(\bar{p}^T \gamma) + \beta^2 \gamma^T \gamma} \geq \bar{\lambda},$$

and the equality holds if and only if $\alpha = 0$. Thus,

$$\lambda_{\min}(\bar{M}) = \frac{\gamma^T \bar{M} \gamma}{\gamma^T \gamma} = \bar{\lambda}.$$

Let us verify that the matrix $A^* = \gamma\gamma^T/\gamma^T\gamma$ satisfies conditions of Theorem 3.3.1 for $\xi^* = \bar{\xi}$. Due to the properties of the Chebyshev polynomial, we have

$$\max_{t\in[a,b]} f^T(t)A^* f(t) = \max_{t\in[a,b]} \frac{(f^T(t)\gamma)^2}{\gamma^T\gamma} = \frac{1}{\gamma^T\gamma} = \lambda_{\min}(\bar{M}).$$

Therefore, due to Theorem 3.3.1, $\bar{\xi}$ is an E-optimal design and $\mathcal{P} = \{\alpha\gamma\}$, $\alpha \in R^1$, and dim $\mathcal{P} = 1$.

(c) Assume that $\dim\mathcal{P} = 1$. Let p be the unique vector in $\mathcal{P}$ such that $\|p\| = 1$. By Theorem 3.3.2 we obtain that $A^* = pp^T$. Therefore,

$$f^T(t)A^* f(t) = (p^T f(t))^2,$$

and by Theorem 3.3.1,

$$\max_{t\in[a,b]} \left(p^T f(t)\right)^2 = \lambda^*,$$

$$\left(p^T f(t)\right)^2 = \lambda^*, \quad t \in \text{supp } \xi^*,$$

where ξ^* is an E-optimal design and $\lambda^* = \lambda_{\min}(M(\xi^*))$.

Due to the uniqueness of the Chebyshev polynomial, it follows from here that

$$p = \gamma/\sqrt{\lambda^*}, \quad A^* = \gamma\gamma^T/\gamma^T\gamma \text{ and } \lambda^* = \bar{\lambda}.$$

Additionally, if the Chebyshev points are uniquely determined, then

$$\text{supp } \xi^* = \{\bar{t}_1, \ldots, \bar{t}_n\}$$

for any E-optimal design. Let ω^* be the vector of weight coefficients of an E-optimal design ξ^*; then $\xi^* = \bar{\xi}_{\omega^*} = \begin{pmatrix} \bar{t}_1 & \cdots & \bar{t}_m \\ \omega^* & \cdots & \omega^* \end{pmatrix}$.

Since $\gamma \in \mathcal{P}$, we have

$$M(\xi^*)\gamma = F\omega^* = \lambda^*\gamma = \bar{\lambda}\gamma$$

and this means that

$$\omega^* = F^{-1}\gamma/\gamma^T\gamma = \bar{\omega}, \quad \xi^* = \bar{\xi}.$$

Thus, $\bar{\xi}$ is a unique E-optimal design, $\bar{\lambda} = \lambda_{\min}(M(\bar{\xi}))$, and $\bar{\lambda}$ is a simple eigenvalue of $M(\bar{\xi})$.

Hence,

$$\lambda_2(M(\bar{\xi})) > \lambda_{\min}(M(\bar{\xi})) = \bar{\lambda}.$$

■

Let us return to the system $f_1(t) = 1$, $f_2(t) = t, \ldots, f_m(t) = t^{m-1}$ on arbitrary interval $[a, b]$. This system is a Chebyshev one and it is easy to check that

$$\gamma^T f(t) = T_{m-1}\left(\frac{t-v}{r}\right), \quad r = \frac{b-a}{2}, \quad v = \frac{b+a}{2},$$

where $T_{m-1}(t) = \cos[(m-1)\arccos t]$, the Chebyshev polynomial of the first kind. Additionally, points $\bar{t}_i = \bar{t}_i(r, v)$ are determined uniquely, and

$$\bar{t}_i(r, v) = r\bar{t}_i(1, 0) + v, \tag{3.38}$$

where $\bar{t}_i(1, 0) = \cos\left(\frac{i-1}{m-1}\pi\right)$, $i = 1, \ldots, m$.

Let us call *a design ξ nonsingular* if the matrix $M(\xi)$ is invertible.

The following lemma provides a basis for the application of Theorem 3.3.5.

Lemma 3.3.3 *Consider the polynomial regression model on an interval $[a, b]$. Let either of the following two conditions be fulfilled:*

(i) $m > 2$ and $[a, b] \subset [-1, 1]$.

(ii) $ab \geq 0$.

Then, for an arbitrary nonsingular design ξ, the minimal eigenvalue of $M(\xi)$ is simple.

From this lemma it obviously follows that under condition (i) we have $\dim \mathcal{P} = 1$ and, therefore, due to Theorem 3.3.5(c), the Chebyshev design $\bar{\xi}$ is the unique E-optimal design.

Proof of Lemma 3.3.3. Let condition (i) be fulfilled and ξ be an arbitrary nonsingular design on $[a, b]$. Since $[a, b] \subset [-1, 1]$ ξ is a nonsingular design on $[-1, 1]$. Let us prove that the minimal eigenvalue of $M(\xi)$ is simple. Assume, oppositely, that the multiplicity of the minimal eigenvalue is equal to or more than 2. Then there exists a vector $\bar{q}^T = (q^T \! : \! 0)$, where $q \in R^{m-1}$ such that

$$\bar{q}^T M \bar{q} = \lambda \bar{q}^T \bar{q}, \quad M = M(\xi), \quad \lambda = \lambda_{\min}(M).$$

Set $\tilde{q}^T = (0 \! : \! q^T) \in R^m$. Then

$$\tilde{q}^T M \tilde{q} = \int_{-1}^{1} \left(\tilde{q}^T f(t) \right)^2 \xi(dt)$$

$$= \int_{-1}^{1} \left(\bar{q}^T f(t) \right)^2 t^2 \xi(dt)$$

$$\leq \int_{-1}^{1} \left(\bar{q}^T f(t) \right)^2 \xi(dt)$$

$$= \bar{q}^T M \bar{q} = \lambda \bar{q}^T \bar{q} = \lambda \tilde{q}^T \tilde{q}$$

and equality is achieved if and only if supp $\xi = \{-1, 1\}$. However, any nonsingular design has at least $m > 2$ support points and, therefore,

$$\frac{\tilde{q}^T M \tilde{q}}{\tilde{q}^T \tilde{q}} < \lambda,$$

which is impossible. The obtained contradiction proves the proposition of the lemma in case (i).

Now, let condition (ii) be fulfilled and ξ be an arbitrary nonsingular design on $[a, b]$. Consider the case $b > a \geq 0$. The case $a < b \leq 0$ can be obtained by the substitution $t \to -t$.

Denote $\{t_1, \ldots, t_n\} = $ supp ξ, $t_1 < t_2 < \cdots < t_n$. From nonsingularity of $M(\xi)$, it follows that $n \geq m$. Note that any determinants of the form

$$\left(f_{i_j}(t_{k_l}) \right)_{j,l=1}^{s},$$

$s = 1, 2, \ldots, m$ does not vanish (see, e.g., Gantmacher (1998)). Due to the Binet–Cauchy's formula, it follows from here that any minor of matrix $M(\xi)$ is positive. For matrices with this property (such matrices are called strictly positive), all eigenvalues are simple (Gantmacher, 1998). ∎

Denote by $\bar{\xi} = \bar{\xi}(r, v)$, $r = (b - a)/2$, $v = (b + a)/2$, the Chebyshev design for the polynomial model (with m parameters) on the interval $[a, b]$,

$$\bar{\xi} = \bar{\xi}(r, v) = \begin{pmatrix} \bar{t}_1 & \cdots & \bar{t}_m \\ \bar{\omega}_1 & \cdots & \bar{\omega}_m \end{pmatrix},$$

where $\bar{t}_i = \bar{t}_i(r, v)$, $\bar{\omega}_i = \bar{\omega}_i(r, v)$ are defined by formulas (3.38) and (3.37), respectively.

Due to Theorem 3.3.5(c), this design is an unique E-optimal design for intervals defined in Lemma 3.3.3.

Consider now the case of symmetrical design intervals $[-r, r]$ and show that the Chebyshev design is not an E-optimal design for such intervals with sufficiently large r. First, let us show that support points of an E-optimal design for symmetrical design intervals are located symmetrically with respect to the origin.

Denote by $\xi^* = \xi^*(r)$,

$$\xi^*(r) = \begin{pmatrix} x_1^* & \cdots & x_n^* \\ \omega_1^* & \cdots & \omega_n^* \end{pmatrix},$$

$x_i^* = x_i^*(r)$, $\omega_i^* = \omega_i^*(r)$, $x_1^* < x_2^* < \cdots < x_n^*$, $i = 1, \ldots, n$, an E-optimal design on $[-r, r]$,

$$\lambda^* = \lambda^*(r) = \lambda_{\min}(M(\xi^*(r)).$$

Due to Theorem 3.3.4, an E-optimal design is unique and $n = m$ for $m > 2$.

Lemma 3.3.4 *For $m > 2$, an E-optimal design for polynomial regression on symmetrical intervals is unique and located in points symmetrical with respect to the origin:*

$$-x_i^* = x_{2k+1-i}^*, \ i = 1, \ldots, k \text{ with } m = 2k,$$

$$-x_i^* = x_{2k+2-i}^*, \ i = 1, \ldots, k, \ x_{k+1}^* = 0 \text{ with } m = 2k + 1.$$

Weight coefficients of the design have also symmetrical values:

$$\omega^* = \omega_{2k+1-i}^*, \ i = 1, \ldots, k \text{ with } m = 2k,$$

$$\omega_i^* = \omega_{2k+2-i}^*, \ i = 1, \ldots, k, \text{ with } m = 2k + 1$$

Moreover, for symmetrical intervals, $\dim \mathcal{P} = 1$ or 2.

Proof. Let

$$\xi^* = \begin{pmatrix} x_1^* & \cdots & x_m^* \\ \omega_1^* & \cdots & \omega_m^* \end{pmatrix}$$

be an E-optimal design on the design interval $[-r, r]$. Assume that $m = 2k$ (the proof for $m = 2k + 1$ is similar). Consider the design

$$\tilde{\xi} = \begin{pmatrix} \tilde{x}_1 & \cdots & \tilde{x}_m \\ \tilde{\omega}_1 & \cdots & \tilde{\omega}_m \end{pmatrix}, \ \tilde{x}_i = -x_{2k+1-i}^*, \ \tilde{\omega}_i = \omega_{2k+1-i}^*, \ i = 1, \ldots, k.$$

The matrix $M(\xi)$ for the design $\xi = (\xi^* + \tilde{\xi})/2$ with rows and columns moved to the first positive assumes the form

$$\begin{pmatrix} M_1 & 0 \\ 0 & M_2 \end{pmatrix}.$$

It is known (see, e.g., Gantmacher (1998)) that the minimal eigenvalue of any non-negative definite matrix M can be represented in the form

$$\min_{||p||=1} p^T M p.$$

Therefore,

$$\lambda_{\min} \begin{pmatrix} M_1 & 0 \\ 0 & M_2 \end{pmatrix} \geq \lambda_{\min} \begin{pmatrix} M_1 & C \\ C^T & M_2 \end{pmatrix}$$

for $M_1, M_2 \geq 0$ and any matrix C, from which

$$\lambda_{\min}(M(\xi) \geq \lambda_{\min}(M(\xi^*)),$$

and ξ is an E-optimal design. If the design ξ^* does not satisfy conditions of the lemma, then $\xi \neq \xi^*$. Since by Theorem 3.3.4 an E-optimal design is unique for $m > 2$, this is impossible.

Let us now prove that dim $\mathcal{P} \leq 2$. Note that the set of eigenvalues of matrix $M(\xi) = M(\xi^*)$ is a conjunction of such sets for M_1 and M_2.

Similar to the proof of the second part of Lemma 3.3.3, it can be checked that all eigenvalues of M_1 and M_2 are simple. Therefore the multiplicitly of any eigenvalue of $M(\xi^*)$ is no more than 2 and it means that dim$\mathcal{P} \leq 2$.
∎

In the following, we will use a lemma on polynomials with fixed absolute values.

Let $p(x) = \sum_{i=1}^{n+1} p_i x^{i-1}$ be a polynomial of degree n such that for $0 \leq x_1 < \cdots < x_{n+1}$

$$p(x_i) = (-1)^i a_i,$$

where $a_i > 0$, $i = 1, 2, \ldots, n+1$. Let

$$\tilde{p}(x) = \sum_{i=1}^{n+1} \tilde{p}_i x^{i-1}$$

be a polynomial such that $|\tilde{p}(x_i)| \leq a_i$, $i = 1, \ldots, n+1$.

Lemma 3.3.5 $|p_i| \geq |\tilde{p}_i|$ and $p_i(-1)^i > 0$, $i = 1, \ldots, n+1$, for the polynomials being considered.

Proof. Represent the polynomial $\tilde{\varphi}(x)$ in the form

$$\tilde{\varphi}(x) = \sum_{i=1}^{n+1} \tilde{p}_i x^{i-1} = \det \begin{pmatrix} 0 & 1 & x & \cdots & x^n \\ \tilde{a}_1 & 1 & x_1 & \cdots & x_1^n \\ \vdots & \vdots & \vdots & \ddots & \vdots \\ \tilde{a}_{n+1} & 1 & x_{n+1} & \cdots & x_{n+1}^n \end{pmatrix} \Big/ \prod_{j<i}(x_i - x_j),$$

where $\tilde{a}_i = \tilde{\varphi}(x_i)$. Note that

$$\tilde{p}_i = (-1)^i \det \left(\tilde{a}_j \ 1 \ \ldots \ x_j^{i-1} x_j^{i+1} \ \ldots \ x_j^n \right)_{j=1}^{n+1} / \delta$$

$$= (-1)^i \sum_{s=1}^{n+1} \tilde{a}_s (-1)^{s+1} \delta_{i,s} / \delta,$$

where

$$\delta_{i,s} = \det \left(1 \ \ldots \ x_j^{i-1} x_j^{i+1} \ \ldots \ x_j^k \right)_{j \neq s, \, j \in 1 : n+1}$$

$$= \prod_{l < j, \, l, j \neq s} (x_j - x_l) \sum x_{j_1} \ldots x_{j_{n-1}} > 0,$$

$$\delta = \prod_{j < i} (x_i - x_j) > 0.$$

Since $\varphi(x_i) = (-1)^i a_i$, $|p_i| = |\sum_{s=0}^k a_s (-1)^s \delta_{i,s} / \delta| \geq |\tilde{p}_i|$, $i = 1, \ldots, n+1$.
∎

Lemma 3.3.6 *The Chebyshev design $\bar{\xi} = \bar{\xi}(r, 0)$ is not an E-optimal design for polynomial regression on design interval $[-r, r]$ with sufficiently large r.*

Proof. Let $\bar{q} = \bar{q}(r)$ be the coefficient vector of the polynomial $T_{m-1}(t/r)$,

$$\bar{q}^T f(t) = T_{m-1}(t/r),$$

where $T_{m-1}(x) = \cos[(m-1) \arccos x]$ is the Chebyshev polynomial of the first kind.

Denote

$$\bar{\lambda}(r) = \left(\bar{q}^T(r) \bar{q}(r) \right)^{-1},$$

$$\bar{\xi} = \bar{\xi}(r) = \bar{\xi}(r, 0) = \begin{pmatrix} \bar{t}_1 & \ldots & \bar{t}_m \\ \bar{\omega}_1 & \ldots & \bar{\omega}_m \end{pmatrix},$$

where $\bar{t}_i = \bar{t}_i(r)$ and $\bar{\omega}_i = \bar{\omega}_i(r)$, $i = 1, \ldots, m$, are defined by formulas (3.38) and (3.37).

Consider the case $m = 2k + 1$, $k = 1, 2, \ldots$. Let us show that in this case,

$$\bar{\omega}_i(r) = \mathbf{O}(r^{-4}), \quad i = 1, \ldots, k-1, k+1, \ldots; m, \, r \to \infty.$$

It is easy to check that $M = M(\bar{\xi}(r))$ is of the form

$$M = F \Lambda F^T,$$

where $\Lambda = \text{diag}\{\bar{\omega}_1, \ldots, \bar{\omega}_m\}$ and $F = \left(f_i(t_j)(-1)^j \right)_{i,j=1}^m$.

Also, it can be checked that rows of F^{-1} (denote them by $l_{(1)}, \ldots, l_{(m)}$) consists of coefficients of the Lagrange interpolation polynomials constructed by points $\bar{t}_1, \ldots, \bar{t}_m$:

$$l_{(i)}^T f(t) = e_i^T F^{-1} f(t) = \prod_{j \neq i} \frac{t - \bar{t}_j}{\bar{t}_i - \bar{t}_j}, \ i = 1, \ldots, m,$$

where $e_1^T = (1, 0, \ldots, 0), \ldots, e_m^T = (0, 0, \ldots, 0, 1)$.

The vector $\bar{q}(r)$ is

$$\bar{q}(r) = \sqrt{\bar{\lambda}(r)}(q_1, 0, q_3 r^{-2}, 0, \ldots, 0, q_{2k+1} r^{-2k})^T,$$

where $(q_1, 0, q_3, 0, \ldots, 0, q_{2k+1})^T = \bar{q}(1)$.

Since $\bar{t}_i(r) = r\bar{t}_i(1)$,

$$l_{(i)j} = b_j^{(i)} r^{-j+i}, \ i, j = 1, \ldots, m,$$

where $b_j^{(i)}$ does not depend on r.

Therefore,

$$\bar{\omega}_i = \bar{\omega}_{m+1-i} = \bar{\lambda}(r) \sum_{j=1}^{k} b_{2j}^{(i)} q_{2j+1} r^{-4j} = \mathbf{O}\left(\frac{1}{r^4}\right)$$

with $r \to \infty$ since $\bar{\lambda}(r) \to 1$ with $r \to \infty$.

Thus, we obtain that

$$\lambda_{\min}\left(M(\bar{\xi}(r))\right) = \min_{\|p\|=1} p^T M(\xi(r)) p \leq \left(M(\bar{\xi}(r))\right)_{22}$$

$$= \sum_{i=1}^{m} \bar{t}^2(r)\bar{\omega}_i(r)$$

$$= \sum_{i \neq k} \bar{t}_i^2(0) r^2 \bar{\omega}_i(r) = \mathbf{O}\left(\frac{1}{r^2}\right).$$

Therefore,

$$\lambda_{\min}\left(M(\bar{\xi}(r)\right) \to 0$$

with $r \to \infty$ and it means that $\bar{\xi}(r)$ cannot be an E-optimal design for sufficiently large r.

Let now $m = 2k$. Note that, in this case, the vector $\bar{q} = \bar{q}(r)$ has the form $\bar{q} = (\bar{q}_1, \bar{q}_2, \ldots, \bar{q}_{2k})^T$, where $\bar{q}_{2i+1} = 0$, $i = 0, 1, \ldots, k-1$. Therefore,

$$\bar{\lambda}(r) = \left(\sum_{i=1}^{k} (\bar{q}(0))_{2i}^2 \, r^{-2(2i-1)}\right)^{-1} \to \infty$$

with $r \to \infty$.

Suppose that $\lambda^*(r) = \lambda_{\min}\left(M(\bar{\xi}(r))\right)$.

Due to Lemma 3.3.1, $\lambda^*(r) < 1$ for any $r > 0$. Let us move rows and columns of the matrix $M(\bar{\xi}(r))$ to the first positions; then the matrix assumes the form

$$\begin{pmatrix} M_1 & 0 \\ 0 & M_2 \end{pmatrix} = \begin{pmatrix} M_1(\bar{\xi}(r)) & 0 \\ 0 & M_2(\bar{\xi}(r)) \end{pmatrix}.$$

By the arguments in the proof of Lemma 3.3.1, all eigenvalues of M_1 are simple and, therefore,

$$\bar{\lambda}(r) = \min\{\lambda_{\min}\left(M_1(\bar{\xi}(r))\right), \lambda_{min}\left(M_2(\bar{\xi}(r))\right)\}.$$

Hence, $\lambda^*(r)$ is a simple eigenvalue of $M(\bar{\xi}(r))$ for $r > \bar{r}$, where $\bar{r}$ is a sufficiently large number, providing that $\bar{\lambda}(r) > 1$.

Let q^* be the eigenvector corresponding to $\lambda^*(r)$ such that $||q^*|| = 1$.

With $r > \bar{r}$, we obtain by Theorems 3.3.1 and 3.3.2 that $A^* = q^*q^{*T}$ and

$$\left|q^{*T}f(\bar{t}_i)\right|^2 = \lambda^*(r) < \bar{\lambda}(r), \ i = 1, \ldots, m.$$

At the same time, we have

$$\left(\bar{q}^T(r)f(\bar{t}_i)\right)^2 = 1, \ i = 1, \ldots, m,$$

and, therefore,

$$\left|q^T f(\bar{t}_i)\right|^2 = \bar{\lambda}(r), \ i = 1, \ldots, m,$$

where $q = \bar{q}(r)/\sqrt{\bar{\lambda}(r)}$, $||\alpha|| = 1$.

Due to the lemma on polynomials with fixed absolute values (Lemma 3.3.5), we obtain

$$||q|| \geq ||q^*||\frac{\bar{\lambda}(r)}{\lambda^*(r)} > 1$$

with $r > \bar{r}$, but $||q|| = 1$. The obtained contradiction proves that $\bar{\xi}(r)$ is not an E-optimal design on $[-r, r]$ for sufficiently large r in the case $m = 2k$.

∎

3.3.4   A boundary equation

The following lemma provides a necessary and sufficient condition for E-optimality of the Chebyshev design in the case of symmetrical intervals. Remember the notation $\lambda_2(A) = \lambda_{\min}(A_-)$, see Section 3.3.3.

Lemma 3.3.7 *For polynomial regression on an arbitrary symmetrical interval $[-r, r]$, the design ξ_r is E-optimal if and only if*

$$\lambda_2\left(M(\bar{\xi}_r)\right) \geq \bar{\lambda}(r).$$

Proof. Let us denote for the sake of brevity $\lambda_2(r) = \lambda_2(M(\bar{\xi}_r))$. It follows from Theorem 3.3.5 that the condition $\lambda_2(r) > \bar{\lambda}(r)$ is a sufficient condition for E-optimality of the design $\bar{\xi}_r$.

The same arguments show that the condition $\lambda_2(r) \geq \bar{\lambda}(r)$ implies E-optimality of the design $\bar{\xi}_r$. From the arguments at the end of the proof of Lemma 3.3.6, it follows that under the condition $\lambda_2(r) < \bar{\lambda}(r)$, the design $\bar{\xi}_r$ is not E-optimal design. Thus, the condition $\lambda_2(r) \geq \bar{\lambda}(r)$ is not only sufficient but also necessary for E-optimality of the design $\bar{\xi}_r$. ∎

It follows from Lemma 3.3.6 that

$$\lambda_2\left(M(\bar{\xi}_r)\right) = \bar{\lambda}(r) \tag{3.39}$$

has at least one positive solution. This equation will be called the *boundary equation*. It can be solved numerically.

In Section 3.4.4, we will point out an explicit form for this equation for the cases $m = 3, 4$, and 5.

Denote by r^* the minimal positive root of this equation. Due to Lemma 3.3.7 with $r \leq r^*$, the design $\bar{\xi}_r$ is the unique E-optimal design for polynomial regression on $[-r, r]$. The case $r > r^*$ will be studied in the next section.

3.4   Non-Chebyshev E-Optimal Designs

The present section is devoted to constructing and studying E-optimal designs for the polynomial regression model on sufficiently large symmetrical segments. In this case, as it was proved in the previous section, there exists a unique E-optimal design and this design does not coincide with the Chebyshev one. In fact, for $m = 3, 4$, and 5 we will construct E-optimal designs for all symmetrical segments. The study is based on the functional approach as well as auxiliary results that are of some independent interest.

We will introduce a basic equation determining support points and weights of E-optimal designs as implicitly given functions of the length of the segment. The limit values (under some normalization) of these functions will be found with the length tends to infinity. It will be also proved that these functions are real analytic for sufficiently large r. Also, Taylor expansions will be constructed on the basis of general formulas introduced in Chapter 2.

3.4.1   Basic equation

Consider the polynomial regression model $(f_i(x) = x^{i-1}, i = 1, \ldots, m)$ on a symmetrical interval $\mathfrak{X} = [-r, r]$. The case $m = 1$ is of no interest and the case $m = 2$ was already studied in Theorem 3.3.3.

Let $m > 2$ and r^* be a minimal positive root of the boundary equation (3.39). Due to Lemma 3.3.6, for $r \leq r^*$, there exists a unique E-optimal design and this design coincides with the Chebyshev design $\bar{\xi}_r$. For $r > r^*$, the Chebyshev design is not, generally speaking, E-optimal, but an E-optimal design is unique as well.

Let us construct an equation determining points and weights of the E-optimal design as implicit functions of r with $r > r^*$. This equation has the same form for even and odd m, but some details are different.

First, consider the case of odd m, $m = 2k + 1$. Set $z = 1/r^2$. Due to Lemma 3.3.4, an E-optimal design has the form

$$\xi^* = \xi^*(z) = \begin{pmatrix} -\frac{1}{\sqrt{z}} & -\frac{t_2^*}{\sqrt{z}} & \cdots & -\frac{t_k^*}{\sqrt{z}} & 0 & \frac{t_k^*}{\sqrt{z}} & \cdots & \frac{t_2^*}{\sqrt{z}} & \frac{1}{\sqrt{z}} \\ \nu_1^* & \nu_2^* & \cdots & \nu_k^* & \nu_{k+1}^* & \nu_k^* & \cdots & \nu_2^* & \nu_1^* \end{pmatrix},$$

where $1 > t_2^* > \cdots > t_k^* > 0$, $2\sum_{i=1}^k \nu_i^* + \nu_{k+1}^* = 1$, $\nu_i^* > 0$, $t_i^* = t_i^*(z)$, and $\nu_i^* = \nu_i^*(z)$ are some numbers.

Denote

$$\lambda^*(z) = \lambda_{\min}(M(\xi^*(z)));$$

$\lambda_{\min}(M)$ denotes, as earlier, the minimal eigenvalue of M.

For any arbitrary design ξ, set

$$M_1(\xi) = \int f_{(1)}(t) f_{(1)}^T(t) \xi(dt),$$

$$M_2(\xi) = \int f_{(2)}(t) f_{(2)}^T(t) \xi(dt),$$

where $f_{(1)}(t) = (1, t^2, \ldots, t^{2k})^T$ and $f_{(2)}(t) = (t, t^3, \ldots, t^{2k-1})^T$.

Due to Lemma 3.3.4, for any nonsingular design ξ the multiplicity of the minimal eigenvalue of $M_1(\xi)$ and $M_2(\xi)$ is equal to 1.

Let $p^* = (p_1^*, \ldots, p_k^*, 1)^T$ be an eigenvector of $M_1(\xi)$, corresponding to its minimal eigenvalue, and $q^* = (q_1^*, \ldots, q_{k-1}^*, 1)^T$ a similar vector for $M_2(\xi^*)$. As was shown in the proof of Lemma 3.3.4 both matrices are strictly positive. Therefore (see, e.g. , Gantmacher (1998)), the vectors p^* and q^* are determined uniquely, and all of their elements are not zero and have interlacing signs:

$$\text{sign}(p_{k-i}^*) = (-1)^{i+1}, \quad i = 0, \ldots, k-1,$$

$$\text{sign}(q_{k-i}^*) = (-1)^i, \quad i = 1, \ldots, k-1,$$

Introduce a vector τ,

$$\tau = (\tilde{p}_1, \ldots, \tilde{p}_k, \tilde{q}_1, \ldots, \tilde{q}_k, \omega_1, \ldots, \omega_k, y_2, \ldots, y_k)^T$$

$$= (z^k p_1, \ldots, z p_k, \sqrt{z} z^{k-1} q_1, \ldots, \sqrt{z} q_k,$$

$$2\nu_1/z, \ldots, 2\nu_k/z, t_2^2, \ldots, t_k^2)^T,$$

where $0 < t_k < \cdots < t_2 < 1$, $\nu_i > 0$, $\sum_{i=1}^k \nu_i < 1/2$, $p_1, \ldots, p_k$, and $q_1, \ldots, q_k$ are some numbers.

Let us introduce also the vector

$$\tau^* = \tau^*(z) = (z^k p_1^*, \ldots, z p_k^*, \alpha^* \sqrt{z} z^{k-1} q_1^*, \ldots, \alpha^* \sqrt{z} z q_{k-1}^*, \alpha^* \sqrt{z},$$

$$2\nu_1^*/z, \ldots 2\nu_k^*/z, z t_2^{*2}, \ldots, z t_k^{*2})^T,$$

$$(3.40)$$

where $\alpha^* = \alpha(z)$ is a positive number.

Let

$$\xi_{\tau,z} = \left(\begin{array}{ccccccccc} -r & -rt_2 & \ldots & -rt_k & 0 & rt_k & \ldots & rt_2 & r \\ \nu_1 & \nu_2 & \ldots & \nu_k & \nu_{k+1} & \nu_k & \ldots & \nu_2 & \nu_1 \end{array} \right), \quad r = 1/\sqrt{z},$$

where $\nu_{k+1} = 1 - 2\sum_{i=1}^{k} \nu_k$, be an experimental design corresponding to a vector τ,

$$\varphi(\tau, z) = \frac{p^T M_1(\xi_{\tau,z}) p + q^T M_2(\xi_{\tau,z}) q}{p^T p + q^T q},$$

where $p = (p_1, \ldots, p_k, 1)^T$ and $q = (q_1, \ldots, q_k)^T$.

Consider the equation

$$\frac{\partial}{\partial \tau} \varphi(\tau, z) = 0. \tag{3.41}$$

Since the function $\varphi(\tau, z)$ is obviously differentiable by τ, this equation is well defined.

Lemma 3.4.1 *With arbitrary z, $0 < z < z^*$, where $z^* = 1/r^{*2}$ and r^* is the minimal positive root of the boundary equation (3.39), there exist $\alpha^*(z) > 0$ such that the vector $\tau^*(z)$ determined by relation (3.40) is a solution of (3.41), and*

$$\varphi(\tau^*(z), z) = \lambda^*(z).$$

Proof. By immediate differentiation, we obtain

$$\frac{\partial}{\partial p_i} \varphi(\tau, z) = 2 \left[\frac{e_i^T M_1 p}{p^T p + q^T q} - \frac{\varphi(\tau, z) e_i^T p}{p^T p + q^T q} \right], \quad i = 1, \ldots, k,$$

$$\frac{\partial}{\partial q_i} \varphi(\tau, z) = 2 \left[\frac{e_i^T M_2 q}{p^T p + q^T q} - \frac{\varphi(\tau, z) e_i^T q}{p^T p + q^T q} \right], \quad i = 1, \ldots, k-1,$$

$$(3.42)$$

where $M_1 = M_1(\xi_{\tau,z})$, $M_2 = M_2(\xi_{\tau,z})$, $e_1 = (1, 0, \ldots, 0)^T, \ldots, e_{k+1}^T = (0, \ldots, 0, 1)$, $p = (p_1, \ldots, p_k, 1)^T$, and $q = (q_1, \ldots, q_k)$.

In addition, since p^* and q^* are eigenvectors, we have the relations

$$M_1(\xi_{\tau^*(z),z}) p^* = \lambda^*(z) p^*,$$

$$M_2(\xi_{\tau^*(z),z}) q^* = \lambda^*(z) q^*,$$

$$\varphi(\tau^*(z), z) = \lambda^*(z).$$

Therefore, the right-hand sides of equalities (3.42) vanishe with $\tau = \tau^*(z)$ and we obtain

$$\frac{\partial}{\partial p_i} \varphi(\tau, z)\big|_{\tau = \tau^*(z)} = 0,$$

$$\frac{\partial}{\partial q_i} \varphi(\tau, z)\big|_{\tau = \tau^*(z)} = 0,$$

for $i = 1, 2, \ldots, k$.

In order to prove the equalities

$$\frac{\partial}{\partial w_i} \varphi(\tau, z) = 0, \ \frac{\partial}{\partial y_i} \varphi(\tau, z) = 0, \ i = 1, \ldots, k, \ j = 2, \ldots, k$$

with $\tau = \tau^*(z)$, let us study some properties of the extremal polynomials, defined in Theorem 3.3.1.

We will need the following auxiliary result.

Lemma 3.4.2 *Consider the polynomial regression model on the interval $[-r, r]$ with $r > r^*$, where r^* is the minimal positive root of the boundary equation (3.39). Let $m = 2k + 1 > 2$. Then the extremal polynomial has the unique representation of the form*

$$\rho(t) = (p^T \bar{f}(y))^2 + y(q^T \bar{f}_{(1)}(y))^2, \ y = t^2, \tag{3.43}$$

where $\bar{f}(y) = (1, y, \ldots, y^k)^T$ and $\bar{f}_{(1)}(y) = (1, y, \ldots, y^{k-1})^T$. Moreover, there exist positive values β and β' such that $p = \sqrt{\beta} p^*$ and $q = \sqrt{\beta'} q^*$, where p^* and q^* are defined earlier.

Proof of Lemma 3.4.2. Due to the definition of the extremal polynomial, we have

$$\rho(t) = f^T(t) A^* f(t),$$

where $f(t) = (1, t, \ldots, t^{m-1})^T$ and A^* is defined in Theorem 3.3.1. Due to Theorem 3.3.4 and Lemma 3.3.1, it is a unique polynomial of degree $2(m-1)$ attaining its maximal value equal to $\lambda^*(z)$ in points $0, \pm r t_i^*$, $i = 1, \ldots, k$.

Therefore, $\rho(t)$ is an even polynomial and $\rho(t) = \bar{\rho}(y)$ and $y = t^2$, where $\bar{\rho}(y)$ is a unique polynomial of degree $m+1$ attaining its maximal eigenvalue $\lambda^*(z)$ in points $0, y_1^*/z, \ldots, y_k^*/z, y_i^* = t_i^{*2}, i = 1, \ldots, k$. It follows from here that

$$\sum_{i=1}^{j} (A^*)_{ij+1-i} = 0, \ \ j = 2, 4, \ldots, 2k. \tag{3.44}$$

Denote

$$\bar{p}^* = (p_1^*, 0, p_2^*, 0, \ldots, 0, p_{k+1}^*)^T, \ p_{k+1}^* = 1,$$

$$\hat{p} = \bar{p}^* / \|\bar{p}^*\|,$$

$$\bar{q} = (0, q_1^*, 0, q_2^*, \ldots, 0, q_k^*)^T, \ q_k^* = 1,$$

$$\hat{q} = \bar{q}^* / \|\bar{q}^*\|.$$

Note that vectors $\hat{p}$ and $\hat{q}$ generate an orthonormal basis in $\mathcal{P}$ since

$$M(\xi_{\tau^*(z),z})\hat{p} = \lambda^*(z)\hat{p},$$

$$M(\xi_{\tau^*(z),z})\hat{q} = \lambda^*(z)\hat{q}.$$

Let us show that this basis is an extremal one (in the sense of Theorem 3.3.1). In fact, due to Theorem 3.3.2,

$$A^* = \kappa p_{(1)}p_{(1)}^T + (1-\kappa)p_{(2)}p_{(2)}^T,$$

where

$$p_{(1)} = \delta\hat{p} + \sqrt{1-\delta^2}\hat{q},$$

$$p_{(2)} = \sqrt{1-\delta^2}\hat{p} - \delta\hat{q}$$

for some κ and δ, where $0 \le \kappa, \delta \le 1$.

Thus, from relations (3.44) it follows that

$$\kappa \sum_{i=1}^{j} p_{(1)i}p_{(1)j+1-i} + (1-\kappa) \sum_{i=1}^{j} p_{(2)i}p_{(2)j+1-i} = 0,$$

$j = 2, 4, \ldots, 2k$. A simple calculation now gives

$$\delta\sqrt{1-\delta^2}(1-2\kappa) \sum_{i=1}^{l} p_i^* q_{l+1-i}^* = 0, \; l = 1, 2, \ldots, k.$$

Note that due to the interlacing property of signs of $\{p_i^*, q_i^*\}$, we have

$$\sum_{i=1}^{l} p_i^* q_{l+1-i}^* \ne 0, \; j = 1, 2, \ldots, k,$$

and, therefore,

$$\delta\sqrt{1-\delta^2}(1-2\kappa) = 0;$$

that is, either $\delta\sqrt{1-\delta^2} = 0$ or $\kappa = 1/2$. In both cases, we receive

$$A^* = \kappa'\hat{p}\hat{p}^T + (1-\kappa')\hat{q}\hat{q}^T$$

with $\kappa' = \kappa$ or $\kappa' = 1-\kappa$ and this means that the vectors $\hat{p}$ and $\hat{q}$ generate an extremal basis of $\mathcal{P}$. Due to last formula, we have

$$\rho(t) = f^T(t)A^*f(t)$$

$$= \kappa'(\hat{p}^T f(t))^2 + (1-\kappa')(\hat{q}^T f(t))^2.$$

Denote $\bar{f}(y) = (1, y, \ldots, y^k)$ and $\bar{f}_{(1)}(y) = (1, y, \ldots, y^{k-1})$ and rewrite the extremal polynomial in the form

$$\rho(t) = \bar{\rho}(y) = \beta \left(p^{*T}\bar{f}(y)\right)^2 + \beta' \left(q^{*T}\bar{f}_{(1)}(y)\right)^2 y, \qquad (3.45)$$

where $\beta, \beta' \geq 0$.

Thus, there exists a representation of the form (3.43), and $p = \sqrt{\beta}p^*$ and $q = \sqrt{\beta'}q^*$.

It is easy to check that such a representation is unique.

In fact, note first that β and β' are determined uniquely. Suppose, oppositely, that for $(\beta_1, \beta_1') \neq (\beta, \beta')$,

$$\rho(t) = \bar{\rho}(y) = \beta_1(p^{*T}\bar{f}(y))^2 + \beta_1'(q^{*T}\bar{f}_{(1)}(y))^2 y.$$

Then it should be that

$$(\beta_1 - \beta)(p^{*T}\bar{f}(y))^2 \equiv (\beta' - \beta_1')(q^{*T}\bar{f}_{(1)}(y))^2 y,$$

which is impossible.

Note that $\beta \neq 0$, otherwise $\bar{\rho}(y)$ has degree $m - 2$, which is impossible.

Suppose that $\beta' = 0$. Then $\rho(t) = \text{const} \, (T_{m-1}(t/r))^2$ since the Chebyshev polynomial of the first kind of degree $m-1$ is the unique polynomial of degree $m - 1$ with leading coefficient 2^{m-2} achieving its maximal absolute value in the interval $[-1, 1]$ in m points (see, e.g., Rivlin (1974)). It is easy to check that the design $\xi^*(z)$ should coincide with $\bar{\xi}_r$ which is impossible due to Lemma 3.3.7 for $r > r^*$.

Thus, β and $\beta' > 0$, and from (3.45) we obtain

$$\bar{\rho}(y) = \frac{\left(p^{*T}\bar{f}(y)\right)^2 + \alpha^{*2}\left(q^{*T}\bar{f}_1(y)\right)y}{p^{*T}p^* + \alpha^{*2}q^{*T}q^*}$$

for some $\alpha^* > 0$ ($\alpha^{*2} = \beta'/\beta$).

Note that for $0 < z < z^*$,

$$\left(\|p\|^2 + \|q\|^2\right)\varphi(\tau, z) = \sum_{i=1}^{k}\left[\left(\bar{f}^T(y_i)p\right)^2 + y\left(\bar{f}_{(1)}^T(y_i)q\right)^2\right]\omega_i$$
$$+ (\bar{f}(0)^T p)^2(1 - \omega_1 - \cdots - \omega_k).$$

Therefore, with $\tau = \tau^*(z)$,

$$\frac{\partial}{\partial y_i}\varphi(\tau, z) = \left(\bar{\rho}\left(\frac{y_i^*}{z}\right)\right)' \omega_i = 0, \ i = 1, \ldots, k,$$

$$\frac{\partial}{\partial \omega_i}\varphi(\tau, z) = \bar{\rho}\left(\frac{y_i^*}{z}\right) - \bar{\rho}(0) = 0, \ i = q, \ldots, k.$$

Thus, we proved that for any z, $0 < z < z^*$, there exists $\alpha^* = \alpha^*(z)$ such that $\tau = \tau^*(z)$ is a solution of the equation $\frac{\partial}{\partial \tau}\varphi(\tau, z) = 0$. ∎

Remark 3.4.1 Note that with $m = 2k + 1$, the extremal polynomial $\bar{\rho}(y)$ is a nonnegative polynomial of degree $m - 1 = 2k$ for any $y \geq 0$. Also, for arbitrary nonnegative polynomial of degree $2k$, say $Q(y)$, there exists a unique representation of the form

$$Q(y) = \alpha_1 \prod_{i=1}^{k} (y - u_i)^2 + \alpha_2 \prod_{i=1}^{k-1} (y - v_i)^2 y, \quad \alpha_1 > 0, \; \alpha_2 \geq 0,$$

where $u_1 \leq v_1 \leq u_2 \leq \cdots \leq v_{k-1} \leq u_k$ (see Karlin and Studden (1966, Chap.5)). We will call such a representation the Karlin–Shapley representation. Since this representation has the form (3.43) and the representation of such a form is unique for the extremal polynomial due to Lemma 3.4.2, we obtain that

$$p^{*T} \bar{f}(y) = p_{k+1}^{*} \prod_{i=1}^{k} (y - u_i),$$

$$q^{*T} \bar{f}_{(1)}(y) = q_{k}^{*} \prod_{i=1}^{k-1} (y - v_i);$$

that is, both polynomials have the maximal number of positive roots and they are interlacing, as described above.

For $m = 2k$, a similar result holds. This remark will be needed in the following section.

The equation

$$\frac{\partial \varphi(\tau, z)}{\partial \tau} = 0$$

will be called the *basic equation*. This equation determines support points and weights of an E-optimal design as well as the elements of an extremal basis as implicit functions of the length of the design interval.

It proves possible to find a limit of the vector $\tau^{*}(z)$ with $z \to 0$.

Denote by $J(z)$ the Jacobi matrix of the basic equation

$$J(z) = \left(\frac{\partial^2 \varphi(\tau, z)}{\partial \tau_i \partial \tau_j} \right)_{i,j=1}^{s} \Big|_{\tau = \tau^{*}(z)},$$

where $s = 4k - 1$ is the size of the vector τ.

For any (scalar, vector, or matrix) function $Q(z)$ denote

$$Q_{(0)} = \lim_{z \to z_{(0)}} Q(z),$$

$$Q_{(n)} = \lim_{z \to z_{(0)}} \frac{Q^{(n)}(z)}{n!}, \quad n = 1, 2, \ldots.$$

Let

$$U_n(t) = \frac{\sin((n+1) \arccos t)}{\sqrt{1 - t^2}}, \quad t \in [-1, 1],$$

be the Chebyshev polynomial of the second kind.

Lemma 3.4.3 *With $m = 2k + 1 > 2$, there exists the limit of $\tau^*(z)$ for $z \to z_{(0)} = 0$:*

$$\tau^*_{(0)} = (\tilde{p}_{1(0)}, \ldots, \tilde{p}_{k(0)}, \tilde{q}_{1(0)}, \ldots, \tilde{q}_{k(0)}, \tilde{\omega}_{1(0)}, \ldots, \tilde{\omega}_{k(0)}, \tilde{y}_{2(0)}, \ldots, \tilde{y}_{k(0)})^T,$$

where $\tilde{y}_{k-i+1(0)} = t^2_{k+i}$, $i = 1, \ldots, k$, and $0 < t_{k+1} < \cdots < t_{2k} = 1$ are positive extremal points of $T_{2k-1}(t)$ on $[0, 1]$,

$$(1, t^2, \ldots, t^{2k})\tilde{p}_{(0)} = (t^2 - 1)U_{2(k-1)}(t) \big/ 2^{2(k-1)},$$

$$\tilde{p}_{(0)} = (\tilde{p}_{1(0)}, \ldots, \tilde{p}_{k(0)}, 1)^T,$$

$$(t, t^3, \ldots, t^{2k-1})\tilde{q}_{(0)} = T_{2k-1}(t) \big/ 2^{2(k-1)},$$

$$\omega_{(0)} = \frac{\tilde{q}_{1(0)}}{|\tilde{p}_{1(0)}|} F^{-1} e_1, \quad e_1 = (1, 0, \ldots, 0)^T,$$

$$F = \left((-1)^{k-j-1} \sqrt{\tilde{y}_{j(0)}} \tilde{y}_{j(0)}^{i-1} \right)_{i,j=1}^k.$$

A proof of the analog of this lemma for $m = 2k$ will be given in Section 3.4.3.

Consider now the case $m = 2k > 2$. In this case, we will use the following notation:

$$f_{(1)}(t) = (1, t^2, \ldots, t^{2(k-1)})^T, \quad f_{(2)}(t) = t f_{(1)}(t),$$

$$M_1(\xi) = \int f_{(1)}(t) f_{(1)}^T(t) \xi(dt),$$

$$M_2(\xi) = \int f_{(2)}(t) f_{(2)}^T(t) \xi(dt),$$

$$\tau = (\tilde{p}_1, \ldots, \tilde{p}_k, \tilde{q}_1, \ldots, \tilde{q}_{k-1}, \tilde{\omega}_1, \ldots, \tilde{\omega}_{k-1}, \tilde{y}_2, \ldots, \tilde{y}_{k-1}, y_k)^T$$

$$= (z^{k-1}\sqrt{z} p_1, \ldots, \sqrt{z} p_k, z^{k-1} q_1, \ldots, z q_{k-1},$$

$$2\nu_1/z, \ldots, 2\nu_{k-1}/z, t_2^2, \ldots, t_{k-1}^2, t_k^2/z)^T = (\tau_1, \ldots, \tau_s)^T,$$

$$s = 4k - 3, \quad \nu_i > 0, \quad \sum_{i=1}^{k-1} \nu_i < 1/2,$$

$$\xi_{\tau,z} = \begin{pmatrix} -r & -rt_2 & \cdots & -rt_k & rt_k & \cdots & rt_2 & r \\ \nu_1 & \nu_2 & \cdots & \nu_k & \nu_k & \cdots & \nu_2 & \nu_1 \end{pmatrix},$$

$$\nu_k = 1/2 - \sum_{i=1}^{k-1} \nu_i,$$

$$\varphi(\tau, z) = \frac{p^T M_1(\xi_{\tau,z}) p + q^T M_2(\xi_{\tau,z}) q}{p^T p + q^T q},$$

$$p = (p_1, \ldots, p_k)^T, \quad q = (q_1, \ldots, q_{k-1}, 1)^T.$$

Denote by $p^* = (p_1^*, \ldots, p_{k-1}^*, 1)^T$ and $q^* = (q_1^*, \ldots, q_{k-1}^*, 1)^T$ the eigen-vectors of $M_1(\xi^*(z))$ and $M_2(\xi^*(z))$, corresponding to their minimal eigen-value, respectively, where $\xi^*(z)$ is the unique E-optimal design.

Due to Lemma 3.3.4, this design is of the form

$$\xi^*(z) = \begin{pmatrix} -r & -rt_2^* & \cdots & -rt_k^*, & rt_k^* & \cdots & rt_2^* & r \\ \nu_1^* & \nu_2^* & \cdots & \nu_k^* & \nu_k^* & \cdots & \nu_2^* & \nu_1^* \end{pmatrix}, \qquad (3.46)$$

where $\nu_i > 0$, $i = 1, \ldots, k$, and $\sum_{i=1}^{k} \nu_i^* = 1/2$.

Let $\lambda^*(z) = \lambda_{\min}(M(\xi^*(z)))$ and

$$\tau^*(z) = (\alpha^* z^{k-1} \sqrt{z} p_1^*, \ldots, \alpha^* z \sqrt{z} p_{k-1}^*, \alpha^* \sqrt{z}, z^{k-1} q_1^*, \ldots, q_{k-1}^*,$$

$$2\nu_1^*/z, \ldots, 2\nu_{k-1}^*/z, t_2^{*2}, \ldots, t_{k-1}^{*2}, t_k^{*2}/z)^T,$$

where $\alpha^* = \alpha^*(z)$ is a positive number.

As in the case of odd m, it can be proved that the equation

$$\frac{\partial}{\partial \tau} \varphi(\tau, z) = 0$$

with arbitrary z, $0 < z < z^*$ possesses the solution $\tau = \tau^*(z)$ with some $\alpha^* = \alpha^*(z)$.

Let us point out the form of the limit $\tau^*(z)$ with $z \to 0$.

Lemma 3.4.4 *With $m = 2k$ and $z \to z_{(0)} = 0$, there exists the limit*

$$\tau_{(0)}^* = \lim_{z \to 0} \tau^*(z)$$

$$= (\tilde{p}_{1(0)}, \ldots, \tilde{p}_{k(0)}, \tilde{q}_{1(0)}, \ldots, \tilde{q}_{k-1(0)}, \tilde{\omega}_{1(0)}, \ldots, \tilde{\omega}_{k-1(0)},$$

$$\tilde{y}_{2(0)}, \ldots, \tilde{y}_{k-1(0)}, y_{k(0)})^T,$$

where $\tilde{y}_{i(0)} = t_i^2$, $i = 2, \ldots, k$, $y_{k(0)} = 1$, $0 = t_k < \ldots < t_2 < 1$ are nonnegative extremal points of $T_{2k-1}(t)$,

$$\sum_{i=1}^{k} \tilde{p}_{i(0)} t^{2(i-1)} = T_{2k-2}(t)/2^{2k-3},$$

$$\sum_{i=1}^{k} \tilde{q}_{i(0)} t^{2i-1} = (t^2 - 1)U_{2k-3}(t)/2^{2k-3}, \tilde{q}_{k(0)} = 1,$$

$$\tilde{\omega}_{(0)j} = (-1)^k e_j^T F^{-1}\left(|\frac{\tilde{p}_{2(0)}}{\tilde{p}_{1(0)}}|e_1 + e_2\right), \quad j = 1, \ldots, k - 1,$$

$$F = \left(\tilde{y}_{j(0)}^{i-1}(-1)^{j-1}\right)_{i,j=1}^{k}.$$

A proof of this lemma will be given in the next section.

Let $\tau^*(z)$ be such as described above for $m = 2k+1$ or $m = 2k$ with the corresponding determination of the function $\varphi(\tau, z)$. Let us codetermine the function $\tau^*(z)$ in $z = 0$ by relation $\tau^*(0) = \tau^*_{(0)}$ and for $z < 0$, $|z| \in (0, z^*)$ by relation $\tau^*(z) = \tau^*(-z)$.

The following theorem describes properties of functions $\tau^*(z)$ and $J(z)$. It is a typical result of the functional approach.

Theorem 3.4.1 *Let $m > 2$ and the vector function $\tau^*(z)$ be such as it was determined above. Then there exists a number $\hat{z}$, $0 < \hat{z} \leq z^*$ such that the matrix $J(z)$ is nonsingular with $z \in (-\hat{z}, \hat{z})$ and the vector function $\tau^*(z)$ is real analytic with $z \in (-\hat{z}, \hat{z})$.*

The proof will be given in Section 3.4.3.

Denote

$$g(\tau, z) = \left(\frac{\partial}{\partial \tau_i} \varphi(\tau, z) \right)^s_{i=1} .$$

Due to Theorem 3.4.1 and Lemmas 3.4.3 and 3.4.4, we can calculate the coefficients in the Taylor expansion

$$\tau^*(z) = \sum_{n=0}^{\infty} \tau^*_{(n)} z^n$$

by general recurrent formulas from Section 2.4:

$$\tau^*_{(n+1)} = - \left(J_{(1)} \right)^{-1} \left(g(\tau^*_{<n>}(z), z) \right)_{(n+2)} , \tag{3.47}$$

where $n = 0, 1, \ldots$, and $\tau^*_{(0)}$ is determined in Lemmas 3.4.3 and 3.4.4,

$$\tau^*_{<n>}(z) = \sum_{i=0}^{n} \tau^*_{(i)} z^i .$$

As it was already pointed out in Section 2.4, these formulas can be realized in the software package Maple and others.

Consider now the problem of exactness of these expansions. Denote by $\xi_{(n)} = \xi_{(n)}(z)$ the design

$$\xi_{(n)} = \xi_{\tau^*_{<n>}(z), z}$$

obtained by using n coefficients of these expansions.

Let for $m = 2k + 1$,

$$\bar{\rho}_{(n)}(y, z) = \frac{((1, y, \ldots, y^k)p)^2 + ((1, y, \ldots, y^{k-1})q)^2 y}{p^T p + q^T q} ,$$

and for $m = 2k$,

$$\bar{\rho}_{(n)}(y, z) = \frac{((1, y, \ldots, y^{k-1})p)^2 + ((1, y, \ldots, y^{k-1})q)^2 y}{p^T p + q^T q} ,$$

where

$$p = \tilde{p}_{<n>}(z) = \sum_{i=0}^{n} \tilde{p}_{(i)} z^i,$$

$$q = \tilde{q}_{<n>}(z) = \sum_{i=0}^{n} \tilde{q}_{(i)} z^i,$$

Denote

$$\lambda^*(z) = \lambda_{\min}\left(M(\xi_{\tau^*(z),z})\right),$$

$$\lambda_{[n]}(z) = \lambda_{\min}\left(M(\xi_{(n)}(z))\right),$$

and note that $\lambda_{[n]}(z) \neq \lambda_{<n>}(z)$, generally speaking.

Lemma 3.4.5 *With $m > 2$, the following inequalities are satisfied:*

$$\lambda_{[n]}(z) \leq \lambda^*(z) \leq \max_{0 \leq y \leq 1} \bar{\rho}_{(n)}(y, z).$$

This lemma is an obvious corollary of Theorem 3.3.1. Thus, we have

$$0 \leq \lambda^*(z) - \lambda_{[n]}(z) \leq \max_{0 \leq y \leq 1}\left(\bar{\rho}_n(y, z) - \lambda_{[n]}(z)\right).$$

3.4.2   Limiting designs

In the present section, we will prove Lemma 3.4.4 (note that Lemma 3.4.3 can be proved in a quite similar way). The following proof is based in the representations for extremal polynomials.

Let $m = 2k > 2$, $z = 1/r^2$, and

$$\xi^*(z) = \left(\begin{array}{cccccccc} -t_1^* & -t_2^* & \cdots & -t_k^* & t_k^* & \cdots & t_2^* & t_1^* \\ \nu_1^* & \nu_2^* & \cdots & \nu_k^* & \nu_k^* & \cdots & \nu_2^* & \nu_1^* \end{array} \right),$$

where $t_i^* = t_i^*(z)$, $\nu_i^* = \nu^*(z)$, $i = 1, \ldots, k$, $t_1^* = r = 1/\sqrt{z}$, is the unique E-optimal design for polynomial regression on the interval $[-r, r]$.

Consider the equality

$$\lambda + \gamma(y - 1/z) \prod_{i=2}^{k} (y - y_i)^2 = \frac{(p^T \bar{f}(y))^2 + y(q^T \bar{f}(y))^2}{p^T p + q^T q}, \qquad (3.48)$$

where $\bar{f}(y) = (1, y, \ldots, y^{k-1})^T$, $0 \leq y_k < \ldots < y_2 \leq 1/z$, $\lambda \leq 1$.

Due to Lemmas 3.3.1 and 3.3.4, the left-hand side of this equality with $\lambda = \lambda^*(z)$, $y_i = t_i^{*2}$, $i = 2, \ldots, k$, and a positive $\gamma > 0$ coincides with $\bar{\rho}(y) = \rho(\sqrt{y})$, where $\rho(t)$ is the extremal polynomial defined in Theorem 3.3.1. Additionally, $\lambda^*(z) < 1$ due to Lemma 3.3.1, and by Lemma 3.4.2, the right-hand side with $p = \sqrt{\alpha^*}p^*$, $q = q^*$ is also equal to $\bar{\rho}(y)$.

Thus, equality (3.48) holds for $\lambda = \lambda^*(z)$, $y_i = y_i^*$, $i = 2, \ldots, k$, and $p = \sqrt{\alpha^*} p^*$, $q = q^*$ for some $\gamma > 0$. Equating coefficients under y^{2k-1} in both sides of the equality, we obtain the relation

$$\gamma = \frac{q_k^2}{p^T p + q^T q}.$$

Without loss of generality, set $q_k = 1$. Equating the free terms and coefficients under $y^{2(k-1)}$, derive

$$\lambda = \frac{p_1^2 + \prod_{i=2}^{k} y_i^2/z}{p^T p + q^T q}, \tag{3.49}$$

$$p_k^2 + 2q_k q_{k-1} = -1/z - 2\sum_{i=2}^{k} y_i. \tag{3.50}$$

Let us change the variable $y \to y/z$ and use the following notations

$$\tilde{y}_i = z y_i, \; \tilde{p}_i = p_i z^{k-i} \sqrt{z}, \; \tilde{q}_i = q_i z^{k-i}, \; i = 1, \ldots, k,$$

$$\tilde{y}_i^* = z y_i^*, \; \tilde{p}_i^* = \sqrt{\alpha^*} p_i^* z^{k-i} \sqrt{z}, \; \tilde{q}_i^* = q_i^* z^{k-i}, \; i = 1, \ldots, k,$$

$$\Delta = \sum_{i=1}^{k} \tilde{p}_i^2 z^{2(i-1)} + z \sum_{i=1}^{k} \tilde{q}_i^2 z^{2(i-1)}.$$

Now, (3.48)–(3.50) assume the form

$$\lambda + \frac{1}{\Delta}(y-1)\prod_{i=2}^{k}(y-\tilde{y}_i)^2 = \frac{1}{\Delta}\left[\left(\tilde{p}^T \bar{f}(y)\right)^2 + y\left(\tilde{q}^T \bar{f}(y)\right)^2\right], \tag{3.51}$$

$$\lambda = (\tilde{p}_1^2 + \prod_{i=2}^{k} \tilde{y}_i^2)/\Delta, \tag{3.52}$$

$$\tilde{p}_k^2 + 2\tilde{q}_{k-1} = -1 - 2\sum_{i=2}^{k} \tilde{y}_i, \tag{3.53}$$

where $0 < \tilde{y}_k < \cdots < \tilde{y}_2 < 1$.

By standard arguments, it follows that $t_i^*, p_i^*, q_i^*, i = 1, \ldots, k$, are continuous functions of z. Since $z t_i^{*2} \leq 1$, values $\tilde{y}_i = z t_i^{*2}$ tend to finite limits; denote these limits by $\tilde{y}_{i(0)}, i = 1, \ldots, k$. Since $\lambda^*(z) \leq 1$, $\lambda^*(z)$ tends to a finite limit (say $\lambda_{(0)}$).

Due to Remark 3.4.1, we have

$$\tilde{q}^{*T} \bar{f}(y) = \prod_{i=1}^{k-1}(y - u_i), \; 0 \leq u_i \leq 1, \; i = 1, \ldots, k-1,$$

$$\tilde{p}^{*T} \bar{f}(y) = \tilde{p}_k^* \prod_{i=1}^{k-1}(y - v_i), \; 0 \leq v_i \leq 1, \; i = 1, \ldots, k-1.$$

Therefore, $\tilde{q}_k^* = 1$ and $\tilde{q}_i^*$ tends to finite limits with $z \to 0$. Note that $\tilde{p}_k^*$ tends to a finite limit due to (3.53). It follows that all $\tilde{p}_i^*$ $(i = 1, \ldots, k)$ tend to finite limits and $\Delta \to 1$.

Denote the limits of $\tilde{p}^*$ and $\tilde{q}^*$ by $\tilde{p}_{(0)}$ and $\tilde{q}_{(0)}$, respectively.

Considering (3.52), we see that $\lambda > 1$ if $\tilde{y}_{k(0)} > 0$. However, $\lambda^*(z) < 1$ and $\lim \lambda^*(z) \leq 1$. Therefore, $\lambda_{(0)} = 1$, $\tilde{y}_{k(0)} = 0$ and we obtain by an immediate calculation that

$$\lambda(z) = 1 + \lambda_{(1)} z + H(z) z^2,$$

where $\lambda_{(1)} = -(\tilde{q}_{1(0)}/\tilde{p}_{1(0)})^2 = -2|\tilde{p}_{2(0)}/\tilde{p}_{1(0)}|$ and $H(z) \leq \text{const} < \infty$ for any z, $0 < z < z^*$.

Denote

$$\lambda(z, \tilde{p}, \tilde{q}) = \left(\tilde{p}_1^2 + \prod_{i=2}^{k} \tilde{y}_i^2 \right) \Big/ \Delta.$$

Note that the minimum of $\lambda(z, \tilde{p}, \tilde{q})$ by all vectors $\tilde{p}$ and $\tilde{q} \in R^k$ such that $\tilde{q}_k = 1$ and

$$\left(\tilde{p}^T \bar{f}(\tilde{y}_i^*) \right)^2 + \tilde{y}_i^* \left(\tilde{q}^T \bar{f}(\tilde{y}_i^*) \right)^2 = C, \; i = 1, \ldots, k, \tag{3.54}$$

where C is a positive constant, is equal to $\lambda^*(z)$. Moreover, this minimum is achieved if and only if $\tilde{p} = \tilde{p}^*$ and $\tilde{q} = \tilde{q}^*$, where $\tilde{p}^*$ and $\tilde{q}^*$ satisfy (3.51) with $\lambda = \lambda^*(z)$ and $\tilde{y}_i = \tilde{y}_i^*$, $i = 2, \ldots, k$.

In fact, multiplying both sides of (3.54) by ν_i^*/Δ and summing with $i = 1, \ldots, k$, we obtain

$$\frac{C}{\Delta} = \frac{p^T M_1(\xi^*(z))p + q^T M_2(\xi^*(z))q}{p^T p + q^T q}$$
$$\geq \lambda_{\min}\left(M(\xi^*(z)) \right) = \lambda^*(z),$$

where $p = (p_1, \ldots, p_k)^T$, $q = (q_1, \ldots, q_k)^T$ and $p_i = \tilde{p}_i/z^{k-i+1/2}$, $q_i = \tilde{q}_i/z^{k-i}$, $i = 1, 2, \ldots, k$. The equality holds if and only if $\tilde{p} = \tilde{p}^*$ and $q = \tilde{q}^*$, which follows from (3.51). Note that the pair (p^*, q^*) is determined uniquely due to Lemma 3.4.2.

Passing to the limit in (3.51) with $z \to 0$, we obtain that $\tilde{p}_{(0)}$ and $\tilde{q}_{(0)}$ give the minimum of

$$\lim_{z \to 0} \frac{\lambda(z) - \lambda(0)}{z} = \lambda'(0) = -2 \left| \tilde{p}_{2(0)}/\tilde{p}_{1(0)} \right|$$

under the condition

$$\varphi^2(y_i) + y_i \psi^2(y_i) = 1, \; i = 1, 2, \ldots, k,$$

where $y_i = \tilde{y}_{i(0)}$, $i = 1, \ldots, k$,

$$\varphi(y) = \tilde{p}_{(0)}^T \bar{f}(y)/|\tilde{p}_{1(0)}|, \; \psi(y) = \tilde{q}_{(0)}^T \bar{f}(y)/|\tilde{p}_{1(0)}|.$$

Let us rewrite the condition in the form

$$|\varphi(y_i)| = \sqrt{1 - y_i \psi^2(y_i)}, \; y_i = \tilde{y}_{i(0)}, \; i = 1, \ldots, k. \tag{3.55}$$

By Lemma 3.3.5, the absolute value of the coefficient of polynomial $\varphi(y)$ under y (equal to $|\tilde{p}_{2(0)}/\tilde{p}_{1(0)}|$) will be maximal under restrictions (3.55) if

$$\varphi(y_i) = (-1)^{i-l}\sqrt{1 - y_i \psi^2(y_i)}, \; i = 1, \ldots, k,$$

and

$$l = 0 \text{ or } 1, \; \psi(y_i) = 0, \; y_i = \tilde{y}_{i(0)}, \; i = 1, \ldots, k-1.$$

From the last equalities, it follows that

$$\psi(y) = \frac{1}{|\tilde{p}_{1(0)}|} \prod_{i=1}^{k-1}(y - \tilde{y}_{i(0)}).$$

Let us introduce the polynomials

$$h_1(t) = \varphi(t^2), \; h_2(t) = \frac{t}{|\tilde{p}_{1(0)}|} \prod_{i=2}^{k-1}(t^2 - \tilde{y}_{i(0)}).$$

Note that

$$y\psi^2(y) - \frac{1}{\tilde{p}_{1(0)}^2}y^2(y-1)\prod_{i=2}^{k-1}(y - \tilde{y}_{i(0)})^2$$
$$= \frac{1}{\tilde{p}_{1(0)}^2}y(1-y)\prod_{i=2}^{k-1}(y - \tilde{y}_{i(0)})^2 = (1-y)h_2^2(\sqrt{y}). \tag{3.56}$$

Let us divide both sides of (3.51) by $\tilde{p}_1^2$ and pass to the limit with $z \to 0$. Taking into account (3.56), we derive

$$1 = h_1^2(t) + (1 - t^2)h_2^2(t).$$

As is known (see, e.g., Karlin and Studden (1966, Section 9.5)), this identity in the class of polynomials holds if and only if $h_1(t)$ and $h_2(t)$ are Chebyshev polynomials of the first and the second kind, respectively, with a sign precision.

Taking into account degrees of $h_1(t)$ and $h_2(t)$ and signs of coefficients p_k and q_k, we obtain that

$$h_1(t) = T_{2k-2}(t), \; h_2(t) = U_{2k-3}(t).$$

From this, we obtain formulas given in the formulation of Lemma 3.4.4 for $\tilde{p}_{(0)}$ and $\tilde{q}_{(0)}$ by simple calculations.

Additionally, since

$$h_1(t) \leq 1,$$

$\tilde{y}_{2(0)}, \ldots, \tilde{y}_{k-1(0)}$ are squares of extremal points of the Chebyshev polynomial $T_{2k-2}(t)$.

Now, let $\omega_i(z) = 2\nu_i^*(z)$, $i = 1, \ldots, k$, and $\tilde{\omega}_i(z) = \omega_i(z)/z$, $i = 1, \ldots, k - 1$. Since $\omega_i(z) > 0$ and $\sum_{i=1}^{k} \omega_i(z) = 1$, there exist the limits $\omega_{i(0)} = \lim_{z \to 0} \omega_i(z)$, $i = 1, \ldots, k$. Let us find these limits. Then we will prove the existence of the limits $y_{(0)} = \lim_{z \to 0} y_k^*(z)$ and $\tilde{\omega}_{i(0)} = \lim_{z \to 0} \tilde{\omega}_i(z)$, $i = 1, \ldots, k - 1$ and find them.

Let

$$\tilde{\xi} = \tilde{\xi}^*(z) = \begin{pmatrix} -t_1^*(z)\sqrt{z} & \cdots & -t_k^*(z)\sqrt{z} & t_k^*(z)\sqrt{z} & \cdots & t_1^*(z)\sqrt{z} \\ \nu_1^*(z) & \cdots & \nu_k^*(z) & \nu_k^*(z) & \cdots & \nu_1^*(z) \end{pmatrix},$$

where $t_1^*(z)\sqrt{z} = 1$.

Consider the matrices

$$\begin{aligned} M_1^*(z) &= M_1(\xi^*(z)), & M_2^*(z) &= M_2(\xi^*(z)), \\ \tilde{M}_1(z) &= M_1(\tilde{\xi}^*(z)), & \tilde{M}_2(z) &= M_2(\tilde{\xi}^*(z)). \end{aligned}$$

A direct calculation shows that

$$\begin{aligned} M_1^* &= \tilde{Z}_1^{-1} \tilde{M}_1 \tilde{Z}_1^{-1}, & \tilde{Z}_1 &= \mathrm{diag}\{1, z, \ldots, z^{k-1}\}, \\ M_2^* &= \tilde{Z}_2^{-1} \tilde{M}_2 \tilde{Z}_2^{-1}, & \tilde{Z}_2 &= \mathrm{diag}\{\sqrt{z}, \sqrt{z}z, \ldots, \sqrt{z}z^{k-1}\}, \end{aligned}$$

Denote

$$Z_1 = \tilde{Z}_1^2, \ Z_2 = \tilde{Z}_2^2.$$

Then from equalities

$$M_1^* p^* = \lambda^*(z) p^*,$$

$$M_2^* q^* = \lambda^*(z) q^*,$$

it follows, that

$$\tilde{M}_1 \tilde{p} = \lambda^*(z) Z_1 \tilde{p}, \qquad (3.57)$$

$$\tilde{M}_2 \tilde{q} = \lambda^*(z) Z_2 \tilde{q}, \qquad (3.58)$$

where $\tilde{p} = z^{k+1/2} \tilde{Z}_1^{-1} p^*$ and $\tilde{q} = z^{k+1/2} \tilde{Z}_2^{-1} q^*$.

Let us pass to the limit in (3.57). Then we have

$$\tilde{M}_{1(0)} \tilde{p}_{(0)} = \tilde{p}_{1(0)} e_1, \ e_1 = (1, 0, \ldots, 0)^T.$$

Remember that

$$\tilde{p}_{(0)}^T \bar{f}(\tilde{y}_{i(0)}) / |\tilde{p}_{1(0)}| = \varphi(\tilde{y}_{i(0)}) = T_{2k-2}(\sqrt{y_{i(0)}}) = (-1)^{i-1}.$$

Therefore,

$$\tilde{p}_{(0)}^T f_{(1)}(\sqrt{y_{i(0)}}) = \tilde{p}_{(0)}^T \bar{f}(\tilde{y}_{i(0)}) = (-1)^{i-1} |\tilde{p}_{1(0)}|. \qquad (3.59)$$

By the definition of $\tilde{M}_1$ we have

$$\tilde{M}_1(z) = \sum_{j=1}^{k} f_{(1)}(\sqrt{\tilde{y}_j}) f_{(1)}^T(\sqrt{\tilde{y}_j}) \omega_j.$$

Using (3.59), we derive

$$\tilde{M}_{1(0)} \tilde{p}_{(0)} = F\omega_{(0)} |\tilde{p}_{1(0)}| = |\tilde{p}_{1(0)}| e_1,$$

where

$$F = \left(\tilde{y}_{j(0)}^{i-1} (-1)^{j-1} \right)_{i,j=1}^{k}.$$

Since $\tilde{y}_{k(0)} = 0$, we have

$$F e_k = e_1$$

and it means that $\omega_{(0)} = e_k$.

Now, using this limit value, we obtain that

$$\tilde{M}_{1(0)} = f_{(1)}(0) f_{(1)}^T(0) = e_1 e_1^T.$$

It also follows from $\omega_{(0)} = e_k$ that

$$\tilde{M}_{2(0)} = f_{(2)}(0) f_{(2)}^T(0) = 0.$$

Let us divide both sides of (3.58) by z and pass to the limit with $z \to 0$. Taking into account that $M_{2(0)}$ is a zero matrix, we derive

$$\tilde{M}_{2(1)} \tilde{q}_{(0)} = \tilde{q}_{1(0)} e_1.$$

Using the definition of $\tilde{M}_2 = \tilde{M}_2(z)$, rewrite the last equality in the form

$$\sum_{i=1}^{k-1} f_{(2)}(\sqrt{\tilde{y}_{i(0)}}) f_{(2)}^T(\sqrt{\tilde{y}_{i(0)}}) \tilde{\omega}_{i(0)} \tilde{q}_{(0)} + y_{k(0)}^* e_1 e_1^T \tilde{q}_{(0)} = \tilde{q}_{1(0)} e_1. \qquad (3.60)$$

Since

$$f_{(2)}^T(\sqrt{y}) \tilde{q}_{(0)} = |\tilde{p}_{1_{(0)}}| \sqrt{y} \psi(y) = |\tilde{p}_{1_{(0)}}| \sqrt{y} \prod_{i=1}^{k} (y - \tilde{y}_{i(0)}),$$

we have

$$f_{(2)}^T(\sqrt{y}) \tilde{q}_{(0)} = 0, \; y = \tilde{y}_{i(0)}, \; i = 1, \dots, k.$$

Inserting these values to (3.60), we obtain that

$$y_{k(0)}^* \tilde{q}_{1(0)} e_1 = \tilde{q}_{1(0)} e_1$$

and $y_{k(0)}^* = 1$.

Now, let us return to equality (3.57) and divide both sides by z. Passing to the limit on the right-hand side, we obtain the expression

$$\tilde{p}_{1(1)}e_1 + \lambda_{(1)}\tilde{p}_{1(0)}e_1,$$

where $\lambda_{(1)} = -2|\tilde{p}_{2(0)}/\tilde{p}_{1(0)}|$. Since this limit is finite, we conclude that there exist finite limits of $\tilde{\omega}_i(z) = \omega_i(z)/z$, $i = 1, \ldots, k$.

Using values $\omega_{(0)} = e_k$, $y^*_{k(0)} = 1$, we find

$$\tilde{M}_{1(1)} = \sum_{i=1}^{k-1} \left[\bar{f}(\tilde{y}_{i(0)})\bar{f}^T(\tilde{y}_{i(0)}) - \bar{f}(\tilde{y}_{k(0)})\bar{f}^T(\tilde{y}_{k(0)}) \right] \tilde{\omega}_{i(0)} \qquad (3.61)$$
$$+ e_2 e_1^T + e_1 e_2^T.$$

In addition, we obtain from (3.57) that

$$\tilde{M}_{1(1)}\tilde{p}_{(0)} + \tilde{M}_{1(0)}\tilde{p}_{(1)} = \tilde{p}_{1(1)}e_1 + \lambda_{(1)}\tilde{p}_{1(0)}e_1,$$

and since $\tilde{M}_{1(0)}\tilde{p}_{(1)} = e_1 e_1^T \tilde{p}_{(1)} = \tilde{p}_{1(1)}e_1$, we have

$$\tilde{M}_{1(1)}\tilde{p}_{(0)} = \lambda_{(1)}\tilde{p}_{1(0)}e_1. \qquad (3.62)$$

Denote $\tilde{\omega}_{(0)} = (\tilde{\omega}_{1(0)}, \ldots, \tilde{\omega}_{k-1(0)}, -\sum_{i=1}^{k-1} \tilde{\omega}_{i(0)})^T$. Then inserting (3.62) on the left-hand side of (3.62), we find

$$F\tilde{\omega}_{(0)}|p_{1(0)}| + e_2\tilde{p}_{1(0)} + e_1\tilde{p}_{2(0)} = -2\left|\frac{\tilde{p}_{2(0)}}{\tilde{p}_{1(0)}}\right|\tilde{p}_{1(0)}e_1.$$

Since sign $\tilde{p}_{1(0)} = (-1)^{k-1}$ and $\frac{\tilde{p}_{2(0)}}{\tilde{p}_{1(0)}} < 0$, we have

$$\tilde{\omega}_{(0)} = F^{-1}\left[(-1)^k e_2 + (-1)^k |\tilde{p}_{2(0)}/\tilde{p}_{1(0)}|e_1\right].$$

This formula completes our proof.   ∎

In order to elucidate the results obtained, let us consider the case $m = 4$. In this case, $k = 2$ and by the above formulas we have

$$\tilde{y}_{1(0)} = 1, \quad \tilde{y}_{2(0)} = 0,$$

$$h_1(t) = T_2(t) = 2t^2 - 1, \quad h_2(t) = U_1(t) = 2t,$$

$$\tilde{p}_{(0)} = (-1/2, 1)^T, \quad \tilde{q}_{(0)} = (-1, 1)^T,$$

$$\lambda_{(1)} = -2|\tilde{p}_{2(0)}/\tilde{p}_{1(0)}| = -4.$$

The limit of (3.51) is now

$$1 + (y - 1)y^2 = (2y - 1)^2 + 4y(y - 1)^2$$

and

$$F = \begin{pmatrix} 1 & -1 \\ \tilde{y}_{1(0)} & -\tilde{y}_{2(0)} \end{pmatrix} = \begin{pmatrix} 1 & -1 \\ 1 & 0 \end{pmatrix}, \quad F^{-1} = \begin{pmatrix} 0 & 1 \\ -1 & 1 \end{pmatrix},$$

$$\tilde{\omega}_{2(0)} = e_2^T F^{-1}(e_2 + 2e_1) = (-1, 1)\begin{pmatrix} 2 \\ 1 \end{pmatrix} = -1.$$

Thus,

$$\tau_{(0)} = (\tilde{p}_{1(0)}, \tilde{p}_{2(0)}, \tilde{q}_{1(0)}, \tilde{\omega}_{1(0)}, y_{2(0)})^T = (-1/2, 1, -1, 1, 1)^T.$$

These results will be used in Section 3.4.4 for constructing non-Chebyshev E-optimal designs.

3.4.3   Proof of the main theorem

In this section, we will prove Theorem 3.4.1 in the case $m = 2k$. A proof for $m = 2k + 1$ can be constructed in a similar way.

Let $m = 2k$, $\mathfrak{X} = [-r, r]$, and $r = 1/\sqrt{z}$, $0 < z < z^*$. Note that the extremal polynomial is positive for all sufficiently small values of z due to Lemmas 3.4.4 and 3.4.5. Denote by $\hat{z}$ the supremum of all such values.

With $z \geq z^*$ the polynomial $\rho(t)$ is the square of the Chebyshev polynomial and it vanishes in its zeros. Thus, $\hat{z} \leq z^*$.

Let us now prove that with $0 < z < \hat{z}$ the determinant of the Jacobi matrix $J(z)$ does not vanish.

Denote $\pi^T = (p^T, q^T)$ and introduce the matrix P_π:

$$P_\pi = \begin{pmatrix} p_1 & p_2 & \cdots & p_k & 0 & 0 & \cdots & 0 \\ 0 & p_1 & p_2 & \cdots & p_k & 0 & \cdots & 0 \\ \vdots & \vdots & \vdots & \ddots & \vdots & \vdots & \ddots & \vdots \\ 0 & \cdots & 0 & p_1 & p_2 & \cdots & p_k & 0 \\ 0 & q_1 & q_2 & \cdots & q_k & 0 & \cdots & 0 \\ 0 & 0 & q_1 & q_2 & \cdots & q_k & \cdots & 0 \\ \vdots & \vdots & \vdots & \vdots & \ddots & \vdots & \ddots & \vdots \\ 0 & 0 & \cdots & 0 & q_1 & q_2 & \cdots & q_k \end{pmatrix}.$$

(The order of matrix P_π is $2k \times 2k$.)

This matrix is the resultant matrix of polynomials $p^T \bar{f}(t^2)$ and $tq^T \bar{f}(t^2)$, $\bar{f}(y) = (1, y, \ldots, y^{k-1})$. It is known (see, e.g., Van der Warden (1967)) that $\det P_\pi \neq 0$ if and only if these polynomials have no common roots.

Denote

$$Z = \mathrm{diag}\{1, z^2, \ldots, z^{2(k-1)}, z, z^3, \ldots, z^{2k-1}\}.$$

An immediate verification shows that the function $\varphi(\tau, z)$ can be written in the form

$$\varphi(\tau, z) = \tilde{\pi}^T P_{\tilde{\pi}} c / \tilde{\pi}^T Z \tilde{\pi},$$

where

$$c = \sum_{i=1}^{k} f(\tilde{y}_i)\omega_i, \quad \omega_i = 2\nu_i, \quad \tilde{\pi}^T = (\tilde{p}^T, \tilde{q}^T), \quad \omega_i = 2\nu_i^*.$$

By immediately calculating the derivatives $\partial^2 \varphi(\tau, z)/\partial \tau_i \partial \tau_j$ $(i, j = 1, \ldots, s)$ at the point $\tau = \tau^*(z)$, we obtain the following formulas:

$$J = J(z) = \frac{2}{\pi^T Z \pi} \begin{pmatrix} \tilde{M} - \lambda Z & P_\pi Y \\ (P_\pi Y)^T & D \end{pmatrix}_{\underline{}}, \tag{3.63}$$

where the symbol "$\underline{}$" right of a matrix means that its $2k$-th row and $2k$-th column are rejected,

$$\tilde{M} = \begin{pmatrix} M_1(\tilde{\xi}) & 0 \\ 0 & M_2(\tilde{\xi}) \end{pmatrix},$$

$$\tilde{\xi} = \begin{pmatrix} -1 & t_2^* & \cdots & t_k^* & t_k^* & \cdots & t_2^* & 1 \\ \nu_1^* & \nu_2^* & \cdots & \nu_k^* & \nu_k^* & \cdots & \nu_2^* & \nu_1^* \end{pmatrix},$$

$$Y = (Y_\nu \dot{:} Y_y), \quad Y_\nu = ((f(\tilde{y}_i) - f(zy_k))z)_{i=1}^{k-1},$$

$$Y_y = \left((f'(\tilde{y}_i)z\tilde{\omega}_i)_{i=2}^{k-1} \dot{:} f'(zy_k)z\omega_k \right), \quad \tilde{y}_1 = 1, \quad \omega_k = 1 - \sum_{i=1}^{k-1} \tilde{\omega}_i z$$

$$D = \begin{pmatrix} 0 & 0 \\ 0 & E \end{pmatrix}, \quad \lambda = \varphi(\tau^*(z), z),$$

$$E = \frac{1}{2}\text{diag}\left\{ \pi^T P_\pi f''(\tilde{y}_2)\tilde{\omega}_2 z, \ldots, \pi^T P_\pi f''(\tilde{y}_{k-1})\tilde{\omega}_{k-1} z, \right.$$

$$\left. \pi^T P_\pi f''(zy_k)\omega_k z^2 \right\}.$$

$$f(y) = (1, y, \ldots, y^{2k-1})^T, \quad \pi^T = \tilde{\pi}^T = (\tilde{p}^T, \tilde{q}^T).$$

In this representation, vanishing terms, e.g., $\pi^T P_\pi f'(\tilde{y}_i)$, are omitted. Set $z^* = 1/r^{*2}$.

Let A be the matrix $-\dfrac{\pi^T Z \pi}{2}J$ $-$ with omitted k-th column and k-th row, and let a be the omitted column with omitted k-th element, $a^* = (\dfrac{\pi^T Z \pi}{2}J)_{kk}$. Note that the matrix A is of the following form:

$$A = \begin{pmatrix} G & H \\ H^T & D \end{pmatrix}, \tag{3.64}$$

where G is the matrix $\tilde{M} - \lambda Z$ with omitted k-th and $2k$-th columns and k-th and $2k$-th rows, H is the matrix $P_{\tilde{\pi}} Y$ with omitted k-th and $2k$-th rows, matrix D is the same as in (3.63) to constant precision.

Since the multiplicity of the minimal eigenvalue of the matrix $Z_1 \tilde{M} Z_1$ ($Z_1 = \mathrm{diag}\{1, 1/z, \ldots, 1/z^{k-1}, 1/\sqrt{z}, \ldots, 1/(\sqrt{z}z^{k-1})\}$) is not more than 2, and there are no zeros in the vector π^*, the matrix G is positive definite.

Since the Vandermonde determinant is not zero, the matrix Y is of full rank. Since the polynomials $\hat{p}^T f(x)$ and $\hat{q}^T f(x)$ have no common factors and P_π is the resultant matrix of these polynomials, $\det P_\pi \neq 0$. Therefore, the matrix H is of full rank.

By the Frobenius formula (see, e.g., Fedorov (1972, Chap.1)),

$$\det A = \det G \det(\mathcal{D} - H^T G^{-1} H).$$

Since the matrix H is of full rank, $H^T G^{-1} H > 0$.

All of the elements of the matrix $\mathcal{D}$ are nonpositive and $-\mathcal{D} \geq 0$, from which it follows that

$$\det A > 0,$$

since the matrices $\mathcal{D}$ and $H^T G^{-1} H$ are of order $(2k-1) \times (2k-1)$.

Let us demonstrate that $\det J(z) \neq 0$ for $z \in (0, \hat{z})$. Multiplying the 2nd, 3rd, ..., and k-th rows of the matrix $J(z)$ by $\tilde{p}_2, \ldots, \tilde{p}_k$, respectively, adding them to the first one, multiplied by $\tilde{p}_1$, gives us the matrix

$$
\begin{pmatrix}
0 & b_{(1)}^T & b_{(2)}^T \\
b_{(1)} & G & H \\
b_{(2)} & H^T & \mathcal{D}
\end{pmatrix},
$$

where $b_{(1)}^T = (0, \ldots, 0)$ and

$$b_{(2)}^T = \left((\tilde{p}^T \bar{f}(\tilde{y}_1))^2 z - (\tilde{p}^T \bar{f}(y_k z))^2 z, \ldots, (\tilde{p}^T \bar{f}(\tilde{y}_{k-1}))^2 z - (\tilde{p}^T \bar{f}(y_k z))^2 z \right.,$$

$$\left. ((\tilde{p}^T \bar{f}(\tilde{y}_2))^2)' \tilde{\omega}_2 z, \ldots, ((\tilde{p}^T \bar{f}(y_k z))^2)' \omega_k z \right).$$

Note that the vector $b_{(2)} \neq 0$, since, otherwise, $p^T f(y)$ would be a Chebyshev polynomial, which is impossible. Therefore, $\det J(z) = b^T A^{-1} b \neq 0$.
∎

Thus, $\det J(z) \neq 0$ with $z \in (0, \hat{z})$.

By a straightforward but rather tedious calculation, it can be checked that the limit of $J(z)$ with $z \to 0$ exists and is equal to the zero matrix. Also, the limiting matrix $J_{(1)} = \lim_{z \to 0} J(z)/z$ exists and $\det J_{(1)} \neq 0$. The last relation can be proved by arguments similar that was already used for proving $\det J(z) \neq 0$.

Now, by the Implicit Function Theorem (see Section 1.8), $\tau^*(z)$ is a real analytic vector function for $z \in (-\hat{z}, 0) \cup (0, \hat{z})$. Also, by the same theorem, $\tau^*(z)$ is real analytic in a vicinity of zero since $\det J_{(1)} \neq 0$.

In the next section, we will construct E-optimal designs for cases $m = 3, 4$, and 5. For $m = 3$ the solution proves to be available in an explicit form. Also, for $m = 4$ and 5, we will construct E-optimal designs with the help of Theorem 3.4.1.

3.4.4 Examples

For the case $m = 2$, E-optimal designs were found in an explicit form for arbitrary design intervals in Section 3.3.2. Here, we consider the case of the symmetrical design interval for $m = 3$, 4, and 5. In the case of quadratic model the $(m = 3)$, E-optimal designs are found in an explicit form. For the cubic model $(m = 4)$ and the model of the fourth degree $(m = 5)$, we will construct E-optimal designs with the help of the functional approach.

Example 3.4.1 Quadratic model on symmetrical interval

Let

$$\eta(t, \theta) = \theta_1 + \theta_2 t + \theta_3 t^2, \quad t \in [-r, r].$$

In this case, an explicit form of E-optimal designs can be obtained by a direct analysis of the characteristic equation and is given by the following theorem.

Theorem 3.4.2 *For the quadratic model on symmetrical intervals $\mathfrak{X} = [-r, r]$, an E-optimal design is unique and it is*

$$\xi^* = \left(\begin{array}{ccc} -r & 0 & r \\ \mu & 1 - 2\mu & \mu \end{array} \right),$$

where

$$\mu = \begin{cases} \dfrac{1}{4 + r^4} & \text{for} \quad r \leq \sqrt{2} \\[2mm] \dfrac{r^2 - 1}{2r^4} & \text{for} \quad r > \sqrt{2}. \end{cases}$$

In the first, case $\lambda^* = \frac{r^4}{4 + r^4}$, and in the second case, $\lambda^* = \frac{r^2 - 1}{r^2}$.

Proof. From Lemma 3.3.4, it follows that an E-optimal design is unique and is of the form

$$\xi_\mu = \left(\begin{array}{ccc} -r & 0 & r \\ \mu & 1 - 2\mu & \mu \end{array} \right), \quad 0 < \mu < 1/2.$$

Thus, the problem is reduced to maximization by $\mu \in [0, 1/2]$, the value

$$\lambda_{\min} \left(M(\xi_\mu) \right).$$

This value can be found by a direct calculation using the characteristic equation. In fact, eigenvalues of the matrix

$$M(\xi_\mu) = \left(\begin{array}{ccc} 1 & 0 & 2\mu r^2 \\ 0 & 2\mu r^2 & 0 \\ 2\mu r^2 & 0 & 2\mu r^4 \end{array} \right)$$

are the roots of the characteristic equation

$$\det(M - \lambda I) = 0,$$

where I is the identity matrix.

Since

$$\det(M(\xi^*) - \lambda I) = (2\mu r^2 - \lambda)\left[(1 - \lambda)(2\mu r^4 - \lambda) - 4\mu^2 r^4\right],$$

we obtain the following expressions for the eigenvalues,

$$\lambda_1 = 2\mu r^2,$$

$$\lambda_{2,3} = \frac{1 + 2\mu r^4 \pm \sqrt{(1 + 2\mu r^4)^2 + 16\mu^2 r^4 - 8\mu r^4}}{2},$$

and $\lambda_3 \leq \lambda_2$.

Let us find a stationary point of the function $\lambda_3 = \lambda_3(\mu)$ by the equation $(\lambda_3(\mu))' = 0$:

$$2r^4 - \frac{4r^4(1 + 2\mu r^4) + 32\mu r^4 - 8r^4}{2\sqrt{(1 + 2\mu r^4)^2 + 16\mu^2 r^4 - 8\mu r^4}} = 0.$$

Note that the denominator is not equal to zero and the equation is reduced to

$$\sqrt{(1 + 2\mu r^4) + 16\mu^2 r^4 - 8\mu r^4} = 1 + 2\mu r^4 + 8\mu - 2.$$

Squaring both sides and collecting similar terms, we obtain

$$\mu\left((r^4 + 1)\mu - 1\right) = 0.$$

Denote by $\mu^* = \frac{1}{r^4 + 4}$ one of the two roots of this equation.

The root $\mu = 0$ is not appropriate since $\lambda_3(0) = 0$. In addition, $\lambda_3(1/2) = 0$ and, therefore,

$$\sup_{0 < \mu < 1/2} \lambda_3(\mu) = \max_{0 \leq \mu \leq 1/2} \lambda_3(\mu) = \lambda_3(\mu^*) = \frac{r^4}{r^4 + 4}.$$

This value is the minimal eigenvalue of $M(\xi(\mu^*))$ if $\lambda_3(\mu^*) \leq \lambda_1(\mu^*)$; that is, if

$$\frac{r^4}{r^4 + 4} \leq \frac{2r^2}{r^4 + 4}.$$

Thus, with $r \leq \sqrt{2}$,

$$\sup_{0 < \mu < 1/2} \lambda_{\min}\left(M(\xi(\mu))\right) = \lambda_3(\mu^*) = r^4/(r^4 + 4).$$

For $r > \sqrt{2}$, the value μ for an E-optimal design should satisfy the condition $\lambda_1(\mu) = \lambda_3(\mu)$; that is,

$$4\mu r^2 = 1 + 2\mu r^4 - \sqrt{(1 + 2\mu r^4)^2 + 16\mu^2 r^4 - 8\mu r^4},$$

$$\sqrt{(1 + 2\mu r^4)^2 + 16\mu^2 r^4 - 8\mu r^4} = 1 + 2\mu r^4 - 4\mu r^2.$$

Squaring and collecting similar terms, we have

$$-8\mu r^4 = -8\mu r^2 - 16\mu^2 r^6.$$

From this it follows $r^2 = 1 + 2\mu r^4$ and, therefore,

$$\mu = \frac{r^2 - 1}{2r^4}, \quad \lambda^* = \lambda_{1,3}(\mu) = \frac{r^2 - 1}{r^2},$$

which completes the proof.   ■

Note that in the present case, dim $\mathcal{P} = 1$ with $r < \sqrt{2}$ and dim $\mathcal{P} = 2$ with $r \geq \sqrt{2}$. The Chebyshev design is

$$\bar{\xi}(r) = \begin{pmatrix} -r & 0 & r \\ \mu^* & 1 - 2\mu^* & \mu^* \end{pmatrix},$$

where

$$\mu^* = \frac{1}{r^4 + 4} = \frac{z^2}{1 + 4z^2}, \quad z = 1/r^2,$$

and it coincides with an E-optimal design with $r \leq \sqrt{2}$. The value $r^* = \sqrt{2}$ is the minimal positive root of the boundary equation

$$\lambda_2 \left(M(\bar{\xi}(r)) \right) = \bar{\lambda}(r),$$

which assumes the form

$$\frac{2z}{1 + 4z^2} = \frac{1}{1 + 4z^2}, \quad z = 1/r^2,$$

and reduces to

$$2z - 1 = 0.$$

This equation has, in fact, the unique solution $z^* = 1/2$ and $r^* = 1/\sqrt{z^*} = \sqrt{2}$.

With $r > \sqrt{2}$, the E-optimal design is also located in Chebyshev points, but it is not a Chebyshev design in the sense introduced in Section 3.3.

For $m > 3$ it seems impossible to find non-Chebyshev E-optimal designs explicitly and we will apply the functional approach developed above.

Example 3.4.2 Cubic model
Let $m = 4$;

$$\eta(t, \theta) = \theta_1 + \theta_2 t + \theta_3 t^2 + \theta_4 t^3, \quad t \in [-r, r].$$

A direct calculation shows that the boundary equation (3.39) in this case assumes the form

$$64z^5 - 32z^4 + 60z^3 - 30z^2 + 11z - 3 = 0.$$

This equation has the unique positive root $z^* = z^*(4) \approx 0.381425$.

Therefore, due to Theorem 3.4.2 with $r \le r^*(4) = 1/\sqrt{z^*(4)}$, the E-optimal design coincides with the Chebyshev design and has the form

$$\xi^* = \begin{pmatrix} -r & -rt & rt & r \\ \mu & \frac{1}{2} - \mu & \frac{1}{2} - \mu & \mu \end{pmatrix},$$

where t is the unique positive root of the polynomial

$$(T_3(t))' = (4t^3 - 3t)' = 12t^2 - 3,$$

$t = 1/2$, and μ can be found from (3.37) and is equal to

$$\mu = \frac{3 + 16z^2}{6(9 + 16z^2)},$$

$$\lambda_{\min}(M(\xi^*)) = \bar{\lambda}(r) = \frac{1}{9z + 16z^3}, z = 1/r^2.$$

For $r > r^*(4)$, an E-optimal design due to Lemma 3.3.1 has the same form. Due to Lemma 3.4.4, the limiting vector $\tau_{(0)}$ is equal to

$$\tau_{(0)} = (\tilde{p}_{1(0)}, \tilde{p}_{2(0)}, \tilde{q}_{1(0)}, \tilde{\omega}_{1(0)}, y_{2(0)})^T = (-1/2, 1, -1, 1, 1)^T.$$

Using this vector and the explicit form of matrix $\bar{J}(z)$, we obtain $\bar{J}_{(0)} = 0$,

$$\bar{J}_{(1)} = \begin{pmatrix} 4 & 2 & 0 & 1 & 1 \\ 2 & 1 & 0 & 1/2 & -1/2 \\ 0 & 0 & 1 & 0 & -1 \\ 1 & 1/2 & 0 & 0 & 0 \\ 1 & -1/2 & -1 & 0 & 0 \end{pmatrix}, \quad g_{(2)} = \begin{pmatrix} 3 \\ 1/2 \\ -2 \\ 0 \\ -1 \end{pmatrix}.$$

Therefore, $\tau_{(1)} = -\bar{J}_{(1)}^{-1} g_{(2)} = (1, -2, 1, -2, 1)^T$. Coefficients $\tau_{(n)}$ with $n = 0, 1, \ldots, 9$ calculated by (3.47) are given in Table 3.4.

Table 3.4: Table of coefficients $(m = 4)$

n	0	1	2	3	4	5	6	7	8	9
$\tilde{p}_1$	$-\frac{1}{2}$	1	$-\frac{1}{2}$	$\frac{5}{2}$	$-\frac{19}{4}$	8	9	$-\frac{239}{4}$	$\frac{3533}{16}$	$-\frac{2645}{8}$
$\tilde{p}_2$	1	-2	1	-3	$\frac{1}{2}$	2	-38	$\frac{167}{2}$	$-\frac{1297}{8}$	$-\frac{931}{4}$
$\tilde{q}_1$	-1	1	-2	5	-8	4	31	-147	362	-348
$\tilde{\omega}_1$	1	-2	1	0	8	-38	78	2	-579	2064
y_2	1	-1	0	1	-2	6	-13	11	58	-350
λ	1	-4	8	-8	-4	24	-8	-132	404	-364

In Table 3.4 the coefficients $\lambda_{(n)}$, $n = 0, 1, \ldots, 10$, for the function

$$\lambda(z) = \lambda_{\min}\left(M(\xi_{\tau(z),z})\right)$$

are also given.

Note that all coefficients are exact. With the help of Table 3.4, we can construct the design

$$\xi_{(n)}(z) = \xi_{\tau_{(n)}(z),z}.$$

From general formulas (3.46), we obtain

$$rt = \sqrt{y_1}, \quad \mu = z\omega_1/2.$$

Also from Table 3.4 we have

$$y_1 = 1 - z + z^3 - 2z^4 + \cdots,$$
$$\omega_1 = 1 - 2z + z^2 + 8z^4 + \cdots.$$

Thus, the values x and μ in the E-optimal design can be approximately calculated. In order to check the quality of the approximation, consider the case $n = 3$, $r = 2$, $z = 1/r^2 = 1/4$.

In this case, we have

$$y_{1<3>}(0.25) = 1 - \frac{1}{4} + \left(\frac{1}{4}\right)^3 = 0.766$$

$$\omega_{1<3>}(0.25) = 1 - \frac{1}{2} + \frac{1}{16} = 0.563$$

$$rx \approx \sqrt{y_{1<3>}} \approx 0.875, \quad \mu \approx \omega_{1<3>}/8 = 0.07$$

$$\xi_{(3)}(0.25) \approx \left(\begin{array}{cccc} -2 & -0.88 & 0.88 & 2 \\ 0.07 & 0.43 & 0.43 & 0.07 \end{array}\right).$$

Note that due to Lemma 3.3.5 we have

$$\lambda_1 < \lambda(z) < \lambda_2,$$

where $\lambda_1 = \lambda_{(n)}$ and $\lambda_2 = \max_{0 \leq y \leq 1} \rho_n(y)$. The values of λ_1 and λ_2 for different n and z are given in Table 3.5.

From Table 3.5 we conclude that for $r = 2$, the first three coefficients secure an acceptable efficiency of the design $\xi = \xi_{(3)}$:

$$\frac{\lambda_{\min}(M(\xi_{(3)}))}{\lambda_{\min}(M(\xi_{\tau(z),z}))} = \frac{0.37407}{0.37722} \approx 0.99.$$

For r greater than 2, the efficiency of the design $\xi_{(3)}(z)$ will be even greater, as can be seen from Table 3.5. However, numerical calculations, omitted for the sake of brevity showed that the series constructed in a

Table 3.5: Table of efficiency ($m = 4$)

z	n	λ_1	λ_2
0,01	1	0,96010	0,96117
0,01	2	0,96079	0,96081
0,01	3	0,96079	0,96079
0,15	3	0,55219	0,56539
0,15	5	0,55247	0,55302
0,15	10	0,55257	0,55257
0,25	3	0,37407	0,46476
0,25	5	0,37403	0,38293
0,25	10	0,37705	0,37744
0,25	20	0,37721	0,37726
0,25	30	0,37722	0,37722

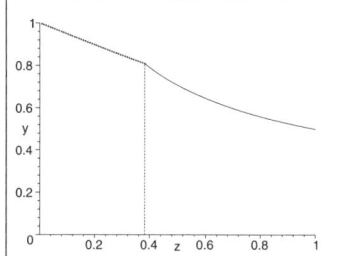

Figure 3.5: The dependence of $x(z) = rt$ for the case $m = 4$

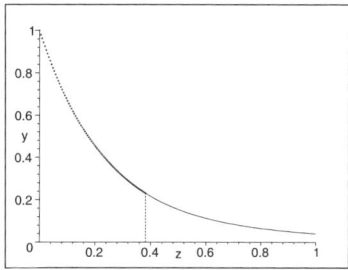

Figure 3.6: The dependence $\lambda^*(z)$ for the case $m = 4$

vicinity of point $z = 0$ converge very slowly if z is close to the critical value $z^*(4) \approx 0.38$.

The behavior of the point $x(z) = rt$ is shown in Figure 3.5. We can see from this figure that the function is monotonic. Note that all other components of the vector $\tau(z)$ also have a monotonic behavior. However, it is difficult to prove this in a strong way. The behavior of $\lambda^*(z)$ is shown in Figure 3.6.

If we need an E-optimal design for points close to $r^*(4)$, it proves reasonable to make calculations in the two stage. First, with the help of expansions in vicinity of $z = 0$ we can approximate the vector $\tau(z)$ with $z = 0.2$. Then we can construct the expansion at the point $z = 0.2$ and use it for calculating the E-optimal design. We will demonstrate this approach for the model of the fourth degree ($m = 5$) in the next example.

Example 3.4.3 Model of fourth degree

Now let $m = 5$,

$$\eta(t, \theta) = \theta_1 + \theta_2 t + \theta_3 t^2 + \theta_4 t^3 + \theta_5 t^4, \; t \in [-r, r].$$

The boundary equation in this case assumes the form

$$64z^5 - 32z^4 + 96z^3 - 44z^2 + 33z - 12 = 0.$$

This equation has the unique positive root

$$z^* = z^*(5) \approx 0.396787.$$

With $r \leq r^*(5) = 1/\sqrt{z^*(5)}$ due to Theorem 3.3.4, the E-optimal design coincides with the Chebyshev design:

$$\xi^* = \begin{pmatrix} -r & -rt & 0 & rt & r \\ \nu_1 & \nu_2 & \nu_3 & \nu_2 & \nu_1 \end{pmatrix},$$

where $\nu_3 = 1 - 2\nu_1 - 2\nu_2$, $t = 1/\sqrt{2}$,

$$\nu_1 = \frac{4z^2(1 + 2z^2)}{64z^4 + 64z^2 + 1}, \quad \nu_2 = \frac{16z^2(1 + z^2)}{64z^4 + 64z^2 + 1},$$

and

$$\lambda_{\min}(M(\xi^*)) = (64z^4 + 64z^2 + 1)^{-1}.$$

Consider the case $r > r^*(5)$. In this case, by Lemma 3.3.6, the E-optimal design is unique and has the same form, but an explicit form for x_1, ν_1, and ν_2 is not available. Let us apply the functional approach.

For $m = 5$, the vector τ has the form

$$\tau = (\tilde{p}_1, \tilde{p}_2, \tilde{q}_1, \tilde{q}_2, \tilde{\omega}_1, \tilde{\omega}_2, zt^2)^T.$$

Applying Lemma 3.4.3, we obtain that

$$\tau_{(0)} = (1/4, -5/4, -3/4, 1, 1, 8, 1/4)^T,$$

$$\lambda_{(0)} = \lim_{z \to 0} \lambda_{\min} \left(M(\xi_{\tau(z), z}) \right) = 1.$$

Matrix $J_{(0)}$ is a zero matrix and $J_{(1)}$ is equal to

$$\begin{pmatrix} 9 & 3 & 0 & 0 & -1/4 & -1/4 & -6 \\ 3 & 3/2 & 0 & 0 & 0 & 0 & -3/2 \\ 0 & 0 & 2 & 3/2 & 1/4 & -1/8 & -2 \\ 0 & 0 & 3/2 & 9/8 & 1/4 & -1/32 & -3/2 \\ -1/4 & 0 & 1/4 & 1/4 & 0 & 0 & 0 \\ -1/4 & 0 & -1/8 & -1/32 & 0 & 0 & 0 \\ -6 & -3/2 & -2 & -3/2 & 0 & 0 & -3/2 \end{pmatrix}.$$

Table 3.6: Coefficients for initial point $z_0 = 0$

	0	1	2	3	4
$\tilde{p}_1$	1/4	−5/12	5/18	7/27	−146/81
$\tilde{p}_2$	−5/4	5/6	−5/9	−26/27	532/81
$\tilde{q}_1$	−3/4	5/4	5/24	−31/72	4717/864
$\tilde{q}_2$	1	−5/3	−5/18	31/54	−151/24
$\tilde{\omega}_2$	8	−688/9	45632/81	−2691136/729	149801216/6561
$\tilde{\omega}_1$	1	−41/9	1888/81	−96560/729	4945072/6561
$\tilde{y}_2$	1/4	5/6	−5/9	−2/27	52/81
λ	1	−9	56	−320	1808

First coefficients of the Taylor expansion of the vector function $\tau(z)$ in a vicinity of $z = 0$ are represented in Table 3.6. Similar coefficients for $\lambda(z) = \varphi(\tau(z), z)$ are also given in the table.

From Table 3.5 we see that the coefficients for weights and for the minimal eigenvalue increase much faster than others. A numerical study shows that the radius of convergency for $\tilde{\omega}_1 = 2\nu_1/z$ and $\tilde{\omega}_2 = 2\nu_2/z$ is approximately 0.16, whereas for other components of $\tau_{(0)}$, it is approximately 0.35.

Note that λ, $\tilde{\omega}_1$ and $\tilde{\omega}_2$ can be expressed by the rest elements of τ.

For arbitrary $m = 2k + 1$, we have from (3.48) that

$$\lambda = \frac{\left(\sum_{i=1}^{k+1} \tilde{p}_i\right)^2 + \left(\sum_{i=1}^{k} \tilde{q}_i\right)^2}{\sum_{i=1}^{k+1} \tilde{p}_i^2 z^{2i} + \sum_{i=1}^{k} \tilde{q}_i^2 z^{2i}}, \quad \tilde{p}_{k+1} = 1,$$

and for $k = 2$, a direct calculation gives

$$\tilde{\omega}_2 = \frac{\lambda(\tilde{q}_1 - \tilde{q}_2 z^2)}{\tilde{y}_2(1 - \tilde{y}_2)(\tilde{q}_1 + \tilde{q}_2 \tilde{y}_2)},$$

$$\tilde{\omega}_1 = \frac{\lambda(\tilde{q}_2 z^2 - \tilde{q}_1 \tilde{y}_2)}{(1 - \tilde{y}_2)(\tilde{q}_1 + \tilde{q}_2)}.$$

Thus, we can approximate $\tilde{p}_1$, $\tilde{p}_2$, $\tilde{q}_1$, $\tilde{q}_2$, and $\tilde{y}_2$ by the Taylor expansions and calculate λ, $\tilde{\omega}_1$, and $\tilde{\omega}_2$ by the above formulas. This method allows one to find support points and weights of the E-optimal design with a great precision for $z \leq 0.3$ ($r \geq 1/\sqrt{0.3}$).

In order to solve the problem in the intermediate case $0.3 < z < 0.396$, we constructed re-expansion of the function $\tau(z)$ in a vicinity of point $z = 0.2$.

By the method described above, using 15 coefficients of the expansion at zero point, we calculated the first column in Table 3.7. Other columns are calculated by the recurrent formulas.

To show the efficiency of designs $\xi_{(n)}(z)$ obtained by the Taylor expansion at $z = 0.2$, we calculated the bounds λ_1 and λ_2 for different values of

Table 3.7: Coefficients for initial point $z_1 = 0.2$

	0	1	2	3	4	5
$\tilde{p}_1$	0.17804	−0.30698	0.22962	−0.20355	−0.02974	0.36925
$\tilde{p}_2$	−1.10688	0.60901	−0.43501	0.55512	−0.10179	−1.11856
$\tilde{q}_1$	−0.48819	1.41862	1.00063	3.72679	11.81191	40.91730
$\tilde{q}_2$	0.65139	−1.88521	−1.31832	−5.06216	−16.14959	−54.73347
$\tilde{\omega}_2$	2.21396	−9.66086	36.65304	−121.46442	374.99630	−1108.795
$\tilde{\omega}_1$	0.52635	−1.33370	2.49681	−6.18865	16.62481	−45.22451
$\tilde{y}_2$	0.39473	0.61899	−0.48325	0.25700	0.20932	−0.34837
λ	0.23768	−1.44590	5.21544	−15.39351	41.72046	−109.9943

z and n. The results are presented in Table 3.8. For example, consider the case $z = 0.35$. With $n = 2$ the efficiency of design $\xi_{(n)}(z)$ is grater than

$$\frac{0.10124}{0.10132} = 0.999.$$

Table 3.8: Table of efficiency

z	n	λ_1	λ_2
0.05	10	0.65880	0.65887
0.05	15	0.65880	0.65885
0.05	20	0.65880	0.65881
0.35	2	0.10124	0.10487
0.35	5	0.10132	0.10217
0.35	10	0.10132	0.10142
0.35	15	0.10132	0.10133
0.35	18	0.10132	0.10132
0.39	3	0.08185	0.08887
0.39	5	0.08185	0.08554
0.39	10	0.08185	0.08347
0.39	19	0.08185	0.08246
0.39679	2	0.07896	0.08649
0.39679	5	0.07897	0.08355
0.39679	10	0.07897	0.08143
0.39679	20	0.07897	0.08022

In the rest of this section, let us point out an efficient way of calculating $\xi_{(n)}(z)$. A numerical verification shows that the series for $\tilde{p}_1$, $\tilde{p}_2$, and $\tilde{y}_2$ have a very quick convergency. This can also be seen immediately from Table 3.7.

The weights for the E-optimal design and the minimal eigenvalue can be calculated from $\tilde{p}_1$, $\tilde{p}_2$, and $\tilde{y}_2$ by the formulas

$$2\nu_i = \tilde{\omega}_i z = \lambda A_i \quad i = 1, 2,$$

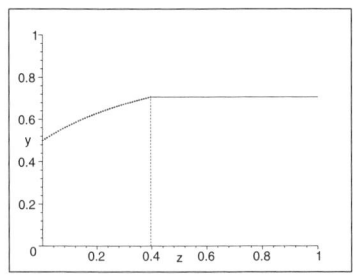

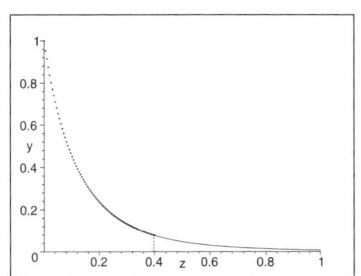

Figure 3.7: The dependence of normalized support point $x = x(z)$ for the model of the fourth degree

Figure 3.8: The behavior of the minimal eigenvalue for the case $m=5$

$$\lambda = 1/(A_1 + A_2 + A_3)$$
$$A = (A_1, A_2, A_3)^T = F^{-1}\text{diag}\{1, z^2, z^4\}\tilde{p},$$

where $F = \left(\hat{p}_j, \tilde{y}_j^{i-1}\right)_{j,i=1}^3$, $\hat{p}_i = (1, \tilde{y}_i, \tilde{y}_i^2)\tilde{p}$, $\tilde{p} = (\tilde{p}_1, \tilde{p}_2, 1)^T$, $\tilde{y}_1 = 1$, $\tilde{y}_3 = 0$.

Table 3.9: Coefficients for functions $\tilde{p}_1$, $\tilde{p}_2$, and $\tilde{y}_2$ in point $z = 0.2$

	0	1	2	3
$\tilde{p}_1$	0.17804	−0.307	0.230	−0.204
$\tilde{p}_2$	−1.10688	0.609	−0.435	0.555
$\tilde{y}_2$	0.39473	0.619	−0.483	0.257

For the convenience of the readers we provide a simplified table of coefficients for $\tilde{p}_1$, $\tilde{p}_2$, and $\tilde{y}_2$ (see Table 3.9). Using these coefficients, the reader can calculate an E-optimal design for any given design interval $[-r, r]$, $r > r^*(5) = 1.5875$. The behavior of the normalized support point $x = x(z)$ is represented in Figure 3.7. Again, we have a monotony dependence. It is worth mentioning that the Taylor expansions provide the construction of the figures and such a construction would be more difficult and not so perfect by a merely numerical approach.

Chapter 4

Trigonometrical Models

Trigonometrical models can be considered as approximations of unknown continuous functions by segments of their Fourier series. Such models have a wide field of actual and potential applications.

In the present chapter, we consider trigonometrical regression models of one variable on arbitrary intervals. We study D- and E-optimal designs for this class of models with the help of the functional approach. In comparison with polynomial models, the main differences consist in the following. First, support points of D-optimal designs for arbitrary intervals cannot be obtained by the scale transformation of such points for the standard design interval. Second, information matrices of E-optimal designs can have a minimal eigenvalue of an arbitrary multiplicity.

4.1 Introduction

Trigonometric regression models of the form

$$y = \beta_0 + \sum_{j=1}^{m} \beta_{2j-1} \sin(jt) + \sum_{j=1}^{m} \beta_{2j} \cos(jt) + \varepsilon, \quad t \in [c, d], \tag{4.1}$$

$-\infty < c < d < \infty$, are widely used to describe periodic phenomena (see, e.g., Mardia (1972), Graybill (1976), or Kitsos, Titterington, and Torsney (1988)) and the problem of designing experiments for Fourier regression models has been discussed by several authors (see, e.g., Hoel (1965), Karlin and Studden (1966, p. 347), Fedorov (1972, p. 94), Hill (1978), Lau and Studden (1985), Riccomagno, Schwabe, and Wynn (1997)). Most authors concentrate on the design space $[-\pi, \pi]$. For the D-criterion, it was established that optimal designs are concentrated in equidistant points and have one and the same weights for these points. Pukelsheim (1980) and Melas (1982) showed that these designs are also E-optimal ones.

However, Hill (1978) and Kitsos, Titterington, and Torsney (1988) pointed out that in many applications, it is impossible to take observations on the full circle $[-\pi, \pi]$. We refer to Kitsos, Titterington, and Torsney (1988) for a concrete example; they investigated a design problem in rhythmometry involving circadian rhythm exhibited by peak expiratory flow, for which the design region has to be restricted to a partial cycle of the complete 24-hour period.

In the present chapter, we address the question of designing experiments in trigonometric models, where the design space is not necessarily the full circle but an arbitrary interval $[c, d] \subset \mathbb{R}$. Recently, Dette and Melas (2003) considered optimal designs for estimating individual coefficients in this model and gave a partial solution to this problem. In the present chapter, we consider the D- and E-optimal designs.

The D-criterion is a reasonable criterion if efficient estimates of all parameters in the model are desired (see Section 1.5). D-Optimal designs are investigated in Section 4.2*. Some preliminary results are given in Section 4.2.1. It is demonstrated that the structure of the D-optimal design depends only on the length $a = (c - d)/2$ of the design space and that there only exist two types of D-optimal design (this result seems to be unknown for the complete circle). Our main result of Section 4.2.2 proves that the support points (and weights) of the D-optimal design are analytic functions of the parameter a and that an appropriately scaled version of the D-optimal design converges weakly as $a \to 0$ to a nondegenerate discrete distribution on the interval $[0, 1]$. Following the functional approach developed in Chapter 2, these results are applied to obtain Taylor expansions for the support points of the D-optimal design (considered as a function of the parameter $a = (d - c)/2$), which allows a complete solution of the D-optimal design problem in the trigonometric regression model (4.1) on the interval $[c, d]$.

Note that it proves possible to calculate explicitly as roots of a classical polynomial the limit of the support points of D-optimal designs normed by the length of the design interval when this length tends to zero. The technique is based on a differential equation for the polynomial having roots in the support points. This equation is similar to that was introduces by Stilties (see Karlin and Studden (1966, Chap. X)) for maximization of Vandermonde determinants. In Section 4.2.3, a similar equation is derived for D-optimal designs on arbitrary intervals. This allows one to derive, in the next section, an algebraic equation for the vector of coefficients of the polynomial with roots in the support points of D-optimal designs. The

---

combination of the functional and the algebraic approaches allows one to make calculations much more quickly.

Finally, some examples are given in Section 4.2.5, and in the linear and quadratic trigonometric regression model on the interval $[-a, a]$, D-optimal designs are determined explicitly.

In Section 4.3, similar results are obtained for E-optimal designs[†]. In Section 4.3.1, we introduce two auxiliary results. The first is the equivalence theorem on E-optimal designs for general linear models, discussed earlier in Section 3.2 and repeated here for the convenience of the reader. The second result shows that E-optimal designs for the interval $[c, d]$ can be obtained from such results on the interval $[-a, a]$, $a = (d - c)/2$, simply by adding $c + a$ to all of the support points. This result is similar to that for D-optimal designs proved in Section 4.2.

As was already mentioned for the full interval $[-\pi, \pi]$ E-optimal designs are D-optimal and vice versa. In Section 4.3.2, we show that this result remains true for intervals $[-a, a]$ if and only if $a \geq \pi(1 - 1/(2m + 1))$. In Section 4.3.3, E-optimal designs are explicitly found for intervals $[-a, a]$ with $a \leq \bar{a}_m$, where $\bar{a}_m > \pi/2$ is a critical value. This result is based on the equivalence theorem and on the proposition that for $a < \bar{a}_m$, the minimal eigenvalue of the information matrix for Chebyshev designs (introduced in Section 3.3) is simple. Section 4.3.4 includes the full solution on the basis of the functional approach. Here, E-optimal designs are studied for the intervals $[-a, a]$ with $\bar{a}_m < a < \pi - 1/(2m + 1)$. Results of Section 4.3 are based on Dette and Melas (2002).

In Section 4.4, we perform a numerical comparison of D- and E-optimal designs. We prove that D-optimal designs have a high E-efficiency and E-optimal designs have a high D-efficiency. However, an advantage of E-optimal design is that for $a \leq \bar{a}_m$, such designs can be found explicitly.

4.2 D-Optimal Designs

4.2.1 Preliminary results for D-optimal designs

Consider the trigonometric regression model (4.1); define $\beta = (\beta_0, \beta_1, \ldots, \beta_{2m})^T$ as the vector of parameters and let

$$f(t) = (1, \sin t, \cos t, \ldots, \sin(mt), \cos(mt))^T = (f_0(t), \ldots, f_{2m}(t))^T \quad (4.2)$$

be the vector of basic regression functions. An approximate design is a probability measure ξ on the design space $[c, d]$ with finite support (see Chapter 1). The support points of the design ξ give the locations where observations are taken, whereas the weights give the corresponding proportions of

---

[†]In this section materials (theorems, tables, and figures) are taken from Dette, H., Melas, V.B. (2002). *E*-Optimal designs in Fourier regression models on a partial circle. *Math. Methods Statist.*, **11**(3), pp. 259–296. ©2003 Allerton Press, Inc. with permission of Allerton Press, Inc.

total observations to be taken at these points. Due to the 2π-periodicity of the regression functions, we restrict ourselves without loss of generality to design spaces with length $d - c \leq 2\pi$. For uncorrelated observations (obtained from an approximate design), the covariance matrix of the least squares estimator for the parameter β is approximately proportional to the matrix

$$M(\xi) = \int f(t)f^T(t)d\xi(t) \in \mathbb{R}^{2m+1 \times 2m+1} , \tag{4.3}$$

which is called the *information matrix* in the design literature. An optimal design minimizes (or maximizes) an appropriate convex (or concave) function of the information matrix and there are numerous criteria proposed in the literature that can be used for the discrimination between competing designs (see Section 1.5).

In this section, we are interested in D-optimal designs for the trigonometric regression model (4.1) on the interval $[c, d]$, which maximize the determinant

$$\det M(\xi)$$

of the Fisher information matrix in the space of all approximate designs on the interval $[c, d]$. Note that a D-optimal design minimizes the (approximate) volume of the ellipsoid of concentration for the vector β of the unknown parameters in the model (4.1) (see Section 1.5) and that optimal designs in the trigonometric regression model (4.1) for the full circle $[c, d] = [-\pi, \pi]$ have been determined by numerous authors (see, e.g., Karlin and Studden (1966), Fedorov (1972), Lau and Studden (1985), Pukelsheim (1993), or Dette and Haller (1998) among many others).

Our first preliminary result demonstrates that for the solution of the D-optimal design problem on a partial circle, it is sufficient to consider only symmetric design spaces. To be precise, let

$$\eta = \begin{pmatrix} t_0 & \cdots & t_n \\ \omega_0 & \cdots & \omega_n \end{pmatrix} \tag{4.4}$$

denote a design on the interval $[c, d]$ with different support points $t_0 < \cdots < t_n$ and positive weights $\omega_0, \ldots, \omega_n$ adding to 1 and define its affine transformation onto the symmetric interval $[-a, a]$ by

$$\xi_\eta = \begin{pmatrix} \tilde{t}_0 & \cdots & \tilde{t}_n \\ \omega_0 & \cdots & \omega_n \end{pmatrix}, \tag{4.5}$$

where $a = (d - c)/2$ and $\tilde{t}_i = t_i - (d + c)/2$, $i = 1, \ldots, n$.

Lemma 4.2.1 *Let $M(\eta)$ and $M(\xi_\eta)$ denote the information matrices in the trigonometric regression model (4.1) of the designs η and ξ_η defined by (4.4) and (4.5), respectively; then*

$$\det M(\xi_\eta) = \det M(\eta). \tag{4.6}$$

Proof.   If the number of support points satisfies $n + 1 < 2m + 1$, then both sides of (4.6) vanish and the proof is trivial. Next, consider the case $n = 2m$, for which we have (see, e.g., Karlin and Studden (1966))

$$\det M(\xi_\eta) = (\det F(\xi_\eta))^2 \prod_{i=0}^{2m} \omega_i \, , \qquad (4.7)$$

where the matrix $F(\xi_\eta) \in \mathbb{R}^{2m+1 \times 2m+1}$ is defined by

$$F(\xi_\eta) = \left(f_i(\tilde{t}_j) \right)_{i=0,\ldots,2m}^{j=0,\ldots,2m} \, . \qquad (4.8)$$

Now, it to easy to see that the vector $f(t)$ defined by (4.2) satisfies, for any $\alpha \in \mathbb{R}$,

$$f(t + \alpha) = P f(t),$$

where P is a $(2m + 1) \times (2m + 1)$ diagonal block matrix defined by

$$P = \begin{pmatrix} 1 & & & \\ & Q(\alpha) & & \\ & & \ddots & \\ & & & Q(m\alpha) \end{pmatrix}$$

and $Q(\beta)$ is a 2×2 rotation matrix given by

$$Q(\beta) = \begin{pmatrix} \cos(\beta) & \sin(\beta) \\ -\sin(\beta) & \cos(\beta) \end{pmatrix} .$$

Obviously, we have $\det P = 1$ and obtain from (4.7) and (4.8)

$$\det M(\xi_\eta) = \det M(\eta) \, ,$$

which proves the assertion of the lemma in the case $n = 2m$. Finally, in the remaining case, $n > 2m$, the assertion follows from the Binet–Cauchy formula and the arguments given for the case $n = 2m$. ■

From Lemma 4.2.1, it is clear that it is sufficient to determine the D-optimal designs for symmetric intervals

$$[c, d] = [-a, a], \qquad 0 < a \le \pi,$$

and we will restrict ourselves to this case throughout this section. For fixed $a \in (0, \pi]$, let ξ_a^* denote a D-optimal design for the trigonometric regression model on the interval $[-a, a]$. Note that, in general, the D-optimal design for the trigonometric regression model is not necessarily unique (see, e.g., Fedorov (1972), who considered the case $a = \pi$). However, it is known that the optimal information matrix $M(\xi_a^*)$ is unique and nonsingular (see, e.g.,

Pukelsheim (1993, p. 151)). Moreover, due to the equivalence theorem for D-optimality (see Section 1.6), the design ξ_α^* satisfies

$$d(t, \xi_\alpha^*) \le 0 \quad \text{for all } t \in [-a, a], \tag{4.9}$$

with equality at the support points, where

$$d(t, \xi) = f^T(t) M^{-1}(\xi) f(t) - (2m + 1). \tag{4.10}$$

Let $\Xi_a^{(1)}$ denote the set of all designs of the form

$$\xi = \xi(a) = \begin{pmatrix} -t_m & \cdots & -t_1 & t_0 & t_1 & \cdots & t_m \\ \frac{1}{2m+1} & \cdots & \frac{1}{2m+1} & \frac{1}{2m+1} & \frac{1}{2m+1} & \cdots & \frac{1}{2m+1} \end{pmatrix}, \tag{4.11}$$

where $0 = t_0 < t_1 < \cdots < t_m = a$, and define

$$\Xi^{(2)} = \left\{ \xi \mid \operatorname{supp}(\xi) \subset [-a, a], \ d(t, \xi) = 0 \text{ for all } t \in [-a, a] \right\} \tag{4.12}$$

as the set of all designs on the interval $[-a, a]$ with a vanishing directional derivative for all $t \in [-a, a]$; then we obtain the following auxiliary result.

Lemma 4.2.2 *Let ξ_a^* denote a D-optimal design on the interval $[-a, a]$; then*

$$\xi_a^* \in \Xi_a^{(1)} \cup \Xi_a^{(2)}.$$

Proof. Due to the equivalence theorem (4.9), any design $\xi \in \Xi_a^{(2)}$ is D-optimal for trigonometric regression model (4.2.1) on the interval $[-a, a]$. Now, assume that

$$\xi = \begin{pmatrix} u_1 & \cdots & u_n \\ \omega_1 & \cdots & \omega_n \end{pmatrix}$$

is D-optimal for the trigonometric regression on the interval $[-a, a]$, where the support points satisfy $-a \le u_1 < \cdots < u_n \le a$. If $\xi \notin \Xi_a^{(2)}$, then $d(t, \xi) \ne 0$, but due the equivalence theorem, we have

$$d(u, \xi) \le 0 \quad \forall\, u \in [-a, a],$$

$$d(u_i, \xi) = 0 \quad \forall\, i = 1, \ldots, n,$$

$$\frac{d}{du} d(u, \xi)\big|_{u=u_i} = 0 \quad \forall\, i = 2, \ldots, n - 1.$$

If $\tilde{\xi}$ denotes the reflection of ξ at the origin, then it is easy to see that $\det M(\xi) = \det M(\tilde{\xi})$ and, consequently, $\tilde{\xi}$ is also D-optimal. Moreover, the concavity of the D-criterion implies that the symmetric design $\xi^* = (\xi + \tilde{\xi})/2$ is also D-optimal in the trigonometric regression (4.1) on the interval $[-a, a]$. Note that there exists a permutation matrix $P \in \mathbb{R}^{2m+1 \times 2m+1}$ such that

$$P M(\xi) P^T = \begin{pmatrix} M_1(\xi) & M_2(\xi) \\ M_2^T(\xi) & M_3(\xi) \end{pmatrix}, \tag{4.13}$$

where

$$M_1(\xi) = \int_{-a}^{a} f_c(t) f_c^T(t) d\xi(t) \in \mathbb{R}^{m+1 \times m+1},$$

$$M_2(\xi) = \int_{-a}^{a} f_c(t) f_s^T(t) d\xi(t) \in \mathbb{R}^{m+1 \times m}, \qquad (4.14)$$

$$M_3(\xi) = \int_{-a}^{a} f_s(t) f_s^T(t) d\xi(t) \in \mathbb{R}^{m \times m},$$

and $f_c(t) = (1, \cos(t), \ldots, \cos(mt))^T$ and $f_s(t) = (\sin(t), \ldots, \sin(mt))^T$. Because the information matrix of the D-optimal design is unique (see Pukelsheim (1993)), we obtain (note that ξ^* is symmetric)

$$M_2(\xi) = M_2(\tilde{\xi}) = M_2(\xi^*) = 0 \in \mathbb{R}^{m+1 \times m},$$

which implies for the directional derivative in (4.10)

$$g(t) = d(t, \xi) = f_c^T(t) M_1^{-1}(\xi) f_c(t) + f_s^T(t) M_3^{-1}(\xi) f_s(t) - (2m+1)$$

$$= \sum_{i=0}^{2m} \gamma_i \cos(it) \qquad (4.15)$$

for appropriate constants $\gamma_0, \ldots, \gamma_{2m}$ (note that the last representation follows by well-known trigonometric formulas). From $\xi \notin \Xi_a^{(2)}$, we obtain that the polynomial $g(t)$ is not identically zero and the equivalence theorem shows that every support point is a zero of the function g. Moreover, the functions $\{1, \cos t, \ldots, \cos(2mt)\}$ form a Chebyshev system on the interval $[0, a]$ and a Chebyshev system on the interval $[-a, 0]$. Consequently, g has at most $2m + 1$ roots in the interval $[0, a]$ and at most $2m + 1$ zeros in the interval $[-a, 0]$ (including counting of multiplicities) (see Section 1.9 and Karlin and Studden (1966), Ch. 10). Consider the case $[0, a]$ and substitute $t = \arccos x$, then it follows, observing the definition of the Chebyshev polynomials of the first kind,

$$T_i(x) = \cos(i \arccos x) \qquad (4.16)$$

(see Rivlin (1974)), that $g(\arccos x)$ is a nonpositive polynomial of degree $2m$ on the interval $[\cos a, 1]$. Consequently, if $g(\arccos x)$ has exactly $2m$ roots (including counting of multiplicities), the boundary points $\cos a$ and 1 have to be roots of $g(\arccos x)$. Note that a similar argument applies to the interval $[-a, 0]$ and, therefore, the nonpositive function g defined in (4.15) has at most $4m$ roots (including counting of multiplicities) in the interval $[-a, a]$. Because the number of regression functions is $2m + 1$, it therefore follows from (4.2.1) that any D-optimal design $\eta \notin \Xi_a^{(2)}$ has exactly $2m + 1$

support points in the interval $[-a, a]$, including the boundary points. A standard argument shows that all weights of the D-optimal design have to be equal i.e. $(\omega_j = 1/(2m + 1),\ j = 1, \ldots, 2m + 1)$. If $\xi \notin \Xi_a^{(1)}$, then $\xi \neq \tilde{\xi}$ and, consequently, $\xi^* = (\xi + \tilde{\xi})/2$ is a D-optimal design for the trigonometric regression model (4.1) on the interval $[-a, a]$ with more than $2m + 1$ support points, which is impossible, by the above discussion. This shows that $\xi \in \Xi_a^{(1)}$ and proves Lemma 4.2.2. ∎

4.2.2 Analytic properties of D-optimal designs

Lemma 4.2.2 motivates the consideration of designs of the form (4.11) and our next lemma gives an explicit representation for the determinant of the information matrix of this type of design.

Lemma 4.2.3 *Let ξ denote a design of the form (4.11) and $x_i = \cos t_i$, $i = 0, \ldots, m$; then*

$$\det M(\xi) = \frac{2^{2m^2}}{(2m + 1)^{2m+1}} \prod_{i=1}^{m} (1 - x_i^2)(1 - x_i)^2 \prod_{1 \leq i < j \leq m} (x_j - x_i)^4. \quad (4.17)$$

Proof. For any design ξ of the form (4.11), we have

$$\det M(\xi) = \det M_1(\xi) \det M_3(\xi),$$

where the matrices $M_1(\xi), M_2(\xi)$, and $M_3(\xi)$ are defined by (4.14). Define the design η_ξ by

$$\eta_\xi = \left(\begin{array}{cccc} x_0 & x_1 & \cdots & x_m \\ \frac{1}{2m+1} & \frac{2}{2m+1} & \cdots & \frac{2}{2m+1} \end{array} \right);$$

then it is straightforward to see that

$$M_1(\xi) = \left(\int_{-1}^{1} T_i(x) T_j(x) \, d\eta_\xi(x) \right)_{i,j=0}^{m},$$

$$M_3(\xi) = \left(\int_{-1}^{1} (1 - x^2) U_i(x) U_j(x) \, d\eta_\xi(x) \right)_{i,j=0}^{m-1},$$

where $T_i(x)$ is the Chebyshev polynomial of the first kind defined in (4.16) and

$$U_i(x) = \frac{\sin((i + 1) \arccos x)}{\sin(\arccos x)}$$

is the Chebyshev polynomial of the second kind (see Rivlin (1974)). Because $T_i(x)$ is a polynomial of degree i with leading coefficient 2^{i-1}, it follows that

$M_1(\xi)$ is essentially a Vandermonde determinant; that is,

$$\det M_1(\xi) = 2^{m(m-1)} \frac{2^m}{(2m+1)^{m+1}} \left(\det \left((x_j^i)_{i=0,\ldots,m}^{j=0,\ldots,m} \right) \right)^2$$

$$= \frac{2^{m^2}}{(2m+1)^{m+1}} \prod_{i=1}^{m} (1-x_i)^2 \prod_{1 \le i < j \le m} (x_j - x_i)^2$$

(note that $x_0 = 1$). Note that the support point x_0 of η_ξ has a vanishing contribution to the matrix $M_3(\xi)$ and that the leading coefficient of $U_i(x)$ is 2^i. Therefore, we have, by similar arguments,

$$\det M_3(\xi) = \frac{2^{m^2}}{(2m+1)^m} \prod_{i=1}^{m} (1-x_i^2) \prod_{1 \le i < j \le m} (x_j - x_i)^2$$

and a combination of these formulas yields (4.17), which proves the assertion of Lemma 4.2.3. ∎

We now study the function

$$\phi(x,a) = \prod_{i=1}^{m} (1-x_i^2)(1-x_i)^2 \prod_{1 \le i < j \le m} (x_j - x_i)^4 \qquad (4.18)$$

as a function of the length a of the design space. To this end, we note that $x_m = \cos(a)$ and introduce the sets

$$T = \{(\tau_1, \ldots, \tau_{m-1})^T \mid 0 < \tau < \cdots < \tau_{m-1} < 1\} \qquad (4.19)$$

$$\mathcal{X} = \{(x_1, \ldots, x_{m-1})^T \mid x_i = \cos(a\tau_i),\ i=1,\ldots,m-1,\ (\tau_1, \ldots, \tau_{m-1})^T \in T\}.$$

Note that any design $\xi \in \Xi_a^{(1)}$ of the form (4.11) is uniquely determined by a point $\tau = (\tau_1, \ldots, \tau_{m-1})^T \in T$ or its corresponding function $x = (x_1, \ldots, x_{m-1})^T \in \mathcal{X}$ by the transformation $t_i = a\tau_i = \arccos x_i$, $i = 1, \ldots, m-1$ (note that $t_0 = 0, t_m = a$), and by Lemma 4.2.3 the determinant of $M(\xi)$ is proportional to the function ϕ given in (4.18). By standard arguments, it can now be verified that for fixed $a \in (0, \pi]$, the function ϕ in (4.18) is a strictly concave function of $x = (x_1, \ldots, x_{m-1})^T \in \mathcal{X}$. Therefore (for fixed a) the function $\phi(x, a)$ has a unique maximum in $\mathcal{X}$, which will be denoted by $x^*(a)$ (because of its dependence on the length of the design space). The function ϕ is obviously differentiable and $x^*(a)$ can be obtained as the unique solution of the equations

$$\frac{\partial}{\partial x} \phi(x, a) = 0 \in \mathbb{R}^{m-1}. \qquad (4.20)$$

Moreover, for any $x \in \mathcal{X}$, the matrix of the second partial derivatives

$$G(x, a) = \left(\frac{\partial^2}{\partial x_i \partial x_j} \phi(x, a) \right)_{i,j=1}^{m-1}$$

is positive definite and, in particular, the matrix

$$J(a) = G(x^*(a), a)$$

is positive definite for all $a \in (0, \pi]$. It, therefore, follows from the Implicit Function Theorem (see Section 1.8) that the function

$$x^* : \begin{cases} (0, \pi] & \to & \mathcal{X} \\ a & \to & x^*(a) \end{cases},$$

defined as the solution of (4.20), is real analytic. In other words, for any point $a_0 \in (0, \pi]$ there exists a neighborhood U_0 of a_0, such that the function $x^*|_{U_0}$ can be expanded in a convergent Taylor series. Observing the symmetry $\phi(x, a) = \phi(x, -a)$, it therefore follows that the function

$$\tau^* : \begin{cases} [-\pi, \pi] \backslash \{0\} \to T \\ a \to \tau^*(a) = \left(\dfrac{\arccos x_1^*(|a|)}{a}, \ldots, \dfrac{\arccos x_{m-1}^*(|a|)}{a} \right)^T \end{cases} \quad (4.21)$$

is also real analytic. The following result shows that the function τ^* can be extended to a real analytic function on the full circle $[-\pi, \pi]$.

Lemma 4.2.4 *The function τ^* defined by (4.21) can be extended to a real analytic function on the interval $[-\pi, \pi]$, where*

$$\tau^*(0) = \lim_{a \to 0} \tau(a) = (\tau_1^*, \ldots, \tau_{m-1}^*)^T ,$$

$\tau_1^* < \cdots < \tau_{m-1}^*$, *are the positive roots of the polynomial*

$$P_{m-1}^{(1,1/2)}(2x^2 - 1) = \frac{1}{2x} P_{2m-1}^{(1,1)}(x) = \frac{1}{(2m+1)x} P'_{2m}(x)$$

and $P_i^{(\alpha, \beta)}(x)$ denotes the i-th Jacobi polynomial orthogonal with respect to the measure $(1-x)^\alpha (1+x)^\beta dx$ and $P_{2m}(x)$ is the $2m$-th Legendre polynomial orthogonal with respect to the Lebesgue measure on the interval $[-1, 1]$.

Proof.   The assertion of Lemma 4.2.4 follows if we prove the existence of $\lim_{a \to 0} \tau^*(a)$ and the claimed form of its components. Let $x_\tau = (\cos(a\tau_1), \ldots, \cos(a\tau_{m-1}))^T$; then the expansions $\sin t = t + o(t)$ and $\cos t = 1 - t^2/2 + o(t^2)$ show that for $a \to 0$,

$$\phi(x_\tau, a) = \frac{a^{2m(2m+1)}}{2^{2m^2}} \prod_{i=1}^m \tau_i^6 \prod_{1 \le i < j \le m} (\tau_i^2 - \tau_j^2)^4 (1 + o(a))$$

$(\tau_m = 1)$ and, consequently, the limit $\lim_{a \to 0} \tau^*(a)$ exists and can be obtained by maximizing the function

$$\bar{\phi}(\tau) = \prod_{i=1}^m \tau_i^3 (1 - \tau_i^2)^2 \prod_{1 \le i < j \le m-1} (\tau_i^2 - \tau_j^2)^2$$

over the set T defined in (4.19). Note that standard arguments show the strict concavity of the function $\bar{\phi}$ and, consequently, the point $\tau^* = (\tau_1^*, \ldots, \tau_{m-1}^*)^T$, where the maximum is obtained, is unique. Taking partial derivatives of the logarithm of $\bar{\phi}$ yields the system

$$\frac{3}{\tau_i} + \frac{4\tau_i}{\tau_i^2 - 1} + \sum_{j=1, j \neq i}^{m-1} \frac{4\tau_i}{\tau_i^2 - \tau_j^2} = 0, \ i = 1, \ldots, m-1,$$

and substituting $\tau_i^2 = y_i \in (0, 1)$ gives

$$\frac{3}{y_i} + \frac{4}{y_i - 1} + \sum_{j=1, j \neq i}^{m-1} \frac{4}{y_i - y_j} = 0, \ i = 1 \ldots m-1.$$

Similar arguments as given in Karlin and Studden (1966) or Fedorov (1972) show that the polynomial $\psi(y) = \prod_{i=1}^{m-1}(y - y_i)$ satisfies the differential equation

$$y(1-y)\psi''(y) + (3/2 - 7/2y)\psi'(y) + (m-1)(m+3/2)\psi(y) = 0. \quad (4.22)$$

It is well known (see, e.g., Szegö (1975, Section 4.21)) that the unique polynomial solution of this differential equation is given by the polynomial

$$P_{m-1}^{(1/2,1)}(1 - 2y)$$

and the assertion of the Lemma now follows from transformation $y = \tau^2$ and the equation $P_n^{(\alpha,\beta)}(-x) = (-1)^n P_{m-1}^{(\beta,\alpha)}(x)$ (see Szegö (1975, formula (4.1.3))). The alternative representations of the polynomial $P_{m-1}^{(1,1/2)}(2x^2 - 1)$ are a consequence of $P_n^{(0,0)}(x) = P_n(x)$ and Theorem 4.1 in Szegö (1975). ∎

Table 4.1 shows the polynomial $P_{m-1}^{(1,1/2)}(2y - 1)$ (normalized such that the leading coefficient is 1) and the corresponding values $\tau_i^* = \sqrt{y_i}$ for lower degrees $m = 2, 3, 4, 5$. The following result shows that for small designs space, that is,

$$a \leq \pi(1 - 1/(2m + 1)),$$

the solution of the optimal design problem can be obtained by a Taylor expansion of the function τ^* in (4.21) at the point $a = 0$, where the i-th component $\tau_i^*(0)$ of the vector $\tau^*(0)$ is the i-th positive root of the polynomial $P_{m-1}^{(1,1/2)}(2x^2 - 1)$.

Theorem 4.2.1 *Consider the trigonometric regression model (4.1) with design space $[-a, a]$, where $0 < a \leq \pi$.*

(i) *If $a > \pi(1 - 1/(2m + 1))$, then the design ξ_a^* with equal masses at the $2m + 1$ points,*

$$t_i^* = 2\pi \frac{i - 1 - m}{2m + 1}, \quad i = 1, \ldots, 2m + 1, \quad (4.23)$$

is a D-optimal design.

Table 4.1: Values of the components $\tau_1^*(0), \ldots, \tau_{m-1}^*(0)$ of the vector $\tau^*(0)$ defined in Lemma 4.3.2 and the polynomial solution of the differential equation (4.22) for various values of m

m	$\psi(y)$ and $\tau_j(0)$
2	$\psi(y) = y - 3/7$
	$\tau_1^*(0) = \sqrt{3/7} \approx 0.6546$
3	$\psi(y) = y^2 - 10/11y + 5/33$
	$\tau_1^*(0) \approx 0.4688, \tau_2^*(0) \approx 0.8302$
4	$\psi(y) = y^3 - 7/5y^2 + 7/13y - 7/143$
	$\tau_1^*(0) \approx 0.3631, \tau_2^*(0) \approx 0.6772, \tau_3^*(0) \approx 0.8998$
5	$\psi(y) = y^4 - 36/19y^3 + 378/323y^2 - 84/323y + 63/4199$
	$\tau_1^*(0) \approx 0.2958, \tau_2^*(0) \approx 0.5652, \tau_3^*(0) \approx 0.7845, \tau_4^*(0) \approx 0.9340$

(ii) If $a < \pi(1 - 1/(2m + 1))$, the D-optimal design (denoted by ξ_a^*) is unique and is of the form

$$
\begin{pmatrix}
\dfrac{-a}{\frac{1}{2m+1}} & \dfrac{-a\tau_{m-1}^*}{\frac{1}{2m+1}} & \cdots & \dfrac{-a\tau_1^*}{\frac{1}{2m+1}} & \dfrac{0}{\frac{1}{2m+1}} & \dfrac{a\tau_1^*}{\frac{1}{2m+1}} & \cdots & \dfrac{a\tau_{m-1}^*}{\frac{1}{2m+1}} & \dfrac{a}{\frac{1}{2m+1}}
\end{pmatrix}, \quad (4.24)
$$

where τ^* is a real analytic vector function on the interval $[-\pi, \pi]$ defined by (4.21) and Lemma 4.2.4.

Proof. Recall the definition of the set $\Xi_a^{(2)}$ in (4.12) and assume that the design $\xi^* \in \Xi_a^{(2)}$ is D-optimal for the trigonometric regression model (4.2.1) on the interval $[-a, a]$. Because $d(t, \xi^*) = 0$ for all $t \in [-a, a]$, it follows from the Chebyshev property of the functions $\{1, \sin t, \cos t, \ldots, \sin mt, \cos mt\}$ that $d(t, \xi^*)$ also vanishes on the full circle $[-\pi, \pi]$ (see Karlin and Studden (1966, p. 20)). Consequently, ξ^* is also D-optimal for the trigonometric regression on the interval $[-\pi, \pi]$, which implies (by the uniqueness of the D-optimal information matrix)

$$M(\xi^*) = \mathrm{diag}(1, 1/2, \ldots, 1/2),$$
$$\det M(\xi^*) = 2^{-2m}.$$

On the other hand, we have

$$\lim_{a \to 0} \max_{\xi} \det M(\xi) = 0,$$

and, consequently, for sufficiently small a, the D-optimal design cannot be an element of the set $\Xi_a^{(2)}$. From Lemma 4.2.2, it follows that the D-optimal design must belong to the set $\Xi_a^{(1)}$ and the discussion in the first part of this section shows that for sufficiently small a, the D-optimal design is unique and of the form (4.24). Now, let ξ_a^* denote the design defined by (4.24) and

$$a^* = \sup\{a \in (0, \pi] \mid \xi_a^* \text{ is } D\text{-optimal}\}$$

$$= \sup\{a \in (0, \pi] \mid \det M(\xi^*) < 2^{-2m}\} \tag{4.25}$$

(note that the second equality follows by continuity and Lemma 4.2.2). It is well known (see Fedorov (1972) or Pukelsheim (1993)) that the uniform distribution ξ_u at the $2m + 1$ points defined by (4.23) is D-optimal for the trigonometric regression model on the interval $[-\pi, \pi]$. If $\hat{a} = \pi(1-1/(2m+1))$ denotes the largest support point of this design, it follows that $\xi_{\hat{a}}^* = \xi_u$. Consequently, the design $\xi_{\hat{a}}^*$ specified in part (i) of Theorem 4.2.1 is also D-optimal for the trigonometric regression on the interval $[-\hat{a}, \hat{a}]$ and the D-optimality of $\xi_{\hat{a}}^*$ on $[-\pi, \pi]$ shows

$$\xi_{\hat{a}}^* \in \Xi_{\hat{a}}^{(1)} \cap \Xi_{\hat{a}}^{(2)},$$

which implies the inequality $a^* \leq \hat{a}$ for the critical bound in (4.24). Now, for any design of the form

$$\xi = \xi(a) \begin{pmatrix} -t_m & \cdots & -t_1 & t_0 & t_1 & \cdots & t_m \\ \frac{1}{2m+1} & \cdots & \frac{1}{2m+1} & \frac{1}{2m+1} & \frac{1}{2m+1} & \cdots & \frac{1}{2m+1} \end{pmatrix} \tag{4.26}$$

with $0 < t_1 < \cdots < t_m \leq \pi$, it follows from Lemma 4.2.3 that

$$\det M(\xi) = C \prod_{i=1}^{m} (1 - x_i^2)(1 - x_i)^2 \prod_{1 \leq i < j \leq m} (x_j - x_i)^4 =: h(x_\xi)$$

with $C = 2^{2m^2}/(2m + 1)^{2m+1}$, $x_\xi = (x_1, \ldots, x_m)^T$, and $x_i = \cos t_i$ ($i = 1, \ldots, m$). The discussion at the beginning of this section shows that h is strictly concave. Additionally, we have for the design $\xi_{\hat{a}}^*$,

$$h(x_{\xi_{\hat{a}}^*}) = 2^{-2m},$$

and for any other design ξ of the form (4.26),

$$h(x_\xi) < 2^{-2m}$$

(because otherwise a convex combination of $\xi_{\hat{a}}^*$ and ξ_a would have an information matrix with a determinant larger than 2^{-2m}, which is impossible). Consequently, because ξ_u^* is of the form (4.26), it follows for the quantity a^* defined by (4.25) that $a^* = \hat{a}$.

If $a \geq \hat{a}$, the discussion in this proof shows that the design specified by part (i) of Theorem 4.2.1 is D-optimal. If $a < \hat{a}$, definition (4.25) shows that the D-optimal design is in the set $\Xi_a^{(1)}$ and Lemmas 4.2.3 and 4.2.4 (with their corresponding proofs) imply that the D-optimal design for the trigonometric regression on the interval $[-a, a]$ is of the form (4.24), which completes the proof of the theorem.   ∎

Note that Theorem 4.2.1 provides a complete solution of the D-optimal design problem. In case (i) with $a \geq \pi(1 - 1/(2m + 1))$, a D-optimal design for the trigonometric regression model (4.2.1) on the interval $[-a, a]$ is explicitly given by the uniform distribution at the support points specified by (4.23), but is not necessarily unique. If $a < \pi(1 - 1/(2m + 1))$, the D-optimal design is unique and specified by (4.24), where the vector $\tau^*(a) = (\tau_1^*(a), \ldots, \tau_{m-1}^*(a))^T$ can be obtained by means of a Taylor expansion at the point $a = 0$,

$$\tau^*(a) = \sum_{i=0}^{\infty} \tau_{(i)}^* a^i, \tag{4.27}$$

and the vector $\tau_{(0)}^* = \tau^*(0)$ is given in Lemma 4.2.4. It follows from general formulas of Section 2.4 that the coefficients in the above expansion can be calculated by the recursive relations

$$\tau_{(s+1)}^* = -\frac{1}{(s+1)!} J^{-1}(0) \left(\frac{d}{da} \right)^{s+1} g(\tau_{<s>}^*(a), a) \Big|_{a=0},$$

$s = 0, 1, 2, \ldots$, where

$$\tau_{<s>}^*(a) = \sum_{i=0}^{s} \tau_{(i)}^* a^i$$

denotes the Taylor polynomial of degree $s \in \{0, 1, 2, \ldots\}$,

$$J(0) = \left(\frac{\partial^2}{\partial \tau_i \partial \tau_j} \phi(x_\tau, a) \right)_{i,j=1}^{m-1} \Big|_{\tau = \tau^*(0)},$$

and

$$g(\tau, a) = \frac{\partial}{\partial \tau} \phi(x_\tau, a) \in \mathbb{R}^{m-1}.$$

Note that, in general, an exact determination of the radius of convergence for the Taylor expansion (4.27) seems to be intractable. In general, several re-expansions might be needed to obtain the D-optimal design for any $a \in (0, \pi(1 - 1/(2m + 1)))$. However, our numerical calculations in Section 4.2.5 indicate that the expansion at the point $a = 0$ always gives the D-optimal design for the trigonometric regression model (4.1) on the interval $[-a, a]$ for any $a \in (0, \pi(1 - 1/(2m + 1)))$.

4.2.3   The differential equation and the eigenvalue problem

Note that due to formula (4.17), the support points of the D-optimal design on the interval $[-a, a]$, $t_i(a)$, $i = 1, \ldots, m$, can be written in the form

$$t_i(a) = \arccos(x_i^*),$$

where $x^* = (x_1^*, \ldots, x_m^*)$ is the unique point of maximum of the function $\phi(x, a)$ on the set

$$\mathcal{X} = \{x = (x_1, \ldots, x_m);\ 0 < x_1 < \cdots < x_m = \cos(a)\}.$$

Calculating the first partial derivatives of $\phi(x, a)$, we obtain

$$\frac{1}{1 + x_i} - \frac{3}{1 - x_i} + \frac{4}{x_i - 1 + \alpha} + \sum_{j=1, j \neq i}^{m-1} \frac{4}{x_i - x_j} = 0,$$

$i = 1, \ldots, m - 1$, with $x_i = x_i^*$, where $\alpha = 1 - \cos(a)$. Consider the supporting polynomial

$$\psi(z) = \prod_{i=1}^{m-1} (z - x_i^*) = z^{m-1} + \sum_{i=0}^{m-2} \psi_i z^i.$$

Applying the following well-known equality (see, for instance, Fedorov, (1972))

$$\sum_{j=1, j \neq i}^{m-1} \frac{1}{x_i^* - x_j^*} = \frac{1}{2} \frac{\psi''(x_i^*)}{\psi'(x_i^*)}, \quad i = 1, \ldots, m - 1,$$

we obtain the relations

$$\frac{-1 - 2z}{1 - z^2} + \frac{2}{z - 1 + \alpha} + \frac{\psi''(z)}{\psi'(z)} = 0$$

for $z = x_1^*, \ldots, x_{m-1}^*$. Multiplying the equation by the common denominator, we obtain

$$(1 - z^2)(z - 1 + \alpha)\psi''(z) + (-4z^2 + (1 - 2\alpha)z + 3 - \alpha)\psi'(z) = 0$$

again for $z = x_1^*, \ldots, x_{m-1}^*$.

Since on the left-hand side there is a polynomial of degree m vanishing at the $m - 1$ points x_i^*, $i = 1, \ldots, m - 1$, we can equate this to the polynomial ψ multiplied by a linear factor, so that the problem turns out to be one of solving a second-order differential equation

$$P(z) := (1 - z^2)(z - 1 + \alpha)\psi''(z) + (-4z^2 + (1 - 2\alpha)z + 3 - \alpha)\psi'(z)$$
$$- (\vartheta_0 z + \lambda)\psi(z) \equiv 0, \tag{4.28}$$

where $\vartheta_0 = -(m - 1)(m + 2)$ is obtained by comparing coefficients of z^m and λ is an unknown real constant.

Since the solution ψ^* of the differential equation is supposed to be a polynomial of degree $m - 1$, we can rewrite $P(z)$ in the form

$$P(z) = (z^m, \ldots, z, 1)\, A(\lambda, \alpha)\, \psi, \tag{4.29}$$

where $\psi = (\psi_{m-1}, \ldots, \psi_0)^T$ and $A = A(\lambda, \alpha)$ is some $(m + 1) \times m$-matrix. Note that the first row of A consists of zeros. Let $B = B(\lambda, \alpha)$ be the matrix obtained from A by deleting the first row with elements $b_{i,j} = (B(\lambda, \alpha))_{i,j}$, $i, j = 1, \ldots, m$. Comparing the coefficients of the monomials z^j, $j = 0, \ldots, m$, in (4.29) yields

$$
b_{i,j} = \begin{cases}
(m - j)(m - j + 3) - \vartheta_0, & j - i = 1 \\
(m - j)((1 - \alpha)(m - j + 1) - 1) - \lambda, & j - i = 0 \\
(m - j)(m - j - \alpha + 2), & j - i = -1 \\
(m - j)(m - j - 1)(\alpha - 1), & j - i = -2 \\
0, & \text{otherwise.}
\end{cases} \tag{4.30}
$$

Note that the matrix B is of the form $B = B(\lambda, \alpha) = \tilde{B}(\alpha) - \lambda\, I_m$, and λ is an eigenvalue of the matrix $\tilde{B}(\alpha)$. Therefore, we can rewrite (4.28) in the form

$$
(\tilde{B}(\alpha) - \lambda\, I_m)\, \psi = 0. \tag{4.31}
$$

For known λ, we conclude from (4.30) that the vector ψ can be calculated by the following recursive relations

$$
\psi_{m-1} = 1
$$

$$
\psi_\nu = - \sum_{j=\nu+1}^{m-1} b_{m-\nu-1, m-j}\, \psi_j \,/\, b_{m-\nu-1, m-\nu}, \tag{4.32}
$$

where $\nu = m - 2, m - 3, \ldots, 0$.

A method to calculate the eigenvalue of interest will be described in the following subsection. Our approach based on the algebraic equation (4.32) will be called an algebraic approach. Note that a similar method was suggested in Dette, Haines, Imhof (1999) and Melas (1999) for studying (locally) D-optimal designs for rational models. Here, we will combine this approach with the functional approach.

4.2.4   A functional-algebraic approach

Consider the function

$$
g(\lambda, \alpha) = \det(\tilde{B}(\alpha) - \lambda\, I_m).
$$

The unknown value λ in (4.28) is a function of α ($\lambda^*(\alpha)$, say), to be explicitly given by

$$
g(\lambda, \alpha) = 0.
$$

Since λ is a simple eigenvalue of $\tilde{B}(\alpha)$ (recursive formula (4.31) shows that the corresponding normalized eigenvector is unique), the following equation holds:

$$
\frac{d}{d\lambda} g(\lambda, \alpha)\Big|_{\lambda = \lambda^*(\alpha)} \neq 0.
$$

Due to the Implicit Function Theorem (see Section 1.9), $\lambda^*(\alpha)$ is a real analytic function on the interval $(0, \hat{\alpha})$, where $\hat{\alpha} = 1 - \cos(\hat{a})$ and $\hat{a}$ is defined in (4.5). This also follows from the fact that simple eigenvalues of a matrix are real analytic (see Lancaster (1969)). Consequently, the function $\lambda^*(\alpha)$ can be expanded into a Taylor series on this interval. To expand this function in a neighborhood of the origin, we must continue it to the interval $(-\hat{\alpha}, \hat{\alpha})$. So our aim is to find the limit of $\lambda^*(\alpha)$ when $\alpha \to 0$, which can be realized by taking the limit in (4.28). Since all of the points in the D-optimal design tend to zero, it follows that $x_i^* \to 1$, $i = 1, \ldots, m-1$, and for the supporting polynomial, $\psi(z) \to (1 - z)^{m-1}$. By direct calculations, we obtain

$$\lim_{\alpha \to 0} \lambda^*(\alpha) = 1 - m^2.$$

Hence, the function

$$\hat{\lambda}(\alpha) = \begin{cases} \lambda^*(\alpha), & 0 < \alpha < \hat{\alpha} \\ \lambda^*(-\alpha), & 0 > \alpha > -\hat{\alpha} \\ 1 - m^2, & \alpha = 0 \end{cases}$$

is real analytic on the interval $(-\hat{\alpha}, \hat{\alpha})$. Consider its Taylor expansion

$$\hat{\lambda}(\alpha) = \sum_{i=0}^{\infty} \lambda_{(i)} \, \alpha^i, \quad \lambda_{(0)} = 1 - m^2, \tag{4.33}$$

and let

$$\lambda_{<n>}(\alpha) = \sum_{i=0}^{n} \lambda_{(i)} \, \alpha^i,$$

$$(g(\lambda_{<n>}(\alpha), \alpha))_{(n)} = \frac{1}{n!} \frac{\partial^n}{\partial \alpha^n} g(\lambda_{<n>}(\alpha), \alpha) \Big|_{\alpha=0}.$$

To determine the coefficients $\lambda_{(i)}$ in this expansion, we will use the following recursive formulas, which are the particular case of general formulas from Section 2.4:

$$\lambda_{(n+1)} = -J^{-1}(0) \, (g(\lambda_{<n>}(\alpha), \alpha))_{(n+1)}, \quad n = 0, 1, \ldots,$$

$$J(\alpha) = \frac{\partial}{\partial \lambda} \, g(\lambda, \alpha).$$

The first values of the scaled coefficients $\bar{\lambda}_{(i)} = \lambda_{(i)} 2^i$ are given in Table 4.2.

Note that since the eigenvectors of a matrix are real analytic functions (see Lancaster (1969)), the coefficients $\psi_j = \psi_j(\alpha)$, $j = m-2, \ldots, 0$ are real analytic functions on the interval $(0, \hat{\alpha})$. So the problem of determining the components of the (normalized) eigenvector can be dealt with analogously to that of calculating the eigenvalue. By the relations

$$\hat{\psi}_j(\alpha) = \begin{cases} \psi_j(\alpha), & 0 < \alpha < \hat{\alpha} \\ \psi_j(-\alpha), & 0 > \alpha > -\hat{\alpha} \\ \psi_j(0), & \alpha = 0, \end{cases}$$

Table 4.2: Coefficients $\bar{\lambda}_{(i)} = 2^i \lambda_{(i)}$ in the expansion (4.33) of the eigenvalue and coefficients $\bar{\psi}_{j(i)}$ in the expansion (4.34) of the components of the corresponding eigenvector $(\hat{\psi}_0, \ldots, \hat{\psi}_{m-1})$.

	i	0	1	2	3	4	5
$m = 2$	$\bar{\psi}_{0(i)}$	-1	0.85714	-0.06997	-0.02856	-0.00886	-0.00133
	$\bar{\lambda}_{(i)}$	-3	-0.57143	-0.27988	-0.11424	-0.03544	-0.00533
$m = 3$	$\bar{\psi}_{0(i)}$	1	-1.81818	0.67618	-0.02666	-0.00204	0.00190
	$\bar{\psi}_{1(i)}$	-2	1.81818	-0.07012	-0.03345	-0.01963	-0.01248
	$\bar{\lambda}_{(i)}$	-8	-1.09091	-0.42074	-0.20068	-0.11779	-0.07489
$m = 4$	$\bar{\psi}_{0(i)}$	-1	2.80000	-2.22277	0.48317	-0.00808	0.00222
	$\bar{\psi}_{1(i)}$	3	-5.00000	2.29169	-0.05781	-0.01309	-0.00261
	$\bar{\psi}_{2(i)}$	-3	2.80000	-0.06892	-0.03375	-0.02060	-0.01400
	$\bar{\lambda}_{(i)}$	-15	-1.60000	-0.55138	-0.27001	-0.16480	-0.11199
$m = 5$	$\bar{\psi}_{0(i)}$	1	-3.78947	4.74901	-2.23711	0.32001	-0.00380
	$\bar{\psi}_{1(i)}$	-4	11.36842	-9.56600	2.36016	-0.03433	-0.00019
	$\bar{\psi}_{2(i)}$	6	-11.36842	4.88488	-0.08948	-0.02490	-0.00831
	$\bar{\psi}_{3(i)}$	-4	3.78947	-0.06792	-0.03357	-0.02072	-0.01430
	$\bar{\lambda}_{(i)}$	-24	-2.10526	-0.67923	-0.33566	-0.20720	-0.14296

where $\psi_j(0) = (-1)^{m-j-1}(m-1)!\,/\,(j\,!(m-j-1)!)$, these functions can be analytically expanded on the interval $(-\hat{\alpha}, \hat{\alpha})$. The Taylor expansions

$$\hat{\psi}_j(\alpha) = \sum_{i=0}^{\infty} \bar{\psi}_{j(i)} \, \alpha^i / \, 2^i \tag{4.34}$$

can be constructed using the recursive formulas (4.30). The first coefficients are listed in Table 4.2.

Using the values of the $\psi_i(\alpha)$ for the components of the eigenvector, the Taylor expansions of the functions (which give the support points of the D-optimal design), $t_i(a)$, $i = 1, \ldots, m-1$, can be constructed as follows. Note that these functions are real analytic because the roots of a polynomial are real analytic functions of its coefficients.

Let us define the polynomial $\rho(u, \alpha)$ by the relation

$$\rho(u, \alpha) = \alpha^{1-m} \, \psi(1 - \alpha u).$$

Denote by $u_i(0)$, $i = 1, \ldots, m-1$, the roots of $\rho(u, 0) = \text{const } P_{m-1}^{(1,1/2)}(2u-1)$, where $P_{m-1}^{(\beta,\gamma)}$ is the Jacobi polynomial with parameters (β, γ) of degree $m - 1$. Construct expansions of the solutions $u_i(\alpha) = u(\alpha)$ of the equation $\rho(u, \alpha) = 0$ with the initial condition $u(0) = u_i(0)$ by the functional approach described earlier and return to the original variables $t_i(a) = \arccos(x_i(\alpha)) = \arccos(1 - \alpha u_i(\alpha))$, $\alpha = 1 - \cos(a)$, $i = 1, \ldots, m-1$.

Proceeding as described earlier, we obtained the first coefficients of the Taylor expansions for the support points $t_i(a)$, $i = 1, \ldots, m - 1$, of the D-optimal design for the trigonometric regression model (4.2.1) on the interval $[-a, a]$ (if $a < \hat{a}$). Note that the present approach appears to be preferable with regard to computer time and memory compared to the direct functional approach. This is not surprising, since the algebraic-analytical approach takes into account the special structure of the problem at hand.

4.2.5 Examples

Example 4.2.1 Our first example considers the linear trigonometric regression model ($m = 1$) on the interval $[-a, a]$, for which the solution is rather obvious. If $a \geq 2\pi/3$, the design

$$\xi_a^* = \begin{pmatrix} -\frac{2\pi}{3} & 0 & \frac{2\pi}{3} \\ \frac{1}{3} & \frac{1}{3} & \frac{1}{3} \end{pmatrix}$$

is D-optimal, whereas for $a < 2\pi/3$, the D-optimal design for the linear trigonometric regression model on the interval $[-a, a]$ is given by

$$\xi_a^* = \begin{pmatrix} -a & 0 & a \\ \frac{1}{3} & \frac{1}{3} & \frac{1}{3} \end{pmatrix}.$$

This follows directly from Theorem 4.2.1.

Example 4.2.2 In the quadratic regression model, the situation is more complicated. If $a \geq 4\pi/5$, then part (i) of Theorem 4.2.1 shows that the design

$$\xi_a^* = \begin{pmatrix} -\frac{4\pi}{5} & -\frac{2\pi}{5} & 0 & \frac{2\pi}{5} & \frac{4\pi}{5} \\ \frac{1}{5} & \frac{1}{5} & \frac{1}{5} & \frac{1}{5} & \frac{1}{5} \end{pmatrix}$$

is D-optimal. If $a < 4\pi/5$, the D-optimal design can be obtained by means of a Taylor expansion, as indicated in the second part of Theorem 4.2.1. However, in this particular case, an explicit solution is possible by a careful inspection of the arguments given in Section 4.2.2. Part (ii) of Theorem 4.2.1 shows that the D-optimal design in the quadratic trigonometric regression model is in the set $\Xi_a^{(1)}$ whenever $a < 4\pi/5$ and, consequently, only one support point $t_1^* = t_1^*(a)$ has to be determined. This can be done by a direct differentiation of the function $\phi(x, a)$ in (4.18). Note that $m = 2$, $x_2 = \cos a$, and, therefore, $\phi(x, a)$ is a function of only one variable, say $x_1 \in (-1, 1)$. Elementary calculus yields that the derivative of ϕ has zeros at the points $x_1 = \cos a$, $x_2 = 1$ and

$$x_{3,4} = \frac{1}{8}\left[2\cos(a) - 1 \mp \sqrt{33 + 12\cos(a) + 4\cos(a)^2}\right].$$

It is easy to see that only one of these two points yields a solution in the interval $[\cos a, 1]$ and, consequently, the D-optimal design for the quadratic

Table 4.3: Coefficients in the expansion (4.35). The D-optimal design in the trigonometric regression model (4.2.1) on the interval $[-a, a]$ with $0 < a < \pi(1 - 1/(2m+1))$ has equal masses at the points $-a, -t_{m-1}, \ldots, -t_1$, $0, t_1, \ldots, t_{m-1}, a$, where $t_i = a\tau_i^*(a)$, $i = 1, \ldots, m-1$.

	i	0	2	4	6	8	10
$m = 2$	$\tau_{1(i)}^*$	0.65465	-0.21977	-0.07747	0.04852	0.06118	-0.02116
$m = 3$	$\tau_{1(i)}^*$	0.46885	-0.19145	-0.00875	0.02584	-0.00184	-0.00283
	$\tau_{2(i)}^*$	0.83022	-0.13502	-0.10286	-0.05465	-0.00161	0.03946
$m = 4$	$\tau_{1(i)}^*$	0.36312	-0.15556	0.00820	0.01117	-0.00368	-0.00011
	$\tau_{2(i)}^*$	0.67719	-0.18093	-0.07349	0.00094	0.02393	0.01100
	$\tau_{3(i)}^*$	0.89976	-0.08456	-0.07603	-0.06025	-0.03806	-0.01256
$m = 5$	$\tau_{1(i)}^*$	0.29576	-0.12851	0.01204	0.00501	-0.00238	0.00036
	$\tau_{2(i)}^*$	0.56524	-0.18316	-0.03971	0.01585	0.01178	-0.00245
	$\tau_{3(i)}^*$	0.78448	-0.14366	-0.08805	-0.03360	0.00483	0.01980
	$\tau_{4(i)}^*$	0.93400	-0.05677	-0.05431	-0.04874	-0.03965	-0.02762

trigonometric regression model on the interval $[-a, a]$ with $0 < a \leq 4\pi/5$ is given by

$$\xi_a^* = \begin{pmatrix} -a & -t_1^*(a) & 0 & t_1^*(a) & a \\ \frac{1}{5} & \frac{1}{5} & \frac{1}{5} & \frac{1}{5} & \frac{1}{5} \end{pmatrix},$$

where

$$t_1^*(a) = \arccos\left(\frac{1}{8}\left[2\cos(a) - 1 + \sqrt{33 + 12\cos(a) + 4\cos(a)^2}\right]\right).$$

Example 4.2.3 In the general case $m \geq 3$, the second part of Theorem 4.2.1 has to be applied if $a \leq \pi(1 - 1/(2m+1))$ (note that in the remaining case, a D-optimal design is explicitly given in part (i) of Theorem 4.2.1). From Table 4.1, we obtain the values of $\tau_i^*(0)$, $i = 1, \ldots, m-1$ (provided $m \leq 5$), and the nontrivial support points $\tau_i^*(a)$ for $0 < a < \pi(1 - 1/(2m+1))$ can now be calculated by means of a Taylor expansion, as indicated at the end of Section 4.2.2. Using the functional algebraic approach, we calculated the values of the first coefficients in the expansion

$$\tau_i^*(a) = \sum_{l=0}^{\infty} \tau_{i(l)}^* \left(\frac{a}{\pi}\right)^l, \quad i = 1, \ldots, m-1 \qquad (4.35)$$

for $m = 2, 3, 4, 5$. These coefficients are collected in Table 4.3. It can easily be shown that $\tau_i^*(a)$ is an even function of the parameter a and, consequently, the odd coefficients vanish and only the even coefficients are displayed.

Consider as a concrete example the case $m = 3$. If $a \geq 6\pi/7$, a D-optimal design for the cubic trigonometric regression model on the interval

$[-a, a]$ is given by part (i) of Theorem 4.2.1; that is,

$$\xi_a^* = \left(\begin{array}{ccccccc} -\frac{6\pi}{7} & -\frac{4\pi}{7} & -\frac{2\pi}{7} & 0 & \frac{2\pi}{7} & \frac{4\pi}{7} & \frac{6\pi}{7} \\ \frac{1}{7} & \frac{1}{7} & \frac{1}{7} & \frac{1}{7} & \frac{1}{7} & \frac{1}{7} & \frac{1}{7} \end{array} \right).$$

If $0 < a < 6\pi/7$, the D-optimal design can be calculated from the expansion (4.21) and Table 4.3. For example, if $a = 1$, we obtain that the D-optimal design for the cubic trigonometric regression model on the interval $[-1, 1]$ is given by

$$\xi_a^* = \left(\begin{array}{ccccccc} -1 & -0.8154 & -0.4494 & 0 & 0.4494 & 0.8154 & 1 \\ \frac{1}{7} & \frac{1}{7} & \frac{1}{7} & \frac{1}{7} & \frac{1}{7} & \frac{1}{7} & \frac{1}{7} \end{array} \right).$$

4.3 E-Optimal Designs

4.3.1 Preliminary results and E-optimal designs on large design spaces

Consider the common regression model

$$y = \sum_{j=0}^{k} \theta_j f_j(x) + \varepsilon, \quad x \in \mathcal{X}, \tag{4.36}$$

where the explanatory variable varies in the compact design space $\mathcal{X}$, $f_0, \ldots, f_k$ are continuous and linearly independent regression functions, and observations at different points are assumed to be independent.

In the present section, we are interested in the E-optimality criterion, which is given by

$$\Phi(\xi) = \lambda_{\min}(M(\xi)) ,$$

where $\lambda_{\min}(A)$ denotes the minimum eigenvalue of a symmetric matrix $A \in \mathbb{R}^{k+1 \times k+1}$. Note that maximizing Φ is equivalent to minimizing the function

$$\frac{1}{\Phi(\xi)} = \lambda_{\max}(M^{-1}(\xi)) = \max_{\|a\|_2=1, a \in \mathbb{R}^{k+1}} a^T M^{-1}(\xi) a.$$

The expression $a^T M^{-1}(\xi) a$ is proportional to the variance of the least squares estimate for the linear combination $a^T \theta$ ($a \in \mathbb{R}^{k+1}$) and, therefore, an E-optimal design minimizes the worst variance over all possible (normalized) linear combinations.

It follows by standard arguments (see, e.g., Pukelsheim (1993)) that an E-optimal design exists. For the convenience of the reader, we will repeat here the equivalence theorem already formulated in Section 3.2 (Theorem 3.3.1).

For an E-optimal design ξ_E, we define $\mathcal{P}_{\xi_E}$ as the eigenspace corresponding to the minimal eigenvalue $\lambda_{\min}(M(\xi_E))$ and

$$\mathcal{P} = \bigcap_{\xi_E \text{ is } E\text{-optimal}} \mathcal{P}_{\xi_E} \tag{4.37}$$

as the intersection of all eigenspaces corresponding to E-optimal designs. It can easily be verified that $\mathcal{P} \neq \emptyset$ and the following lemma gives a characterization for E-optimal designs.

Lemma 4.3.1 (Equivalence Theorem) *A design ξ^* is E-optimal for the regression model (4.36) if and only if there exists a non-negative definite matrix $A^* \in \mathbb{R}^{k+1 \times k+1}$ such that $\operatorname{tr} A^* = 1$ and*

$$\max_{x \in \mathcal{X}} f^T(x) A^* f(x) \leq \lambda_{\min}(M(\xi^*)). \tag{4.38}$$

Moreover, if x^* is a support point of ξ^*, there is equality in (4.38), that is,

$$f^T(x^*) A^* f(x^*) = \lambda_{\min}(M(\xi^*)),$$

and the matrix A^* can be represented as

$$A^* = \sum_{i=1}^{s} \alpha_i z_i z_i^T ,$$

where $z_1, \ldots, z_s$ is an orthonormal basis of the set $\mathcal{P}$ defined in (4.37), $s = \dim \mathcal{P}$, and $\alpha_1, \ldots, \alpha_s \geq 0$ with $\sum_{i=1}^{s} \alpha_i = 1$.

In the specific situation of the trigonometric regression model (4.1), we have $\mathcal{X} = [c,d]$, $f_0(t) = 1/\sqrt{2}$, $f_{2j}(t) = \cos(jt)$ $(j = 1, \ldots, m)$, and $f_{2j-1}(t) = \sin(jt)$ $(j = 1, \ldots, m)$. Note that we use a slightly different parameterization of the intercept, but most of our results are also valid for the trigonometric regression model with $f_0(t) = 1$. Our first result shows that the E-optimal design in the trigonometric regression model is essentially invariant with respect to transformations of the design space by an additive shift.

Lemma 4.3.2 *Let*

$$\eta = \left(\begin{array}{ccc} t_1 & \cdots & t_n \\ w_1 & \cdots & w_n \end{array} \right)$$

denote a design on the interval $[c,d]$, $a = (c+d)/2$, and ξ_η be the design obtained by the linear transformation $t \to t - a$, that is,

$$\xi_\eta = \left(\begin{array}{ccc} t_1 - a & \cdots & t_n - a \\ w_1 & \cdots & w_n \end{array} \right);$$

then the information matrices $M(\eta)$ and $M(\xi_\eta)$ in the trigonometric regression model (4.1) have the same eigenvalues, in particular

$$\lambda_{\min}(M(\eta)) = \lambda_{\min}(M(\xi_\eta)).$$

Proof. Let $f(t) = (1/\sqrt{2}, \sin t, \cos t, \ldots, \sin(mt), \cos(mt))^T$; then we have for any $\alpha \in \mathbb{R}$,

$$f(t + \alpha) = P(\alpha)f(t),$$

where $P(\alpha)$ is a $(2m+1) \times (2m+1)$ (block) matrix given by

$$P(\alpha) = \begin{pmatrix} \frac{1}{\sqrt{2}} & & & \\ & Q(\alpha) & & \\ & & \ddots & \\ & & & Q(m\alpha) \end{pmatrix},$$

with

$$Q(\beta) = \begin{pmatrix} \cos\beta & \sin\beta \\ -\sin\beta & \cos\beta \end{pmatrix} \in \mathbb{R}^{2\times 2}.$$

Because $P(\alpha)$ is orthogonal, the matrices $M(\eta)$ and

$$M(\xi_\eta) = \int_{-a}^{a} f(t)f^T(t)\, d\xi_\eta(t) = \int_{c}^{d} f(t-a)f^T(t-a)\, d\eta(t)$$
$$= P(-a)M(\eta)P^T(-a)$$

have the same eigenvalues and the assertion of the lemma has been established. ∎

From the proof of Lemma 4.3.2 it follows that for any Φ_p-criterion in the sense of Pukelsheim (1993), the solution of the ϕ_p-optimal design problem for the trigonometric regression model (4.1) on the interval $[c, d]$ can be obtained from the solution of the corresponding problem on the interval $[-a, a]$ and a linear transformation. For this reason, we will restrict our subsequent investigations on E-optimal designs to symmetric intervals of the form $[-a, a]$, where $0 < a \le \pi$. Note that, in general, an E-optimal design for the trigonometric regression model (4.1) on the interval $[-a, a]$ is not necessarily unique. For example, it follows from Lemma 4.3.1 that for the full circle $[-a, a] = [-\pi, \pi]$, any design with information matrix

$$M^* = I_{2m+1} = \text{diag}\left(\frac{1}{2}, \frac{1}{2}, \frac{1}{2}, \ldots, \frac{1}{2}\right) \in \mathbb{R}^{2m+1 \times 2m+1} \tag{4.39}$$

is E-optimal. In particular, any design of the form

$$\xi_n^* = \begin{pmatrix} t_1 & \cdots & t_n \\ \frac{1}{n} & \cdots & \frac{1}{n} \end{pmatrix} \tag{4.40}$$

with $n \ge 2m + 1$ and

$$t_j = -\pi + \frac{2j-1}{n}\pi, \qquad j = 1, \ldots, n,$$

has information matrix M^* (see Pukelsheim (1993)) and is therefore E-optimal for the trigonometric regression model on the interval $[-\pi, \pi]$. In the following, we will prove that the E-optimal design for the trigonometric regression model is unique, provided that the design space is sufficiently small.

To this end, let $\Xi_{(a)}^{(1)}$ denote the set of all designs of the form

$$\xi = \xi(a) = \begin{pmatrix} -t_m & \cdots & -t_1 & t_0 & t_1 & \cdots & t_m \\ \frac{w_m}{2} & \cdots & \frac{w_1}{2} & w_0 & \frac{w_1}{2} & \cdots & \frac{w_m}{2} \end{pmatrix},$$

where $0 = t_0 < t_1 < \cdots < t_{m-1} < t_m = a$ and $w_j > 0\,(j = 0, \ldots, m)$ such that $\sum_{j=0}^m w_j = 1$. Furthermore, define

$$\Xi_{(a)}^{(2)} = \Big\{ \xi \mid \operatorname{supp}(\xi) \subset [-a, a], \exists\, A^* \in PD(2m + 1) : \operatorname{tr} A^* = 1,$$

$$= f^T(t) A^* f(t) = \lambda_{\min}(M(\xi)) \;\; \forall\, t \in [-a, a] \Big\}, \tag{4.41}$$

where $PD(2m+1)$ denotes the set of all positive definite $(2m+1) \times (2m+1)$ matrices. A straightforward calculation shows $\xi_{2m+1}^* \in \Xi_{(a)}^{(2)}$, and with the aid of Lemma 4.3.2, it is easy to see that the design ξ_{2m+1}^* defined in (4.40) is E-optimal for the trigonometric regression model on the interval $[-a, a]$, whenever $a > \bar{a}$, where

$$\bar{a} = \bar{a}(m) = \pi \left(1 - \frac{1}{2m + 1} \right) \tag{4.42}$$

denotes the largest support point of the design ξ_{2m+1}^*. The following result shows that E-optimal designs for the trigonometric regression model on the interval $[-a, a]$ are either in the set $\Xi_{(a)}^{(1)}$ or in $\Xi_{(a)}^{(2)}$, depending on the sign of the quantity $a - \bar{a}$.

Theorem 4.3.1 *If $a \in [\bar{a}, \pi]$, then any E-optimal design for the trigonometric regression model (4.1) on the interval $[-a, a]$ is contained in the set $\Xi_{(a)}^{(2)}$ defined in (4.41). If $a \in (0, \bar{a})$, then the E-optimal design for the trigonometric regression model on the interval $[-a, a]$ is unique and contained in the set $\Xi_{(a)}^{(1)}$. Moreover, $\xi \in \Xi_{(a)}^{(2)}$ if and only if the information matrix of ξ is of the form (4.39).*

Proof. Let ξ^* denote an E-optimal design for the trigonometric regression model on the interval $[-a, a]\,(0 < a \leq \pi)$, then it follows by similar arguments as given in the proof of Lemma 4.2.2 of Section 4.2 that

$$\xi^* \in \Xi_{(a)}^{(1)} \cup \Xi_{(a)}^{(2)}$$

(we only have to replace the equivalence theorem for D-optimality by Lemma 4.3.1). The same arguments show that if an E-optimal design for the trigonometric regression model (4.1) on the interval $[-a, a]$ belongs to the set $\Xi_{(a)}^{(1)}$; then it is the unique E-optimal design on $[-a, a]$.

We now prove the last assertion of the theorem. If $a > \bar{a}$, then the design ξ_{2m+1}^* defined in (4.40) is E-optimal for the trigonometric regression on the interval $[-a, a]$ and therefore satisfies $M(\xi_{2m+1}^*) = M^*$, where M^* is given in (4.39). Consequently, any design ξ on the interval $[-a, a]$ with $M(\xi) = M^*$ must also be E-optimal and satisfy $\xi \in \Xi_{(a)}^{(2)}$.

Conversely, let

$$\xi = \begin{pmatrix} t_1 & t_2 & \cdots & t_n \\ w_1 & w_2 & \cdots & w_n \end{pmatrix}$$

denote an arbitrary design on the interval $[-a, a]$; then it is easy to see that the information matrix of ξ in the trigonometric regression model satisfies

$$(M(\xi))_{11} = \frac{1}{2}, \quad \mathrm{tr}(M(\xi)) = m + \frac{1}{2}. \tag{4.43}$$

Now, assume additionally that ξ is E-optimal and $a > \bar{a}$; then the E-optimality of the design ξ_{2m+1}^* in (4.40) implies

$$\lambda_{\min}(M(\xi)) = \lambda_{\min}(M(\xi_{2m+1}^*)) = \lambda_{\min}(M^*) = \frac{1}{2}.$$

On the other hand, we have from the well-known estimates $(M(\xi))_{ii} \geq \lambda_{\min}(M(\xi)) = \frac{1}{2}$ and the equations in (4.43) that

$$m + \frac{1}{2} = \sum_{i=1}^{2m+1} (M(\xi))_{ii} \geq \frac{1}{2}(2m+1) = m + \frac{1}{2},$$

which shows

$$(M(\xi))_{ii} = \frac{1}{2}, \quad i = 1, \ldots, 2m+1.$$

In the next step, let $\alpha = (M(\xi))_{ij} = (M(\xi))_{ji}$ denote the element in the position (i, j) of the information matrix of the design ξ, where $1 \leq i \neq j \leq 2m+1$, and define

$$p = \frac{1}{\sqrt{2}}(e_i - \mathrm{sign}(\alpha)e_j),$$

where $e_i \in \mathbb{R}^{2m+1}$ denotes the i-th unit vector. Then $\|p\|_2^2 = 1$ and we obtain

$$\frac{1}{2} = \lambda_{\min}(M(\xi)) \leq p^T M(\xi)p$$

$$= \frac{1}{2}(1, -\mathrm{sign}(\alpha)) \begin{pmatrix} \frac{1}{2} & \alpha \\ \alpha & \frac{1}{2} \end{pmatrix} \begin{pmatrix} 1 \\ -\mathrm{sign}(\alpha) \end{pmatrix}$$

$$= \frac{1}{2} - |\alpha| \leq \frac{1}{2},$$

which implies $\alpha = (M(\xi))_{ij} = 0$, whenever $1 \leq i \neq j \leq 2m + 1$. Consequently, the information matrix of any E-optimal design is diagonal (i.e., $M(\xi) = \frac{1}{2}I_{2m+1}$).

Now, let $a < \bar{a}$; then it follows from the results of Section 4.2 that for any design ξ on the interval $[-a, a]$,

$$\det M(\xi) < \left(\frac{1}{2}\right)^{2m+1}$$

(see the proof of Theorem 4.2.1). Because any E-optimal design ξ^* in $\Xi_a^{(2)}$ satisfies $\det M(\xi^*) = \det M^* = 2^{-2m-1}$, there are no E-optimal designs on the interval $[-a, a]$, which belong to the set $\Xi_a^{(2)}$ (if $a < \bar{a}$). Consequently, by the discussion at the beginning of the proof, the E-optimal design is unique and an element of the set $\Xi_a^{(1)}$. Finally, if $a > \bar{a}$, we have shown that the information matrix of the E-optimal design for the trigonometric regression model is unique and equal to the matrix $M^* = M(\xi^*_{2m+1}) = \frac{1}{2}I_{2m+1}$, where the design ξ^*_{2m+1} is defined by (4.40). Because $\xi^*_{2m+1} \in \Xi_a^{(2)}$, it follows from definition (4.41) that any E-optimal design belongs to the set $\Xi_a^{(2)}$.   ∎

Note that Theorem 4.3.1 provides a solution of the E-optimal design problem in the trigonometric regression model on the interval $[-a, a]$ whenever $a > \bar{a} = \pi(1 - 1/(2m + 1))$. In this case, the solution is not necessarily unique. However, the information matrix corresponding to E-optimal designs is unique although the E-criterion (considered as a mapping on the positive definite matrices) is *not* strictly concave. If $a < \bar{a}$, the E-optimal design on the interval $[-a, a]$ is unique and will be described explicitly in the following subsection, when the parameter a is sufficiently small.

4.3.2 E-optimal designs on sufficiently small intervals

Throughout this chapter let

$$T_k(x) = \cos(k \arccos x) , \qquad k \in \mathbb{N}_0 ,$$

denote the k-th Chebyshev polynomial of the first kind (see Rivlin (1974)), which are orthogonal with respect to the arcsine distribution; that is,

$$\frac{2}{\pi} \int_{-1}^{1} T_i(x)T_j(x)\frac{dx}{\sqrt{1 - x^2}} = \begin{cases} 1 & \text{if} \quad i = j \geq 1, \\ 2 & \text{if} \quad i = j = 0, \\ 0 & \text{if} \quad i \neq j. \end{cases} \tag{4.44}$$

It is well know (see Rivlin (1974)) that $T_k(x)$ is the unique solution of the extremal problem

$$\min_{a_0,\dots,a_{k-1}\in\mathbb{R}} \quad \max_{x\in[-1,1]} |2^{k-1}x^k + a_{k-1}x^{k-1} + \cdots + a_1x + a_0| ,$$

and, in particular, we have equality at the Chebyshev points $s_i = \cos(i\pi/k)$; that is,

$$T_k(s_i) = (-1)^i, \qquad i = 0, \ldots, k.$$

Throughout this chapter, let $\alpha = \cos a \in [-1, 1)$, define

$$x_i = x_i(a) = \frac{1 - \alpha}{2} s_i + \frac{1 + \alpha}{2}, \qquad i = 0, \ldots, m, \tag{4.45}$$

as the extremal points of the Chebyshev polynomial of the first kind

$$T_m\left(\frac{2x - 1 - \alpha}{1 - \alpha}\right) = \frac{q_{a0}}{\sqrt{2}} + \sum_{i=1}^{m} q_{ai} T_i(x) \tag{4.46}$$

on the interval $[\alpha, 1]$, and define

$$t_i = t_i(a) = \frac{1}{a} \arccos x_i, \qquad i = 0, \ldots, m. \tag{4.47}$$

We will consider designs of the form

$$\hat{\xi}_a = \begin{pmatrix} -at_m & \cdots & -at_1 & t_0 & at_1 & \cdots & at_m \\ \frac{\hat{w}_m}{2} & \cdots & \frac{\hat{w}_1}{2} & \hat{w}_0 & \frac{\hat{w}_1}{2} & \cdots & \frac{\hat{w}_m}{2} \end{pmatrix} \tag{4.48}$$

as candidate for the E-optimal design in the trigonometric regression model (4.1) on the interval $[-a, a]$ (note that $\hat{\xi}_a \in \Xi_{(a)}^{(1)}$). The weights in (4.48) are given by

$$\hat{w}_i = \hat{w}_i(a) = \frac{|q_a^T F^{-1} e_i|}{\sum_{j=0}^{m} |q_a^T F^{-1} e_j|}, \qquad i = 0, \ldots, m, \tag{4.49}$$

where $e_i \in \mathbb{R}^{m+1}$ denotes the $(i + 1)$-st unit vector, the vector $q_a^T = (q_{a0}, \ldots, q_{am}) \in \mathbb{R}^{m+1}$ is defined by the representation (4.46), and the matrix $F \in \mathbb{R}^{m+1 \times m+1}$ is given by

$$F = \begin{pmatrix} \frac{1}{\sqrt{2}} & \frac{1}{\sqrt{2}} & \cdots & \frac{1}{\sqrt{2}} \\ T_1(x_0) & T_1(x_1) & \cdots & T_1(x_m) \\ \vdots & \vdots & \ddots & \vdots \\ T_m(x_0) & T_m(x_1) & \cdots & T_m(x_m) \end{pmatrix}. \tag{4.50}$$

The following result specifies some properties of the design defined in (4.48) and (4.49) and is the main tool for proving its E-optimality for sufficiently small design spaces $[-a, a]$.

Lemma 4.3.3 Let $\hat{\xi}_a$ denote the design defined by (4.48) and (4.49); then the following statements are correct.

(i) If $0 < a \leq \pi/2$, then the weights $\hat{w}_i = \hat{w}_i(a)$ can be represented as

$$\hat{w}_i = \lambda_a (-1)^i \frac{2}{\pi} \int_{-1}^{1} \ell_i(x)\, dx,$$

where the constant λ_a is given by

$$\lambda_a = \frac{1}{q_a^T q_a}, \tag{4.51}$$

the vector $q_a^T = (q_{a0}, \ldots, q_{am})$ is defined in the representation (4.46), and

$$\ell_i(x) = \prod_{j \neq i} \frac{x - x_j}{x_i - x_j}$$

denotes the i-th Lagrange interpolation polynomial with knots $x_0, \ldots, x_m$ given by (4.45).

(ii) For all $a \in (0, \pi]$, the quantity λ_a defined in (4.51) is an eigenvalue of the matrix $M(\hat{\xi}_a)$ with corresponding eigenvector $\bar{q}_a = (q_{a0}, 0, q_{a1}, 0, \ldots, 0, q_{am})^T$.

(iii) The support points and weights defined by (4.47) and (4.49), respectively, satisfy

$$\lim_{a \to 0} t_i(a) = \cos\left(\pi \frac{m-i}{2m}\right), \quad i = 0, \ldots, m,$$

$$\lim_{a \to 0} \hat{w}_i(a) = \begin{cases} \dfrac{1}{m} & \text{if } i = 1, \ldots, m-1 \\ \dfrac{1}{2m} & \text{if } i = 0, m. \end{cases}$$

Proof. Let $w = (w_0, \ldots, w_m)^T \in \mathbb{R}_+^{m+1}; \sum_{i=0}^{m} w_i = 1$ and

$$\xi_a(w) = \begin{pmatrix} -at_m & \cdots & -at_1 & t_0 & at_1 & \cdots & at_m \\ \frac{w_m}{2} & \cdots & \frac{w_1}{2} & w_0 & \frac{w_1}{2} & \cdots & \frac{w_m}{2} \end{pmatrix}$$

be an arbitrary design with positive weights at the points $\pm at_i$ ($i = 0, \ldots, m$). It was shown in Theorems 4.1 and 4.3 of Dette and Melas (2001) that for $a \in (0, \pi/2]$, the optimal designs $\xi_{(0)}, \xi_{(2)}, \ldots, \xi_{(2m)}$ for estimating the individual coefficients $\beta_0, \beta_2, \ldots, \beta_{2m}$, respectively, in the trigonometric regression model (4.1) on the interval $[-a, a]$ are of the form

$$\xi_{(2j)} = \xi_a(w_{(j)}), \quad j = 0, \ldots, m,$$

where the weights $w_{(j)} = (w_{(j)0}, \ldots, w_{(j)m})^T$ are given by

$$w_{(j)i} = \frac{B_{(j)i}}{\sum_{s=0}^{m} B_{(j)s}}, \quad i = 0, \ldots, m, \tag{4.52}$$

and

$$B_{(j)i} = (-1)^{m+i-j} \int_{-1}^{1} \ell_i(x) T_j(x) \frac{dx}{\sqrt{1-x^2}}$$
$$= (-1)^{m+i+j} c_j e_i^T F^{-1} e_j, \qquad (4.53)$$

with $c_0 = \pi/\sqrt{2}$ and $c_j = \pi/2\,(j = 1, \ldots, m)$. Note that we use a slightly different notation for the support points t_i and weights $w_{(j)i}$ compared to the cited reference.

A similar argument as given in the proof of Lemma 4.3.2 of Dette and Melas (2001) shows that the design $\eta_{\xi_{(2j)}}$ obtained by the transformation

$$\eta_\xi(\cos x) = \begin{cases} \xi(x) + \xi(-x) & \text{if } 0 < x \le a \\ \xi(0) & \text{if } x = 0 \end{cases} \qquad (4.54)$$

is optimal for estimating the coefficient δ_j in the Chebyshev regression model

$$y = \frac{\delta_0}{\sqrt{2}} + \sum_{j=1}^{m} \delta_j T_j(x) + \varepsilon , \qquad (4.55)$$

and the representation (4.46) in this chapter and Lemma 4.3.1 in the cited reference show

$$q_{aj}^2 = e_j^T M_T^{-1}(\eta_{\xi_{(2j)}}) e_j, \quad j = 0, \ldots, m,$$

where $M_T(\eta)$ denotes the information matrix of the design η in the model (4.55). Moreover, recalling the definition of the matrix F in (4.50), it follows from Lemma 8.9 in Pukelsheim (1993) that

$$q_{aj}^2 = e_j^T M_T^{-1}(\eta_{\xi_{(j)}}) e_j = \left(\sum_{i=0}^{m} |e_i^T F^{-1} e_j| \right)^2 = \left(\sum_{i=0}^{m} (-1)^{m+i+j} e_i^T F^{-1} e_j \right)^2 \qquad (4.56)$$

$(j = 0, \ldots, m)$, where the last equality is obtained by a careful analysis of the sign pattern in the matrix F^{-1} observing that $a \in (0, \pi/2]$. Because the sign of q_{aj} for $a \in (0, \pi/2]$ is $(-1)^{m-j}$, we obtain

$$q_{aj} = \sum_{i=0}^{m} (-1)^i e_i^T F^{-1} e_j, \quad j = 0, \ldots, m, \qquad (4.57)$$

and the second equality in (4.53) gives, for the vector $q_a^T = (q_{a0}, \ldots, q_{am})$,

$$q_a^T F^{-1} e_i = \sum_{j=0}^{m} q_{aj} (e_j^T F^{-1} e_i)$$
$$= \int_{-1}^{1} \ell_i(x) \sum_{j=0}^{m} \frac{q_{aj}}{c_j} T_j(x) \frac{dx}{\sqrt{1-x^2}} \qquad (4.58)$$
$$= \frac{2}{\pi} \int_{-1}^{1} \ell_i(x) \frac{dx}{\sqrt{1-x^2}}$$

$(i = 0, \ldots, m)$, where we have used the representation (4.46) and the fact that $c_j = \pi/2$ $(j = 1, \ldots, m)$ and $c_0 = \pi/\sqrt{2}$. Moreover, observing that for $a \in (0, \pi/2]$ the sign of q_{aj} and $e_j^T F^{-1} e_i$ is $(-1)^{m-j}$ and $(-1)^{m+i+j}$, respectively, we obtain that the sign of $q_a^T F^{-1} e_i$ is $(-1)^i$. Now, the polynomial in (4.46) attains the values $(-1)^i$ at the point x_i $(i = 0, \ldots, m)$ and it follows that

$$T_m \left(\frac{2x - 1 - \alpha}{1 - \alpha} \right) = \frac{q_{a0}}{\sqrt{2}} + \sum_{j=1}^{m} q_{aj} T_j(x) = \sum_{j=0}^{m} (-1)^j \ell_j(x).$$

Combining these arguments yields

$$\sum_{i=0}^{m} |q_a^T F^{-1} e_i| = \sum_{i=0}^{m} (-1)^i q_a^T F^{-1} e_i$$

$$= \frac{2}{\pi} \sum_{i=0}^{m} (-1)^i \int_{-1}^{1} \ell_i(x) \frac{dx}{\sqrt{1 - x^2}} \qquad (4.59)$$

$$= \frac{2}{\pi} \int_{-1}^{1} T_m^2 \left(\frac{2x - 1 - \alpha}{1 - \alpha} \right) \frac{dx}{\sqrt{1 - x^2}} = q_a^T q_a,$$

where the last equation is a consequence of the representation (4.46) and the orthogonality relations (4.44). Assertion (i) of Lemma 4.3.3 now follows from definition (4.49) and the identity (4.58).

In order to prove the second assertion of Lemma 4.3.3, let $P \in \mathbb{R}^{2m+1 \times 2m+1}$ denote a permutation matrix such that

$$P M(\hat{\xi}_a) P^T = \bar{M}(\hat{\xi}_a) := \begin{pmatrix} M_c(\hat{\xi}_a) & 0 \\ 0 & M_s(\hat{\xi}_a) \end{pmatrix}, \qquad (4.60)$$

where the blocks in the matrix $\bar{M}(\hat{\xi}_a)$ are defined by

$$M_c(\xi) = \int_{-a}^{a} f_c(t) f_c^T(t) \, d\xi(t) \in \mathbb{R}^{m+1 \times m+1}, \qquad (4.61)$$

$$M_s(\xi) = \int_{-a}^{a} f_s(t) f_s^T(t) \, d\xi(t) \in \mathbb{R}^{m \times m}, \qquad (4.62)$$

and the vectors $f_c(t) \in \mathbb{R}^{m+1}$ and $f_s(t) \in \mathbb{R}^m$ are given by

$$f_c^T(t) = (1/\sqrt{2}, \cos t, \ldots, \cos(mt)),$$
$$f_s^T(t) = (\sin t, \ldots, \sin(mt)),$$

respectively. Because the matrices $\bar{M}(\hat{\xi}_a)$ and $M(\hat{\xi}_a)$ have the same eigenvalues and its corresponding eigenvectors are related by the transformation $x \to Px$, assertion (ii) of Lemma 4.3.3 follows, if we prove that the vector

$\tilde{q}_a = P\bar{q}_a = (q_a^T, 0^T)^T \in \mathbb{R}^{2m+1}$ is an eigenvector of the matrix $\bar{M}(\hat{\xi}_a)$ with corresponding eigenvalue

$$\lambda_a = (\tilde{q}_a^T \tilde{q}_a)^{-1} = (q_a^T q_a)^{-1}.$$

However, this follows easily, observing that the sign of $q_a^T F^{-1} e_i$ is $(-1)^i$ for $a \in (0, \pi]$ and from the representation of the weights $\hat{w}_i$ in (4.49), which gives

$$M_c(\hat{\xi}_a) q_a = \sum_{i=0}^{m} f_c(a t_i) f_c^T(a t_i) \hat{w}_i q_a$$

$$= \sum_{i=0}^{m} f_c(a t_i) \frac{1}{q_a^T q_a} q_a^T F^{-1} e_i$$

$$= \frac{1}{q_a^T q_a} F F^{-1} q_a = \lambda_a q_a.$$

Consequently, we obtain

$$\bar{M}(\hat{\xi}_a) \tilde{q}_a = \lambda_a \tilde{q}_a,$$

completing the proof of the second assertion of Lemma 4.3.3.

For the proof of the remaining third part, recall that the sign of q_{aj} and $e_j^T F^{-1} e_i$ is $(-1)^{m-j}$ and $(-1)^{m+i+j}$, respectively. Then (4.57) implies for sufficiently small a,

$$|q_{aj}| = \sum_{i=0}^{m} |e_j^T F^{-1} e_i|,$$

and from the first equation in (4.58), we have

$$(-1)^i q_a^T F^{-1} e_i = (-1)^i \sum_{j=0}^{m} |q_{aj}| (-1)^{m-j} |e_j F^{-1} e_i| (-1)^{m+i+j}$$

$$= \sum_{j=0}^{m} |q_{aj}| |e_j F^{-1} e_i|.$$

A summation of these quantities yields for the weights of the design $\hat{\xi}_a$ defined in (4.48),

$$\hat{w}_i = \frac{|q_a^T F^{-1} e_i|}{\sum_{j=0}^{m} |q_a^T F^{-1} e_i|} = \sum_{j=0}^{m} w_{(j)i} \cdot \alpha_j(a), \quad i = 0, \ldots, m,$$

where

$$\alpha_j(a) = \frac{|q_{aj}|^2}{\sum_{s=0}^{m} |q_{as}|^2}, \quad j = 0, \ldots, m, \tag{4.63}$$

and the weights $w_{(j)i}$ are defined in (4.52), corresponding to the optimal design

$$
\xi_{(2j)} = \xi(w_{(j)}) = \begin{pmatrix} -at_m & \cdots & -at_1 & t_0 & at_1 & \cdots & at_m \\ \frac{w_{(j)m}}{2} & \cdots & \frac{w_{(j)1}}{2} & w_{(j)0} & \frac{w_{(j)1}}{2} & \cdots & \frac{w_{(j)m}}{2} \end{pmatrix}
$$

for estimating the individual coefficient β_{2j} in the trigonometric regression model (4.1) on the interval $[-a, a]$, whenever $0 < a < \pi/2$. Note that we use the second representation in (4.53) and (4.56) to find this normalization. In other words, the design $\hat{\xi}_a$ is obtained as a convex combination of the optimal designs for estimating the individual coefficients in the trigonometric regression model on the interval $[-a, a]$ (whenever $0 < a < \pi/2$); that is,

$$
\hat{\xi}_a = \sum_{j=0}^{m} \alpha_j(a) \xi_{(2j)}. \tag{4.64}
$$

If $a \to 0$, the representation (4.46) implies that $(\alpha = \cos a)$

$$
\lim_{a \to 0} (1 - \alpha)^m q_a = f \in \mathbb{R}^{m+1},
$$

where $f = (f_0, \ldots, f_m)^T \neq 0$ denotes the vector in the expansion

$$
2^{2m-1}(x - 1)^m = \frac{f_0}{\sqrt{2}} + \sum_{j=1}^{m} f_j T_j(x). \tag{4.65}
$$

Consequently, we obtain from (4.63) for the weights in the convex combination (4.64),

$$
\lim_{a \to 0} \alpha_j(a) = \alpha_j^* = \frac{|f_j|^2}{\sum_{i=0}^{m} |f_i|^2}, \quad j = 0, \ldots, m.
$$

Finally, Corollary 4.2 in Dette and Melas (2001) shows that for $j = 0, \ldots, m$ the optimal design $\xi_{(2j)}$ for estimating the individual coefficient β_{2j} in the trigonometric regression model on the interval $[-a, a]$ converges weakly in the following sense

$$
\lim_{a \to 0} \xi_{(2j)}([-a, at]) = \zeta([-1, t]), \quad t \in [-1, 1],
$$

where the limiting design ζ is given by

$$
\zeta = \begin{pmatrix} -y_m & -y_{m-1} & \cdots & -y_1 & y_0 & y_1 & \cdots & y_{m-1} & y_m \\ \frac{1}{4m} & \frac{1}{2m} & \cdots & \frac{1}{2m} & \frac{1}{2m} & \frac{1}{2m} & \cdots & \frac{1}{2m} & \frac{1}{4m} \end{pmatrix}
$$

with

$$
y_i = \cos\left(\frac{\pi(m - i)}{2m}\right), \quad i = 0, \ldots, m.
$$

Consequently, equation (4.64) shows that $\hat{\xi}_a$ has the same weak limit, i.e.

$$\lim_{a \to 0} \hat{\xi}_a([-a, at]) = \zeta[-1, t], \ t \in [-1, 1] ,$$

and assumption (iii) of Lemma 4.3.3 follows by rewriting this statement in terms of the support points and weights of the designs $\hat{\xi}_a$ and ζ, respectively.

∎

Theorem 4.3.2 *For sufficiently small $a > 0$, the design $\hat{\xi}_a$ defined in (4.48) and (4.49) is E-optimal for the trigonometric regression model (4.1) on the interval $[-a, a]$. The minimum eigenvalue is given by $\lambda_{\min}(M(\hat{\xi}_a)) = \lambda_a$, where*

$$\lambda_a^{-1} = q_a^T q_a = \frac{2}{\pi} \int_{-1}^{1} T_m^2 \left(\frac{2x - 1 - \alpha}{1 - \alpha} \right) \frac{dx}{\sqrt{1 - x^2}},$$

and the vector $q_a = (q_{a0}, \ldots, q_{a0})^T$ is defined by the expansion (4.46).

Proof. Recalling the definition of the design $\hat{\xi}_a$ in (4.48) and (4.49), we will study the asymptotic behavior of the matrix

$$a^{4m} M^{-1}(\hat{\xi}_a)$$

as $a \to 0$. To this end, let

$$U_k(x) = \frac{\sin((k + 1) \arccos x)}{\sin(\arccos x)}, \quad k \geq 0, \tag{4.66}$$

denote the Chebyshev polynomial of the second kind and define

$$u = u(t) = \frac{2(1 - \cos t)}{a^2} .$$

Obviously, $\cos(kt) = T_k(1 - \frac{a^2}{2} u)$, $\sin(kt)/\sin t = U_{k-1}(1 - \frac{a^2}{2} u)$, and, consequently, there exists an $(m + 1) \times (m + 1)$ matrix $S_{(1)}$ and an $m \times m$ matrix $S_{(2)}$ such that the vector

$$\bar{f}(t) = (f_c^T(t), f_s^T(t))^T \in \mathbb{R}^{2m+1}$$

can be represented as

$$\bar{f}(t) = S A \tilde{f}(u(t)),$$

where

$$\tilde{f}(t) = \left(\frac{1}{\sqrt{2}}, u(t), \ldots, u^m(t), \frac{\sin t}{a}, \frac{\sin t}{a} u(t), \ldots, \frac{\sin t}{a} u^{m-1}(t) \right)^T, \tag{4.67}$$

and the matrices A and S are defined by

$$A = A(a) = \text{diag}\left\{1, \frac{a^2}{2}, \ldots, \left(\frac{a^2}{2}\right)^m, a, a\frac{a^2}{2} \ldots, a\left(\frac{a^2}{2}\right)^{m-1}\right\} \quad (4.68)$$

and

$$S = \begin{pmatrix} S_{(1)} & 0 \\ 0 & S_{(2)} \end{pmatrix},$$

respectively, $A, S \in \mathbb{R}^{2m+1 \times 2m+1}$. It is easy to see that the matrices $S_{(1)}$ and $S_{(2)}$ do not depend on the parameter a and are lower triangular with nonvanishing diagonal elements. Consequently, we obtain an alternative representation for the matrix $\bar{M}(\hat{\xi}_a)$ defined in (4.60):

$$\bar{M}(\hat{\xi}_a) = SA\tilde{M}(\hat{\xi}_a)AS^T, \quad (4.69)$$

where

$$\tilde{M}(\xi) = \int \tilde{f}(t)\tilde{f}^T(t)\, d\xi(t) \in \mathbb{R}^{2m+1 \times 2m+1}.$$

Now, let

$$\hat{f}(t) = \left(\frac{1}{\sqrt{2}}, t^2, \ldots, t^{2m}, t, t^3, \ldots, t^{2m-1}\right)^T$$

and define for any design ξ,

$$\hat{M}(\xi) = \int \hat{f}(t)\hat{f}^T(t)\, d\xi(t) \quad (4.70)$$

as the corresponding information matrix. From the expansions

$$1 - \cos(at) = \frac{(at)^2}{2}(1 + o(a)) \quad \text{and} \quad \sin(at) = at(1 + o(a))$$

and (4.67), it is easy to see that

$$\lim_{a\to 0} \tilde{f}(at) = \hat{f}(t).$$

Consequently, we obtain from the third part of Lemma 4.3.3 and definition (4.68) that

$$\lim_{a\to 0} \tilde{M}(\hat{\xi}_a) = \hat{M}(\zeta),$$

where ζ is the limiting design defined in (4.66). Moreover, from (4.68), we have

$$\lim_{a\to 0} \left(\frac{a^2}{2}\right)^m A^{-1}(a) = \text{diag}(\underbrace{0, \ldots, 0}_{m}, 1, \underbrace{0, \ldots, 0}_{m}), \quad (4.71)$$

which gives, for the matrix $\bar{M}(\hat{\xi}_a)$ in (4.69),

$$\lim_{a\to 0} \left(\frac{a^2}{2}\right)^{2m} \bar{M}^{-1}(\hat{\xi}_a) = (S^T)^{-1}DS^{-1} \quad (4.72)$$

and for the corresponding $(m+1) \times (m+1)$ block,

$$\lim_{a \to 0} \left(\frac{a^2}{2} \right)^{2m} M_c^{-1}(\hat{\xi}_a) = (S_{(1)}^T)^{-1} D_{(1)} S_{(1)}^{-1}, \tag{4.73}$$

where the matrix $D \in \mathbb{R}^{2m+1 \times 2m+1}$ is defined by

$$D = \begin{pmatrix} D_{(1)} & 0 \\ 0 & 0 \end{pmatrix},$$

with

$$D_{(1)} = \Delta e_m e_m^T \in \mathbb{R}^{m+1 \times m+1},$$

$$e_m = (0, \ldots, 0, 1)^T \in \mathbb{R}^{m+1},$$

$$\Delta = e_m^T \hat{M}_*^{-1}(\zeta) e_m.$$

and $\hat{M}_*(\zeta)$ denotes the $m+1 \times m+1$ matrix formed by the first $m+1$ rows and columns of the matrix $\hat{M}(\zeta)$ defined in (4.70). Because the matrices D and $D_{(1)}$ have rank 1, the matrices on the right-hand sides of (4.71) and (4.72) have only one nonvanishing eigenvalue. From the discussion at the end of the proof of Lemma 4.3.3 we have

$$\lim_{a \to 0} \left(\frac{a^2}{2} \right)^m q_a = f \neq 0 \in \mathbb{R}^{m+1},$$

where the vector f is defined by the expansion (4.65). Similarly, it follows for the eigenvalue λ_a^{-1} of $\bar{M}^{-1}(\hat{\xi}_a)$ that

$$\lim_{a \to 0} \left(\frac{a^2}{2} \right)^{2m} \lambda_a^{-1} = \lim_{a \to 0} \left(\frac{a^2}{2} \right)^{2m} q_a^T q_a = f^T f \neq 0.$$

Consequently, the continuous dependence of the eigenvalues of a matrix from its elements (see Lancaster (1969)), formulas (4.72) and (4.73) imply that for sufficiently small a, the matrices

$$\left(\frac{a^2}{2} \right)^{2m} \bar{M}^{-1}(\hat{\xi}_a), \quad \left(\frac{a^2}{2} \right)^{2m} M_c^{-1}(\hat{\xi}_a)$$

have a maximal eigenvalue of multiplicity 1, which is given by

$$\left(\frac{a^2}{2} \right)^{2m} \lambda_a^{-1}.$$

In other words, the minimal eigenvalue λ_a of the matrix $\bar{M}(\hat{\xi}_a)$ has multiplicity 1, provided that a is close to 0.

Now, let $0 < a < \bar{a}$ be sufficiently small such that this property is satisfied. By Lemma 4.3.3(ii), the vector $\bar{q}_a = (q_{a0}, 0, q_{a1}, 0, \ldots, 0, q_{am})^T$

is the eigenvector corresponding to λ_a and we define $A^* = \lambda_a \bar{q}_a \bar{q}_a^T$. With these notations, from (4.46) we have

$$\max_{t \in [-a,a]} f^T(t) A^* f(t) = \max_{t \in [-a,a]} \frac{(q_a^T f_c(t))^2}{q_a^T q_a} = \max_{x \in [\alpha,1]} \frac{T_m(\frac{2x-1-\alpha}{1-\alpha})}{q_a^T q_a}$$

$$= \frac{1}{q_a^T q_a} = \lambda_a = \lambda_{\min}(\bar{M}(\hat{\xi}_a)) = \lambda_{\min}(M(\hat{\xi}_a))$$

and the optimality of the design $\hat{\xi}_a$ follows from the Equivalence Theorem given in Lemma 4.3.1.

Finally, the integral representation of λ_a^{-1} follows from the orthogonality properties (4.44) of the Chebyshev polynomials and the representation (4.46); that is,

$$\lambda_a^{-1} = q_a^T q_a = \sum_{i,j=0}^{m} q_{ai} q_{aj} \frac{2}{\pi} \int_{-1}^{1} b_j T_i(x) T_j(x) \frac{dx}{\sqrt{1-x^2}}$$

$$= \frac{2}{\pi} \int_{-1}^{1} \left(\frac{q_{a0}}{\sqrt{2}} + \sum_{i=1}^{m} q_{ai} T_i(x) \right)^2 \frac{dx}{\sqrt{1-x^2}}$$

$$= \frac{2}{\pi} \int_{-1}^{1} T_m^2 \left(\frac{2x-1-a}{1-a} \right) \frac{dx}{\sqrt{1-x^2}},$$

where $b_0 = 1/2$ and $b_j = 1$ if $j \geq 1$.   ∎

The following corollary is an immediate consequence of Theorem 4.3.2 and its proof.

Corollary 4.3.1 *Let*

$$\underline{a} = \underline{a}(m) = \sup\{a > 0 \mid \lambda_{\min}(M(\hat{\xi}_a)) = \lambda_a, \tag{4.74}$$

where λ_a is defined in (4.51). Whenever $0 < a < \underline{a}$, the E-optimal design for the trigonometric regression model (4.1) on the interval $[-a, a]$ is given by the design $\hat{\xi}_a$ defined in (4.48) and (4.49). Moreover,

$$\underline{a} = \min\{\underline{a}_{(1)}, \underline{a}_{(2)}\},$$

where the quantities $\underline{a}_{(1)}$ and $\underline{a}_{(2)}$ are given by

$$\underline{a}_{(1)} = \underline{a}_{(1)}(m) = \sup\{a > 0 | \lambda_{\min}(M_c(\hat{\xi}_a)) = \lambda_a\},$$
$$\tag{4.75}$$
$$\underline{a}_{(2)} = \underline{a}_{(2)}(m) = \sup\{a > 0 | \lambda_{\min}(M_s(\hat{\xi}_a)) > \lambda_a\}.$$

The quantities $\underline{a}_{(1)}$ and $\underline{a}_{(2)}$ have been calculated numerically for lower-order trigonometric regression models and are listed in Table 4.4. Note that these values are rather close to the upper bound $\bar{a} = \pi(1 - 1/(2m+1))$

obtained in Section 4.2 and, consequently, Theorem 4.3.1 and Corollary 4.3.1 cover a rather large range of the interval $(0, \pi]$ for the parameter a of the design space $[-a, a]$. Moreover, Table 4.4 indicates that both values might be equal in general, and in Section 4.3.4, we will prove that $\underline{a}_{(1)} = \underline{a}_{(2)}$ for all $m \in \mathbb{N}$.

Table 4.4: Bounds $\bar{a}$, $\underline{a} = \min\{\underline{a}_{(1)}, \underline{a}_{(2)}\}$. The E-optimal design for the trigonometric regression model (4.1) on the interval $[-a, a]$ can be found analytically whenever $a \in (0, \underline{a}] \cup [\bar{a}, \pi]$ for various values of m

m	$\underline{a}_{(1)}$	$\underline{a}_{(2)}$	$\bar{a}$
2	0.741π	0.741π	0.8π
3	0.794π	0.794π	0.857π
4	0.827π	0.827π	0.889π
5	0.851π	0.851π	0.909π

Note Table 4.4 does not contain the case $m = 1$, for which a complete analytic solution is presented in the following section. For later purposes, we require the following auxiliary result, which is probably of independent interest.

Lemma 4.3.4 *Let $0 < a < \bar{a} = \pi(1 - 1/(2m + 1))$ and ξ^* denote the E-optimal design for the trigonometric regression model (4.1) on the interval $[-a, a]$. If the minimum eigenvalue of the information matrix $M(\xi^*)$ has multiplicity 1, then*

$$\xi^* = \hat{\xi}_a,$$

where the design $\hat{\xi}_a$ is defined in (4.48).

Proof. From Theorem 4.3.1 we have that the E-optimal design ξ^* is unique and of the form

$$\xi^* = \begin{pmatrix} -t_m^* & \cdots & -t_1^* & t_0^* & t_1^* & \cdots & t_m^* \\ \frac{w_m^*}{2} & \cdots & \frac{w_1^*}{2} & w_0^* & \frac{w_1^*}{2} & \cdots & \frac{w_m^*}{2} \end{pmatrix}.$$

Now, let

$$\lambda^* = \lambda_{\min}(M(\xi^*)) = \min\{\lambda_{\min}(M_c(\xi^*)), \lambda_{\min}(M_s(\xi^*))\}$$

denote the minimum eigenvalue of the matrix $M(\xi^*)$ and consider first the case where $\lambda^* = \lambda_{\min}(M_c(\xi^*))$. Obviously, λ^* is a simple eigenvalue of $M_c(\xi^*)$ and we define $q = (q_0, \ldots, q_m)^T$ as the corresponding eigenvector. With the notation $\bar{q} = (q_0, 0, q_1, 0, \ldots, 0, q_m)^T$ and $A^* = \bar{q}\bar{q}^T/\bar{q}^T\bar{q}$, it follows from Lemma 4.3.1 that (note the ξ^* is E-optimal)

$$\lambda^* = \max_{t \in [-a,a]} f^T(t)A^*f(t) = \max_{t \in [-a,a]} \frac{(q^T f_c(t))^2}{q^T q}.$$

The polynomial

$$\Psi(x) = q^T f_c(\arccos x)$$

attains its maximum absolute value in the interval $[\alpha, 1]$ at the $m+1$ points $x_i^* = \cos t_i^*$ $(i = 0, \ldots, m)$ and must coincide with the polynomial

$$\mp T_m \left(\frac{2x - 1 - \alpha}{1 - \alpha} \right),$$

which implies $\mathrm{supp}(\xi^*) = \mathrm{supp}(\hat{\xi}_a)$, $q = \mp q_a$ and $\lambda^* = \lambda_a = 1/q_a^T q_a$. From the equation

$$M_c(\xi^*)q_a = \lambda_a q_a,$$

it is then easy to see that the weights w_i^* must coincide with the weights of the design $\hat{\xi}_a$ given in (4.49) and it follows that $\xi^* = \hat{\xi}_a$.

Second, if $\lambda^* = \lambda_{\min}(M_s(\xi^*)) < \lambda_{\min}(M_c(\xi^*))$, then a similar argument shows that ξ^* is concentrated at $2m$ points, which is impossible. ∎

4.3.3 Example: The linear trigonometric regression model on a partial circle

In this subsection we study the linear trigonometric regression model on the interval $[-a, a]$, which indicates that even this relatively simple case is not trivial. Our next proposition specifies the E-optimal designs in the linear trigonometric regression model. In this case, it proves that $\underline{a} = \bar{a}$ and we will show in the following subsection that this equality only holds in the linear case.

Proposition 4.3.1 *Consider the linear trigonometric regression model $(m=1)$ on the interval $[-a, a]$.*

(i) If $\bar{a} = 2\pi/3 \le a \le \pi$, then an E-optimal design is given by

$$\xi_3^* = \left(\begin{array}{ccc} \frac{-2\pi}{3} & 0 & \frac{2\pi}{3} \\ \frac{1}{3} & \frac{1}{3} & \frac{1}{3} \end{array} \right).$$

(ii) If $0 < a \le \underline{a} = 2\pi/3$, then the E-optimal design is unique and given by

$$\xi_a^* = \left(\begin{array}{ccc} -a & 0 & a \\ \frac{\mu(a)}{2} & 1 - \mu(a) & \frac{\mu(a)}{2} \end{array} \right), \tag{4.76}$$

where

$$\mu(a) = \frac{4 + 2\cos a}{4 + 2(1 + \cos a)^2}. \tag{4.77}$$

Proof. The first point and the statement of uniqueness in part (ii) follows from Theorem 4.3.1, which also shows that the E-optimal design is of the form (4.76) whenever $0 < a < \bar{a}$. If a is sufficiently small, we can use Theorem 4.3.2 and Corollary 4.3.1 and obtain from the representation of the weights by the first part of Lemma 4.3.3 $[x_0 = 1, x_1 = \alpha = \cos a, q_{a0} = -\sqrt{2}(1 + \alpha)/(1 - \alpha), q_{a1} = 2/(1 - \alpha)]$

$$1 - \mu(a) = \left\{ \frac{4 + 2(1 + \alpha)^2}{(1 - \alpha)^2} \right\}^{-1} \frac{2}{\pi} \int_{-1}^{1} \frac{x - \alpha}{1 - \alpha} \left\{ q_{a1}T_1(x) + \frac{q_{a0}}{\sqrt{2}} \right\} \frac{dx}{\sqrt{1 - x^2}}$$

$$= 2 \frac{1 + \alpha + \alpha^2}{4 + 2(1 + \alpha)^2},$$

where we have used the orthogonality relation for the Chebyshev polynomials of the first kind. The representation (4.77) now follows from a trivial calculation; that is,

$$\mu(a) = \frac{4 + 2\alpha}{4 + 2(1 + \alpha)^2}. \tag{4.78}$$

Note that this formula can also be obtained from the representation $w_0 = q_a^T F^{-1} e_0 / q_a^T q_a$, where $e_0 = (1, 0)^T$ and

$$F = \begin{pmatrix} \frac{1}{\sqrt{2}} & \frac{1}{\sqrt{2}} \\ 1 & \alpha \end{pmatrix}.$$

A straightforward calculation shows that

$$\lambda_a = \frac{(1 - \alpha)^2}{4 + 2(1 + \alpha)^2} \tag{4.79}$$

is the minimum eigenvalue of $M_c(\hat{\xi}_a)$ and has multiplicity 1. Consequently, the critical value $\underline{a}$ can be obtained as

$$\underline{a} = \sup \left\{ a \in (0, \bar{a}) | \lambda_{\min}(M_c(\hat{\xi}_a)) < \lambda_{\min}(M_s(\hat{\xi}_a)) \right\}$$

$$= \inf \left\{ a \in (0, \bar{a}) | \lambda_{\min}(M_c(\hat{\xi}_a)) = \lambda_{\min}(M_s(\hat{\xi}_a)) \right\},$$

which gives the equation

$$\lambda_a = \frac{(1 - \alpha)^2}{4 + 2(1 + \alpha)^2} = \mu(a)(1 - \alpha^2) = \frac{4 + 2\alpha}{4 + 2(1 + \alpha)^2}(1 - \alpha^2)$$

$$= \frac{(1 + \alpha)(4 + 2\alpha)}{1 - \alpha} \lambda_a,$$

where we have used the representation (4.78) and (4.79) for the last equalities. This yields the equation $2\alpha^2 + 7\alpha + 3 = 0$, which gives

$$\cos \underline{a} = \underline{\alpha} = -\frac{1}{2}, \quad \underline{a} = \frac{2\pi}{3}$$

as unique solution in the interval $[-1, 1]$.

By Theorem 4.3.2 and Corollary 4.3.1, the E-optimal design for the linear trigonometric regression model ($m=1$) on the interval $[-a, a]$ is given by (4.76) and (4.77) whenever $a \in (0, \underline{a}]$, which proves part (ii) of the proposition.   ∎

4.3.4 E-Optimal designs on arbitrary intervals

As was shown in Section 4.3.1, we can restrict the discussion of the E-optimal design problem to the case of symmetric intervals $[-a, a]$, $0 < a \leq \pi$. For $0 < a \leq \underline{a} = \underline{a}(m)$ and $\bar{a} = \bar{a}(m) \leq a \leq \pi$, we have already received explicit solutions for the E-optimal design problem in the trigonometric regression model (4.1) on the interval $[-a, a]$. Note that the range $(\underline{a}, \bar{a})$ not covered by these results is rather small (see Table 4.4) and, consequently, explicit solutions of the E-optimal design problem are available for most cases. Moreover, in Section 4.3.3, we have shown that in the linear trigonometric regression model with $m = 1$, we have $\underline{a} = \bar{a} = 2\pi/3$ and a complete analytic solution is available in this case.

Now, we will prove that for $m \geq 2$ it follows that $\underline{a} < \bar{a}$ and elaborate a technique for the case $\underline{a} < a < \bar{a}$ that can be used for the numerical construction of E-optimal designs and is based on the functional approach. The method will be illustrated for the quadratic and cubic trigonometric regression model at the end of this subsection.

We begin with a reformulation of Lemma 4.3.1. To this end, let us introduce the function

$$\Psi(x) = \Psi(x; q, p) = \frac{(q^T f_{(1)}(x))^2 + (1 - x^2)(p^T f_{(2)}(x))^2}{q^T q + p^T p}, \qquad (4.80)$$

where $q = (q_0, \ldots, q_m)^T \in \mathbb{R}^{m+1}$ is an arbitrary vector with $q_m = 1$, $p = (p_0, \ldots, p_{m-1})^T \in \mathbb{R}^m$ is an arbitrary vector, and the functions $f_{(1)}(x)$ and $f_{(2)}(x)$ are defined by

$$f_{(1)}^T(x) = \left(1/\sqrt{2}, T_1(x), \ldots, T_m(x)\right), \qquad (4.81)$$

$$f_{(2)}^T(x) = (U_0(x), \ldots, U_{m-1}(x)). \qquad (4.82)$$

Due to Theorem 4.3.1, we can restrict our consideration to the case $a < \bar{a}$ and designs $\xi \in \Xi_a^{(1)}$. The following result is a refinement of Lemma 4.3.1 for the model at hand.

Lemma 4.3.5 For the trigonometric regression model (4.1) on the interval $[-a, a]$ with $0 < a < \bar{a}$, the design

$$\xi = \begin{pmatrix} -t_m & \cdots & -t_1 & t_0 & t_1 & \cdots & t_m \\ \frac{w_m}{2} & \cdots & \frac{w_1}{2} & w_0 & \frac{w_1}{2} & \cdots & \frac{w_m}{2} \end{pmatrix}, \qquad (4.83)$$

with $t_0 = 0$ *and* $t_m = a$ *is E-optimal if and only if there exist vectors* $q = q(a) = (q_0, \ldots, q_m)^T \in \mathbb{R}^{m+1}$ *with* $q_m = 1$ *and a vector* $p = p(a) \in \mathbb{R}^m$, *such that the inequality*

$$\Psi(x) = \Psi(x; q, p) \leq \lambda_{\min}(M(\xi)) \tag{4.84}$$

holds for all $x \in [\alpha, 1]$, *where the function* $\Psi(x; q, p)$ *is defined in (4.80).*

Moreover, if a design ξ *of the form (4.83) is E-optimal, then these vectors are eigenvectors of the matrices defined by (4.61) and (4.62) corresponding to the minimum eigenvalue* $\lambda = \lambda_{\min}(M(\xi))$ *of the matrix* $M(\xi)$; *that is,*

$$M_c(\xi)q = \lambda q, \; M_s(\xi)p = \lambda p, \tag{4.85}$$

and

$$\Psi(x_i) = \Psi(\cos a), \; i = 1, \ldots, m-1,$$
$$\Psi'(x_i) = 0, \; i = 1, \ldots, m-1, \tag{4.86}$$

where $x_i = \cos t_i$, $i = 0, 1, \ldots, m-1$.

The polynomial $\Psi(x)$ *is uniquely determined. The vectors* p *and* q *can be chosen such that the polynomials*

$$p^T f_{(2)}(x) \quad and \quad q^T f_{(2)}(x)$$

have interlacing roots, and under this additional condition, the vectors p *and* q *are also uniquely determined. If* $a \in [0, \underline{a}]$, *it follows that* $p = 0$.

Proof. Let us prove that the inequality (4.84) is a necessary condition for E-optimality. To this end, assume that a design ξ of the form (4.83) is E-optimal and let A^* be the matrix, defined in Lemma 4.3.1, such that the inequality (4.38) is satisfied.

Consider the function

$$\Psi(x) = h(\arccos x),$$

where $h(t) = f^T(t)A^*f(t)$, $t = \arccos x$. Note that due to Theorem 4.3.1, $\Psi(x) \not\equiv \text{const}$ whenever $0 < a < \bar{a}$. Since

$$\sin(k \arccos x) = \sqrt{1-x^2}U_{k-1}(x)$$

and

$$\cos(k \arccos x) = T_k(x),$$

it follows that $\Psi(x)$ is a polynomial of degree $2m$ [note that $\Psi(x)$ is not constant, and by Lemma 4.3.1, it has $2m-1$ roots counting multiplicities]. Our polynomial is non-negative for $-1 \leq x \leq 1$ due to non-negative definiteness of the matrix A^*. It is known (see Karlin and Studden (1966, Chap. 2) that such a polynomial can be represented in the form

$$\Psi(x) = \varphi_1^2(x) + (1-x^2)\varphi_2^2(x),$$

where $\varphi_1(x)$ is a polynomial of degree m and $\varphi_2(x)$ is a polynomial of degree $m - 1$, that is,

$$\varphi_1(x) = C_1 \prod_{i=1}^{m} (x - \gamma_i), \quad \varphi_2 = C_2 \prod_{i=1}^{m-1} (x - \delta_i),$$

and that the roots of these polynomials are interlacing, that is,

$$\gamma_1 \leq \delta_1 \leq \gamma_2 \leq \cdots \leq \delta_{m-1} \leq \gamma_m,$$

$$\gamma_1 < \gamma_2 < \cdots < \gamma_m, \ \delta_1 < \cdots < \delta_{m-1}.$$

Moreover, this representation is unique. Since the polynomials $T_0(x)$, $T_1(x), \ldots, T_m(x)$ are linearly independent and the same is true for the polynomials $U_0(x), \ldots, U_{m-1}(x)$, we have

$$\varphi_1(x) = Cq^T f_{(1)}(x),$$
$$\varphi_2(x) = Cp^T f_{(2)}(x),$$

where $C > 0$ is a constant and $q = (q_0, \ldots, q_m)^T \in \mathbb{R}^{m+1}$ and $p \in \mathbb{R}^m$ are appropriate vectors with $q_m = 1$. Recalling that the functions $\sqrt{2} f_k(t)$, $k = 0, 1, \ldots, 2m$, are orthonormal with respect to measure $\frac{1}{2\pi} dt$ on the interval $[-\pi, \pi]$, we obtain

$$\frac{1}{\pi} \int_{-\pi}^{\pi} f^T(t) A^* f(t) \, dt = \mathrm{tr} A^* = 1,$$

and, therefore,

$$1 = \frac{2}{\pi} \int_{-1}^{1} \Psi(x) \frac{dx}{\sqrt{1 - x^2}}$$
$$= \frac{2C}{\pi} \int_{-1}^{1} \left[(q^T f_{(1)}(x))^2 + (p^T f_{(2)}(x))^2 (1 - x^2) \right] \frac{dx}{\sqrt{1 - x^2}}$$
$$= C(q^T q + p^T p).$$

Consequently, $C = 1/(q^T q + p^T p)$, and due to Lemma 4.3.1, it follows for all $t \in [-a, a]$ that

$$f^T(t) A^* f(t) \leq \lambda_{\min}(M(\xi)),$$

or, equivalently,

$$\Psi(x) = \Psi(x; q, p) = \frac{(q^T f_1(x))^2 + (1 - x^2)(p^T f_{(2)}(x))^2}{q^T q + p^T p} \leq \lambda \quad (4.87)$$

for all $x \in [\alpha, 1]$, where $\lambda = \lambda_{\min}(M(\xi))$ denotes the minimum eigenvalue of the matrix $M(\xi)$. Therefore, condition (4.84) follows from the E-optimality of the design ξ. Due to Lemma 4.3.1, the left-hand side of the inequality

(4.87) attains its maximal value λ at the support points $x_i = \cos t_i$, $i = 0, \ldots, m$ (since $\Psi(x) = h(t)$, $t = \arccos x$) and the system of equations in (4.86) also provides a necessary condition for E-optimality.

To prove that (4.85) is also a necessary condition for E-optimality, we put $x = \cos t$ and integrate the left-hand side of (4.87) with respect to the measure $\xi(dt)$. We obtain

$$\frac{q^T M_c(\xi)q + p^T M_s(\xi)p}{p^T p + q^T q} \le \lambda, \tag{4.88}$$

where the second term should be replaced by zero if $p = 0$. Since

$$\min_{\tilde{q}} \frac{\tilde{q}^T M_c(\xi)\tilde{q}}{\tilde{q}^T \tilde{q}} = \lambda_{\min}(M_c(\xi)) \ge \lambda,$$

$$\tag{4.89}$$

$$\min_{\tilde{p}} \frac{\tilde{p}^T M_s(\xi)\tilde{p}}{\tilde{p}^T \tilde{p}} = \lambda_{\min}(M_s(\xi)) \ge \lambda,$$

it follows that q is an eigenvector of the matrix $M_c(\xi)$ corresponding to its minimal eigenvalue λ; that is,

$$M_c(\xi)q = \lambda q.$$

Similarly, p is either equal to $0 \in \mathbb{R}^m$ or an eigenvector of the matrix $M_s(\xi)$ corresponding to its minimal eigenvalue λ. In both cases, we have the equation

$$M_s(\xi)p = \lambda p.$$

Finally, we prove that (4.84) is a sufficient condition for E-optimality of the design ξ. To this end, define

$$A = (qq^T + pp^T)/(q^T q + p^T p);$$

then tr $A = 1$ and it follows from (4.84) that for all $t \in [-a, a]$,

$$f^T(t)Af(t) \le \lambda_{\min}(M(\xi)).$$

Due to Lemma 4.3.1, the design ξ is E-optimal.

Note that the polynomial $\Psi(x)$ is uniquely determined by the conditions (4.86) and (4.84). Moreover, we proved earlier that the vectors p and q are uniquely determined under the additional condition of interlacing roots.

Let $0 < a < \underline{a}$; then $\lambda_{\min}(M_s(\xi)) > \lambda$, and from (4.88) and (4.89), it follows that $p = 0$. In the case $a = \underline{a}$, the equality $p = 0$ follows from a continuity argument.   ∎

Lemma 4.3.5 will be used to obtain a representation for the minimal eigenvalue of the information matrix of the E-optimal design. This representation will be essential for the numerical construction of E-optimal designs.

Lemma 4.3.6 *For the trigonometric regression model (4.1) with $m \geq 2$, we have for the quantities $\underline{a}$ and $\bar{a}$ defined in (4.74) and (4.42), respectively,*

$$\underline{a} < \bar{a}.$$

Proof. It is evident that $\underline{a} \leq \bar{a}$. Suppose that $\underline{a} = \bar{a}$; then Theorem 4.3.2 and Corollary 4.3.1 show that for $a \leq \underline{a}$, the design $\hat{\xi}_a$ defined by (4.48) and (4.49) is E-optimal. For $a < \bar{a}$, there exists a unique E-optimal design by Theorem 4.3.1 and a continuity argument shows that there also exists a unique E-optimal design in the case $a = \underline{a} = \bar{a}$, which is of the form

$$\xi^* = \left(\begin{array}{cccccc} -t_m & \cdots & -t_1 & t_0 & t_1 & \cdots & t_m \\ \frac{1}{2m+1} & \cdots & \frac{1}{2m+1} & \frac{1}{2m+1} & \frac{1}{2m+1} & \cdots & \frac{1}{2m+1} \end{array} \right),$$

where the support points are given by

$$t_i = \frac{\pi i}{m}\left(1 - \frac{1}{2m+1}\right), \quad i = 0, \ldots, m.$$

Therefore, $\xi^* = \hat{\xi}_a$ and we obtain the equations

$$\cos t_i = \cos\left[\frac{\pi i}{m}\left(1 - \frac{1}{2m+1}\right)\right] = \frac{1 - \bar{\alpha}}{2}\cos\frac{\pi i}{m} + \frac{1 + \bar{\alpha}}{2}, \quad i = 0, \ldots, m,$$

where $\bar{\alpha} = \cos\bar{a}$ and $\bar{a} = \pi(1 - \frac{1}{2m+1})$. In order to prove that this is impossible, we note that for $0 < a < \pi$, $1/2 < u < 1$, it follows that

$$\cos au > \frac{1 - \cos a}{2}\cos\pi u + \frac{1 + \cos a}{2}. \qquad (4.90)$$

This inequality can be proved observing that for $a = 0$ and $a = \pi$, we have

$$\cos au - \frac{1 - \cos a}{2}\cos\pi u - \frac{1 + \cos a}{2} = 0$$

and verifying that the derivative of the left-hand side has only one zero in the interval $(0, \pi)$ corresponding to an absolute maximum in this region. Substituting $a = \bar{a}$ and $u = i/m$ in (4.90), we obtain a contradiction, which shows that $\underline{a} < \bar{a}$ whenever $m \geq 2$. ∎

Throughout the remaining part of this subsection, we assume $m \geq 2$ (the linear case $m = 1$ was discussed in Section 4.3.3), $\underline{a} < a < \bar{a}$, and define

$$\tilde{p}(a) = p(a) \in \mathbb{R}^m,$$
$$\tilde{q}(a) = (q_0(a), \ldots, q_{m-1}(a))^T \in \mathbb{R}^m,$$
$$x(a) = (x_1(a), \ldots, x_{m-1}(a))^T \in \mathbb{R}^{m-1},$$
$$w(a) = (w_0(a), \ldots, w_{m-1}(a))^T \in \mathbb{R}^m,$$

where $p(a)$ and $q(a) = (q_1(a), \ldots, q_{m-1}(a), 1)^T$ are the vectors defined by Lemma 4.3.5, $x_i(a) = \cos t_i(a)$, $i = 1, \ldots, m - 1$, and $\{t_i(a)\}_{i=1,\ldots,m-1}$,

$\{w_i(a)\}_{i=0,\ldots,m-1}$ correspond to the positive support points and weights of the E-optimal design ξ_a on the interval $[-a, a]$. For arbitrary vectors $\tilde{q} = (q_0, \ldots, q_{m-1})^T$, $\tilde{p} = (p_0, \ldots, p_{m-1})^T$, $x = (x_1, \ldots, x_{m-1})^T$, and $w = (w_0, \ldots, w_{m-1})^T$, with $\alpha = \cos a < x_{m-1} < \cdots < x_1 < 1$, $w_i > 0$, $i = 0, \ldots, m-1$, and $\sum_{i=0}^{m-1} w_i < 1$ we define the vectors

$$\Theta = (\theta_0, \ldots, \theta_{4m-2})^T = (\tilde{p}^T, \tilde{q}^T, x^T, w^T)^T \in \mathbb{R}^{4m-1},$$

and, similarly,

$$\Theta(a) = (\theta_0(a), \ldots, \theta_{4m-2}(a))^T = (\tilde{p}^T(a), \tilde{q}^T(a), x^T(a), w^T(a))^T \in \mathbb{R}^{4m-1}$$

as the vector containing the support points and weights of the E-optimal design and the components of the vectors $q(a)$ and $p(a)$ defined in Lemma 4.3.5. Let us introduce the function

$$\lambda(\Theta, a) = \sum_{i=0}^{m-1} \frac{(q^T f_{(1)}(x_i))^2 + (1 - x_i^2)(p^T f_{(2)}(x_i))^2}{q^T q + p^T p} w_i \tag{4.91}$$
$$+ \frac{(q^T f_{(1)}(\alpha))^2 + (1 - \alpha^2)(p^T f_{(2)}(\alpha))}{q^T q + p^T p}(1 - w_0 - \cdots - w_{m-1}),$$

where $x_0 = 1$ and the vectors q and p are given by $q = (\tilde{q}^T, 1)^T$ and $p = \tilde{p}$. If ξ_a is the E-optimal design on the interval $[-a, a]$, then

$$\lambda(a) := \lambda(\Theta(a), a) = \lambda_{\min}(M(\xi_a)),$$

and an immediate differentiation of the function $\lambda(\Theta, a)$ shows that the conditions

$$\frac{\partial}{\partial \theta_i} \lambda(\Theta, a) \big|_{\Theta = \bar{\Theta}} = 0, \quad i = 0, \ldots, 4m - 2, \tag{4.92}$$

coincide with conditions (4.85) and (4.86) if $\bar{\Theta} = \Theta(a)$. Therefore, by Lemma 4.3.5, these conditions are necessary conditions for the vector $\Theta(a)$, which gives the support points and weights of the E-optimal design. We will call the vector equation (4.92) the *basic equation*. In order to study the Jacobi matrix of this equation, we will present a couple of auxiliary results, which are of independent interest. To this end, denote by

$$\eta = \begin{pmatrix} x_0 & \cdots & x_m \\ w_0 & \cdots & w_m \end{pmatrix}$$

a design on the interval $[\alpha, 1]$ (with $x_0 = 1$) and let

$$\xi_\eta = \begin{pmatrix} -t_m & \cdots & -t_1 & t_0 & t_1 & \cdots & t_m \\ \frac{w_m}{2} & \cdots & \frac{w_1}{2} & w_0 & \frac{w_1}{2} & \cdots & \frac{w_m}{2} \end{pmatrix} \tag{4.93}$$

be the design corresponding to η by the transformation (4.54), where $t_i =$ arc $\cos x_i$, $i = 0, \ldots m$. Similarly, for any symmetric design ξ of the form (4.93) on the interval $[-a, a]$, we denote by

$$\eta_\xi = \left(\begin{array}{ccc} x_0 & \cdots & x_m \\ w_0 & \cdots & w_m \end{array} \right),$$

with $x_i = \cos t_i$, $i = 0, \ldots, m$, the design on the interval $[\alpha, 1]$ obtained by the transformation (4.54). Finally, $v = v(a)$ denotes the multiplicity of the minimum eigenvalue of the matrix $M_c(\xi_a)$ and $u = u(a)$ is the multiplicity of the minimum eigenvalue of the matrix $M_s(\xi_a)$, where ξ_a denotes the E-optimal design for the trigonometric regression model on the interval $[-a, a]$ and the matrices $M_c(\xi_a)$ and $M_s(\xi_a)$ have been defined in (4.61) and (4.62), respectively.

Lemma 4.3.7 *Let $0 < a \le \pi$. A design ξ_a of the form (4.93) is an E-optimal design for the trigonometric regression model (4.1) on the interval $[-a, a]$ if and only if*

$$\xi_a = \xi_{\eta_\alpha},$$

where η_α is an E-optimal design for the Chebyshev regression model (4.55) on the interval $[\alpha, 1]$ and $\alpha = \cos a$.
   Moreover, the quantities $\underline{a}_{(1)}$ and $\underline{a}_{(2)}$ in (4.75) are equal, that is,

$$\underline{a}_{(1)} = \underline{a}_{(2)}$$

and the multiplicities $v(a)$ and $u(a)$ of the minimal eigenvalues of the matrices $M_c(\xi_a)$ and $M_s(\xi_a)$ of the E-optimal design ξ_a satisfy

$$v(a) = u(a) + 1$$

whenever $v(a) > 1$.

Proof. Let us begin with the last assertion; denote with ξ_a the E-optimal design and by $q_{(1)}, \ldots, q_{(v)}$ the eigenvectors of the matrix $M_c(\xi_a)$ corresponding to its minimal eigenvalue $\lambda_{\min}(M_c(\xi_a))$ and define the coordinates of $q_{(j)}$ by $q_{(j)i}$, $i = 0, \ldots, m$, $j = 1, \ldots, v$. Without loss of generality, we can choose $q_{(1)}$ such that $q_{(1)m} = 1$, $q_{(2)}$ such that $q_{(2)m} = 0$, $q_{(3)}$ such that $q_{(3)m} = q_{(3)m-1} = 0$, and so forth. For $v \ge 2$, we introduce the polynomials

$$\varphi_1^2(x) = \frac{\left(q_{(1)}^T f_{(1)}(x) \right)^2}{q_{(1)}^T q_{(1)}},$$

$$\varphi_2^2(x) = \frac{\left(q_{(i)}^T f_{(1)}(x) \right)^2}{q_{(i)}^T q_{(i)}}, \quad i \ne 1,$$

$$g(x) = \frac{\varphi_1^2(x) + \varphi_2^2(x)}{2},$$

where the vectors $f_{(1)}(x)$ and $f_{(2)}(x)$ have been defined in (4.81) and (4.82), respectively. Note that the polynomial g is non-negative of degree m and

$$\int g(\cos t)\xi_a(dt) = \lambda_{\min}(M_c(\xi_a)). \tag{4.94}$$

As in the proof of Lemma 4.3.5, we can find appropriate vectors $q \in \mathbb{R}^{m+1}$ and $p \in \mathbb{R}^m$ such that the polynomial $g(x)$ can be represented in the form

$$g(x) = \bar{\varphi}_1^2(x) + (1 - x^2)\bar{\varphi}_2^2(x), \tag{4.95}$$

where $\bar{\varphi}_1(x) = q^T f_{(1)}(x)$ and $\bar{\varphi}_2(x) = p^T f_{(2)}(x)$. Substituting $x = \cos t$, integrating both sides of (4.95) with respect to the measure $\xi_a(dt)$ and taking into account the identity (4.94), we obtain

$$\lambda_{\min}(M_c(\xi_a)) = q^T M_c(\xi_a)q + p^T M_s(\xi_a)p$$
$$\geq \lambda_{\min}(M_c(\xi_a))q^T q + \lambda_{\min}(M_s(\xi_a))p^T p. \tag{4.96}$$

A further integration of the function $g(\cos t)$ with respect to the uniform distribution $dt/2\pi$ on the interval $[-\pi, \pi]$ yields (observing the representation (4.95))

$$q^T q + p^T p = 1. \tag{4.97}$$

Earlier we proved that

$$\lambda_{\min}(M_s(\xi_a)) \geq \lambda_{\min}(M_c(\xi_a)),$$

and, consequently, (4.96) and (4.97) imply that one of the following conditions holds:

(i) $v = 1$, $p = 0$, $\lambda_{\min}(M_c(\xi_a)) = \lambda_{\min}(M(\xi_a)) < \lambda_{\min}(M_s(\xi_a))$,

(ii) $v > 1$, $p \neq 0$ is an eigenvalue of the matrix $M_s(\xi_a)$, $\lambda_{\min}(M_c(\xi_a)) = \lambda_{\min}(M_s(\xi_a))$.

Part (ii) is an immediate consequence of the previous discussion. For a proof of case (i), assume that

$$\lambda = \lambda_{\min}(M_c(\xi_a)) = \lambda_{\min}(M_s(\xi_a))$$

and let p and q be vectors such that $p \neq 0$ and

$$M_c(\xi_a)q = \lambda q, \quad M_s(\xi_a)p = \lambda p.$$

We introduce the polynomial

$$g(x) = \bar{\varphi}_1^2(x) + (1 - x^2)\bar{\varphi}_2(x),$$

where $\bar{\varphi}_1(x) = q^T f_{(1)}(x)$ and $\bar{\varphi}_2(x) = p^T f_{(2)}(x)$. This polynomial can be represented in the form

$$\left(q_1^T f_{(1)}(x)\right)^2 + \left(q_2^T f_{(1)}(x)\right)^2$$

and a similar calculation to that given in the previous discussion shows that q_1 and q_2 should be eigenvectors, corresponding to $\lambda_{\min}(M_c(\xi_a))$. Therefore, it follows that $v \geq 2$ and this proves that part (i) is correct.

In the first case, $\lambda_{\min}(M(\xi_a))$ is simple. In the second case, $v \geq 2$, and for each eigenvector $q_{(i)}$, there exists an eigenvector $p_{(i)}$ of the matrix $M_s(\xi_a)$. It can be easily checked that the vectors $p_{(i)}$, $i = 2, \ldots, v$, are of the form

$$(p_{(2)0}, \ldots, p_{(2)m-1})^T,$$
$$(p_{(3)0}, \ldots, p_{(3)m-2}, 0)^T,$$
$$\vdots$$
$$(p_{(v)0}, \ldots, p_{(v)m-v+1}, 0, \ldots, 0)^T.$$

Consequently, these vectors are linearly independent, which gives $v(a) \geq u(a) + 1$. In a similar way, we can prove that $v(a) \leq u(a) + 1$ and for the case $v(a) > 1$ we obtain that

$$v(a) = u(a) + 1.$$

From (4.96) and (4.97) it also follows that $v(a) > 1$ in the case $\lambda_{\min}(M_c(\xi_a)) = \lambda_{\min}(M_s(\xi_a))$. Recalling the definition of $\underline{a}_{(1)}$ and $\underline{a}_{(2)}$ in (4.75), it thus follows that

$$\underline{a}_{(1)} = \inf\{a \mid v(a) > 1\},$$
$$\underline{a}_{(2)} = \inf\{a \mid \lambda_{\min}(M_c(\xi_a)) = \lambda_{\min}(M_s(\xi_a))\} ,$$

and the previous remarks yield

$$\underline{a}_{(1)} = \underline{a}_{(2)} = \underline{a}.$$

In order to prove the first assertion of Lemma 4.3.7, let ξ_a be a symmetric E-optimal design of the form (4.93) for the trigonometric regression model on the interval $[-a, a]$; then it follows from the previous discussion that

$$\lambda_{\min}(M(\xi_a)) = \lambda_{\min}(M_c(\xi_a)).$$

From the definition of the transformation (4.54), we have

$$M_c(\xi_a) = M_1(\eta_{\xi_a}),$$

where

$$M_1(\eta) = \int f_{(1)}(x) f_{(1)}^T(x) \eta(dx)$$

denotes the information matrix of the design η in the Chebyshev regression model (4.55). Therefore, a design ξ_a is an E-optimal design for the regression function $f_c(t)$ on the interval $[-a, a]$ if and only if the design η_{ξ_a} is an

E-optimal design in the Chebyshev regression model (4.55) on the interval $[\alpha, 1]$, where $\alpha = \cos a$. Now, it is easy to verify that any E-optimal design of the form (4.93) for the regression function $f_c(t)$ on the interval $[-a, a]$ is also an E-optimal design for the trigonometric regression model on the interval $[-a, a]$ and vice versa. Thus, a design ξ_{η_α} of the form (4.93) is an E-optimal design for the trigonometric regression model on the interval $[-a, a]$ if and only if the corresponding design η_α is an E-optimal design for the Chebyshev regression model (4.55) on the interval $[\alpha, 1]$. ∎

Throughout this chapter we denote by $\tau(a)$ the number of common roots of the polynomials $\varphi_1(x) = q^T f_{(1)}(x)$ and $\varphi_2(x) = p^T f_{(2)}(x)$ defined by Lemma 4.3.5. The following result provides the basis for the implementation of the functional approach.

Theorem 4.3.3 *Consider the trigonometric regression model(4.1) on the interval $[-a, a]$, where $0 < a \leq \pi$ and $m \geq 2$. Then $\underline{a} < \bar{a}$ and there exists a number $\nu \in \mathbb{N}$ and real quantities*

$$\underline{a} = a_1 < a_2 < a_3 < \cdots < a_\nu = \bar{a}$$

such that the vector function

$$\Theta^* : \left\{ \begin{array}{ll} (\underline{a}, \bar{a}) & \to \mathbb{R}^{4m-1} \\ a & \to \Theta(a) \end{array} \right.$$

is uniquely determined, real analytic on the set

$$\bigcup_{j=1}^{\nu-1} (a_j, a_{j+1}), \tag{4.98}$$

and satisfies the system of equations

$$\frac{\partial}{\partial \theta_i} \lambda(\Theta, a) \bigg|_{\Theta=\Theta(a)} = 0, \quad i = 0, \ldots, 4m - 2, \tag{4.99}$$

where the function $\lambda(\Theta, a)$ is defined in (4.91).

Proof. We have already proved that the vector function Θ^* is uniquely determined and satisfies (4.99). It is also obviously continuous. In order to study its analytic properties we define

$$G(\Theta, a) = \left(\frac{\partial^2}{\partial \theta_i \partial \theta_j} \lambda(\Theta, a) \right)_{i,j=0}^{4m-2}$$

as the Jacobi matrix of the system (4.99) and denote by

$$J = J(a) = G(\Theta(a), a)$$

the corresponding value at the point $\Theta = \Theta(a)$. A straightforward but tedious differentiation shows that this matrix is of the form

$$
J = h \begin{pmatrix} S & B_{(1)}^T & B_{(2)}^T \\ B_{(1)} & D & 0 \\ B_{(2)} & 0 & 0 \end{pmatrix}, \tag{4.100}
$$

where $h = 1/(q^T q + p^T p)$, $q = q(a)$, and $p = p(a)$. The matrices in the block matrix (4.100) are given by

$$
S = \begin{pmatrix} M_{(1)} & 0 \\ 0 & M_{(2)} \end{pmatrix},
$$

where

$$
M_{(1)} = M_s(\xi_a) - \lambda I_m.
$$

Similarly, if A_- denotes the matrix A with deleted last row and last column, $M_{(2)}$ is defined by

$$
M_{(2)} = (M_c(\xi_a) - \lambda I_{m+1})_-,
$$
$$
D = \operatorname{diag}\{d_{11}, \dots, d_{m-1,m-1}\},
$$

where the elements of the matrix D are given by

$$
d_{ii} = \left((q^T f_{(1)}(x))^2 + (p^T f_{(2)}(x))^2 (1 - x^2)\right)'' \Big|_{x=x_i(a)}, \quad i = 1, \dots, m-1,
$$

and

$$
B_{(1)}^T = \left(B_{(1)1}^T \vdots B_{(1)2}^T\right),
$$

$$
B_{(1)1}^T = \left((f_{(1)}(x)_- q^T f_{(1)}(x))' w_i \Big|_{x=x_i(a)}\right)_{i=1,\dots,m-1},
$$

$$
B_{(1)2} = \left((f_{(1)}(x)(1 - x^2))' w_i \Big|_{x=x_i(a)}\right)_{i=1,\dots,m-1},
$$

$$
B_{(2)}^T = \left(B_{(2)1}^T \vdots B_{(2)2}^T\right),
$$

$$
B_{(2)1}^T = \left(f_{(1)}(x_i)_- q^T f_{(1)}(x_i)\right)_{i=0,\dots,m-1},
$$

$$
B_{(2)2}^T = \left(f_{(2)}(x_i) p^T f_{(2)}(x_i)(1 - x^2)\right)_{i=0,\dots,m-1},
$$

where b_- denotes the vector b with deleted last element. Let $\tilde{a} \in (\underline{a}, \bar{a})$ such that the following condition is satisfied:

(A) *There exists a neighborhood $\tilde{U}$ of the point $\tilde{a}$ such that for all $a \in \tilde{U}$, we have*

$$
\tau = \tau(\tilde{a}) = \tau(a).
$$

Denote by $\delta_1, \ldots, \delta_\tau$ the common roots of the polynomials

$$\varphi_1(x) = q^T(a)f_{(1)}(x),$$
$$\varphi_2(x) = p^T(a)f_{(2)}(x),$$

by $\gamma_1, \ldots, \gamma_{m-\tau}$ the remaining roots of the polynomial $\varphi_1(x)$, and by $\kappa_1, \ldots, \kappa_{m-1-\tau}$ the remaining roots of the polynomial $\varphi_2(x)$; that is,

$$\varphi_1(x) = \prod_{i=1}^{\tau}(x - \delta_i) \prod_{i=1}^{m-\tau}(x - \gamma_i),$$

$$\varphi_2(x) = \kappa_{m-\tau} \prod_{i=1}^{\tau}(x - \delta_i) \prod_{i=1}^{m-\tau-1}(x - \kappa_i)$$

(recall that it was shown in the proof of Lemma 4.3.5 that φ_1 and φ_2 have simple roots, which are interlacing, and note that $\kappa_{m-\tau}$ denotes *not* a root of the polynomial φ_2 but its leading coefficient). Define the vector

$$\hat{\Theta}(a) = \hat{\Theta} = (\gamma_1, \ldots, \gamma_{m-\tau}, \kappa_1, \ldots, \kappa_{m-\tau}, \delta_1, \ldots, \delta_\tau, \tilde{x}(a), \tilde{w}(a))^T$$
$$= (\hat{\theta}_0, \ldots, \hat{\theta}_{4m-1-\tau})^T.$$

and note that in a neighborhood of the point $\tilde{a}$, there exists essentially a one-to-one correspondence between the points $\hat{\Theta}(a)$ and $\Theta(a)$. Consider the matrix

$$\tilde{J} = H^T J H, \tag{4.101}$$

with

$$H = \left(\partial\hat{\theta}_i/\partial\theta_j\right)_{i=0,j=0}^{4m-1-\tau,4m-2}.$$

We will prove below that the matrix $\tilde{J}$ is nonsingular for any point $\tilde{a}$ satisfying condition (A). Because $\tau(a) \in \{1, 2, \ldots, m\}$, it therefore follows that all points $a \in (\underline{a}, \bar{a})$ except for a finite set denoted by $\{a_1, \ldots, a_\nu\}$ satisfy condition (A). Therefore, the vector function

$$\Theta^+ : a \to \hat{\Theta}(a)$$

is a real analytic vector function on the set (4.98) due to the well-known Implicit Function Theorem (Gunning and Rossi (1965)). Because the coefficients of a polynomial are analytic functions of its zeros, it follows that the vector function Θ^* is also real analytic on the same set.

The proof of the nonsingularity of the matrix $\tilde{J}$ is tedious and we indicate the main steps. Denote by $\mathcal{P}$ the eigenspace of the matrix $M_c(\xi_a)$ corresponding to its minimal eigenvalue $\lambda_{\min}(M_c(\xi_a))$ and by $\mathcal{P}_\tau$ the subspace of all vectors $r = (r_0, r_1, \ldots, r_m)^T$ such that the polynomial $\sum_{j=0}^m r_j x^j$ has the form

$$\prod_{i=1}^{\tau} (x - \delta_i) \sum_{j=0}^{m-\tau} \tilde{r}_j x^j$$

for some vector $\tilde{r} = (\tilde{r}_0, \ldots, \tilde{r}_{m-\tau})^T$ of size $m - \tau + 1$. For the sake of transparency, we introduce the notation $F(x) = (1, x, \ldots, x^m)^T$ and define the following for vectors $r, s \in \mathbb{R}^{m+1}$:

$$< r, s > = \int_{-1}^1 \left(r^T F(x) \right) \left(s^T F(x) \right) \eta_{\xi_a}(dx) \times$$

$$\times \left[\frac{2}{\pi} \int_{-1}^1 \left(r^T F(x) \right) \left(s^T F(x) \right) \frac{dx}{\sqrt{1 - x^2}} \right]^{-1}.$$

A straightforward calculation shows that the condition

$$\frac{\partial \lambda(\hat{\Theta}, a)}{\partial \gamma_i} = 0$$

is equivalent to the condition

$$< q_{\gamma_i}, q > = \lambda(\hat{\Theta}, a),$$

where the vector $q_{\gamma_i} \in \mathcal{P}_\tau$ is defined by

$$q_{\gamma_i}^T F(x) = \frac{1}{x - \gamma_i} q^T F(x) = \frac{d}{dx} q^T F(x) \Big|_{x = \gamma_i}$$

for any $i = 1, \ldots, m - \tau$. This means that

$$q_{\gamma_i} \in \mathcal{P} , \quad i = 1, \ldots, m - \tau$$

(note that the vectors $q_{\gamma_1}, \ldots, q_{\gamma_{m-\tau}}$ are linearly independent). Note that a direct calculation gives

$$\frac{\partial^2}{\partial \gamma_j \partial \gamma_i} \lambda(\hat{\Theta}, a) = \left(\frac{q_{\gamma_i}^T M(\xi_a) q_{\gamma_i}}{q_{\gamma_j}^T q_{\gamma_i}} - \lambda(\hat{\Theta}, a) \right) \frac{1}{q_{\gamma_j}^T q_{\gamma_i}}, \quad j = 1, \ldots, m - \tau.$$

Since $q_{\gamma_j} \in \mathcal{P}$, we obtain

$$\frac{\partial^2}{\partial \gamma_i \partial \gamma_i} \lambda(\hat{\Theta}, a) = 0 , \quad i, j = 1, \ldots, m - \tau,$$

In a similar way, it follows that

$$p_{\kappa_i} \in \mathcal{P}_{(2)}, \quad i = 1, \ldots, m - \tau,$$

where $\mathcal{P}_{(2)}$ is the eigenspace, corresponding to $\lambda_{\min}(M_{(s)}(\xi_a))$, and by the same arguments we obtain

$$\frac{\partial^2}{\partial \kappa_i \partial \kappa_j} \lambda(\hat{\Theta}, a) = 0, \ i, j = 1, \ldots, m - \tau.$$

It is easy to check that for $a \in (\underline{a}, \bar{a})$, it follows that $\tau \geq 1$. Moreover, using the above formulas, we obtain that the matrix $\tilde{J}$ has the structure indicated in Table 4.5, where A is a non-negative definite matrix and D is the negative definite matrix, defined earlier.

Table 4.5: Structure of the matrix $\tilde{J}$ defined in (4.101)

	1	$m - \tau$	$m - 1$	$m - 1$	m
1	0	b^T			
$m - \tau$		0	V^T	B_1^T	C_1^T
$m - 1$		V	A	B_2^T	C_2^T
$m - 1$	b	B_1	B_2	D	0
m		C_1	C_2	0	0

If $b \neq 0$ and the matrices $C = (C_1 : C_2)$, B_2 and B_1 have full rank it follows by similar arguments as given in Section 3.2 with the help of the Frobenius formula that $\det \tilde{J} \neq 0$. The verification of the listed conditions is equivalent to the verification that certain polynomials are not identically zero. This can be done by the standard technique of counting zeros and is left to the reader. Thus $\det \tilde{J} \neq 0$ for any point a satisfying condition (A). ∎

Since the vector function $\Theta(a) = \Theta(\arccos a)$ is real analytic on the set defined by (4.98), it can be expanded into a Taylor series in a neighborhood of any point $\tilde{a} \neq a_j$, $j = 1, \ldots, \nu$, $\underline{a} < \tilde{a} < \bar{a}$, and we obtain for its components an expansion of the form

$$\theta_i(a) = \sum_{k=0}^{\infty} \theta_{i,k}(\alpha - \tilde{\alpha})^k, \ i = 0, \ldots, 4m - 2,$$

where $\tilde{\alpha} = \cos \tilde{a}$ and $\alpha = \cos a$. For the determination of the coefficients $\{\theta_{i,k}\}$, the general recurrent formulas introduced in Section 2.4 can be applied, provided that initial conditions $\theta_{i,0}$, $i = 0, \ldots, 4m - 2$ are known. To find such initial coefficients $\Theta^{(0)} = (\theta_{0,0}, \ldots, \theta_{4m-2,0})^T$, we solve the equation

$$Q(\Theta^{(0)}) := \sum_{i=0}^{4m-2} \left(\frac{\partial}{\partial \theta_i} \lambda(\Theta, \tilde{a}) \Big|_{\Theta=\Theta^{(0)}} \right)^2 = 0$$

for some $\tilde{a}$, which can be done by standard numerical algorithms. To obtain an approximation of the function $\Theta(a)$ with a given precision, we have to find one or several points $\tilde{a}_1, \ldots, \tilde{a}_k$, construct the corresponding Taylor series and verify that the calculated design is E-optimal with sufficient precision (note that $\Theta(a)$ contains also the vectors $p(a)$ and $q(a)$ by Lemma 4.3.5). In the following examples, we will illustrate this approach for the quadratic and cubic trigonometric regression model on the interval $[-a, a]$.

Example 4.3.1 Consider the quadratic trigonometric regression model on the interval $[-a, a]$

$$\beta^T f(t) = \beta_0/\sqrt{2} + \beta_1 \cos t + \beta_2 \sin t + \beta_3 \cos 2t + \beta_4 \sin 2t.$$

By the discussion of Section 4.3.1, it follows that for $\bar{a} = 0.8\pi \le a \le \pi$, an E-optimal design is given by

$$\begin{pmatrix} -\frac{4\pi}{5} & -\frac{2\pi}{5} & 0 & \frac{2\pi}{5} & \frac{4\pi}{5} \\ \frac{1}{5} & \frac{1}{5} & \frac{1}{5} & \frac{1}{5} & \frac{1}{5} \end{pmatrix}.$$

Similarly, Corollary 4.3.1 and Theorem 4.3.2 show that for $0 < a \le \underline{a} \approx 0.741\pi$, the unique E-optimal design is given by

$$\begin{pmatrix} -a & -t(a) & 0 & t(a) & a \\ \frac{w_2}{2} & \frac{w_1}{2} & w_0 & \frac{w_1}{2} & \frac{w_2}{2} \end{pmatrix},$$

where

$$t(a) = \arccos\left(\frac{1 + \cos a}{2}\right)$$

and the weights w_0, w_1, and w_2 can be found by formula (4.49). In the intermediate case

$$\underline{a} = 0.741\pi < a < 0.8\pi = \bar{a},$$

we will construct the E-optimal design by the functional approach. Note that due to Theorem 4.3.1, an E-optimal design is of the form

$$\begin{pmatrix} -t_2 & -t_1 & t_0 & t_1 & t_2 \\ \frac{w_2}{2} & \frac{w_1}{2} & w_0 & \frac{w_1}{2} & \frac{w_2}{2} \end{pmatrix},$$

where $t_0 = 0$ and $t_2 = a$. Since $w_0 + w_1 + w_2 = 1$, it is enough to consider the weights w_1 and w_2 and the point $x_1 = \arccos t_1$. We take $\tilde{a} = 0.77\pi \approx (\bar{a} + \underline{a})/2$. The first Taylor coefficients for the parameters

$$q_0 = q_0(\arccos \alpha),$$
$$q_1 = q_1(\arccos \alpha),$$
$$p_0 = p_0(\arccos \alpha),$$
$$p_1 = p_1(\arccos \alpha),$$
$$1 - x_1 = 1 - x_i(\arccos \alpha),$$
$$w_1 = w_1(\arccos \alpha),$$
$$w_2 = w_2(\arccos \alpha),$$

in the expansion

$$\Theta(\arccos \alpha) = \sum_{n=0}^{\infty} \Theta^{(n)} (\alpha - \cos \tilde{a})^n$$

are listed in Table 4.6. The dependence of the support points and weights of the E-optimal design in the trigonometric regression model from the parameter $a \in (\underline{a}, \bar{a})$ is illustrated in Figure 4.1. In the present case, it follows that $a_1 = \underline{a} < a_2 = \bar{a}$, and for $a_1 < a < a_2$, we have

$$\tau(a) = 1, \ u(a) = 1, \ v(a) = 2.$$

It is also interesting to note that for $0 < a < a_1 = \underline{a}$, we have

$$u(a) = 0, \ v(a) = 1,$$

whereas for the case $\bar{a} = a_2 < a < \pi$, it follows that

$$u(a) = 2, \ v(a) = 3.$$

In other words, if the parameter a is increased from 0 to π, the multiplicity of the minimum eigenvalue of the information matrix of the E-optimal design changes from 1 to 5 by steps of size 2.

Example 4.3.2 Consider the cubic trigonometric regression model on the interval $[-a, a]$ (i.e., $m = 3$). Then, similar to the preceding example, an E-optimal design can be found in an explicit form whenever $0 < a \leq \underline{a} \approx 0.794\pi$ and $\bar{a} \leq a \leq \pi$, $\bar{a} = 6/7\pi \approx 0.857\pi$. In the case $a > \bar{a}$, the design

$$\begin{pmatrix} -\frac{6\pi}{7} & -\frac{4\pi}{7} & -\frac{2\pi}{7} & 0 & \frac{2\pi}{7} & \frac{4\pi}{7} & \frac{6\pi}{7} \\ \frac{1}{7} & \frac{1}{7} & \frac{1}{7} & \frac{1}{7} & \frac{1}{7} & \frac{1}{7} & \frac{1}{7} \end{pmatrix}$$

is E-optimal (but not necessarily unique), whereas in the case $a < \underline{a}$, the support points of the unique E-optimal design are given by

$$\pm a, \ \pm \arccos\left(\frac{3 + \cos a}{4}\right), \ \pm \arccos\left(\frac{1 + 3\cos a}{4}\right)$$

Table 4.6: Coefficients in the Taylor expansion (4.3.1) for the quadratic trigonometric regression model ($m = 2$), where $0.741 < a/\pi < 4/5 = 0.8$

	0	1	2	3	4	5
p_0	0.4771	-0.0781	-1.3312	-1.9692	1.2116	3.8592
p_1	-0.4928	2.0781	-0.5175	0.0124	1.9268	-1.7913
q_0	-0.3532	-2.7276	11.4353	-92.0212	896.9923	-9.90e+03
q_1	-0.3794	-2.5761	15.1122	-109.6045	1.05e+03	-1.15e+04
$1-x_1$	0.7588	-1.2582	1.5801	3.9027	0.3966	-11.1756
w_1	0.1862	0.1994	0.5826	0.2185	-2.1883	-5.1277
w_2	0.2289	-0.4732	-0.3163	0.5386	-0.2008	0.5346

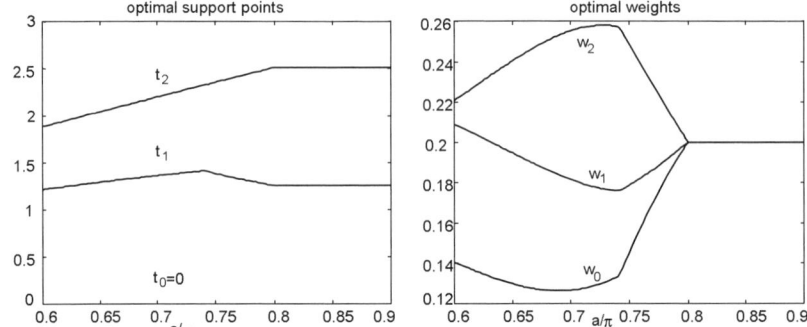

Figure 4.1: Support points and weights of the E-optimal design in a quadratic trigonometric regression models on the interval $[-a, a]$ for various values of a

and the weights are obtained from formula (4.49). It was found numerically that
$$a_1 = \underline{a} < a_2 \approx 0.8113\pi < a_3 = \bar{a} = 6/7\pi \approx 0.857\pi,$$
and for $a \in (a_1, a_2)$, the first coefficients for the Taylor expansion at the point $\bar{a}_1 = 0.81\pi$ are presented in Table 4.7, whereas Table 4.8 contains the corresponding coefficients for the case $a \in (a_2, a_3)$ (for the expansion at the point $\bar{a}_2 = 0.83\pi$). Note that the multiplicities of the minimal eigenvalues of the matrices $M_s(\xi_a)$ and $M_c(\xi_a)$ are given by

$$\begin{aligned}
u(a) &= 0, & v(a) &= 1 \ \text{if} \ a \in (0, a_1), \\
u(a) &= 1, & v(a) &= 2 \ \text{if} \ a \in (a_1, a_2), \\
u(a) &= 2, & v(a) &= 3 \ \text{if} \ a \in (a_2, a_3), & (4.102) \\
u(a) &= 3, & v(a) &= 4 \ \text{if} \ a \in (a_3, \pi),
\end{aligned}$$

where $a_1 = \underline{a}$ and $a_3 = \bar{a}$.

The behavior of the optimal design points and weights is presented in Figure 4.2. It was verified numerically that the points and weights can be

Table 4.7: Coefficients in the Taylor expansion (4.3.1) for the cubic trigonometric regression model $(m = 3)$, where $0.794 < a/\pi < 0.8113$

	0	1	2	3	4	5
p_0	-0.3965	4.0928	-0.3055	-12.3495	-11.4981	761.5907
p_1	0.6477	-1.8835	-3.2233	-5.1845	1.6191	811.6268
p_2	-0.5608	1.6899	3.0650	3.2670	-2.9128	-928.5566
q_0	0.1501	2.0088	-28.2704	430.0770	-8.39e+03	1.96e+05
q_1	0.2599	4.2543	-37.1810	612.9954	-1.27e+04	2.96e+05
q_2	0.2219	3.4819	-34.7987	534.1822	-1.11e+04	2.56e+05
$1-x_1$	0.4047	-1.4247	2.2884	9.9250	15.0767	-48.0838
$1-x_2$	1.3565	0.3121	0.7705	2.0655	2.2545	-12.2375
w_1	0.0966	0.4030	1.2655	1.5036	-9.9210	-81.2046
w_2	0.1397	-0.3048	1.2041	5.8567	-1.1242	-68.7337
w_3	0.2164	-0.4311	-2.8260	-3.6782	21.5715	101.4223

Table 4.8: Coefficients in the Taylor expansion (4.3.1) for the cubic trigonometric regression model $(m = 3)$, where $0.8113 < a/\pi < 6/7 = 0.857$

	0	1	2	3	4	5
p_0	-0.0674	9.6717	12.2522	43.8534	-2.67e+03	6.11e+04
p_1	0.8655	5.9768	-7.3377	1.8819	-2.75e+03	6.44e+04
p_2	-0.7126	-3.5536	9.0726	-65.9985	3.11e+03	-7.76e+04
q_0	0.6670	9.0433	-68.2559	962.3302	-1.87e+04	4.10e+05
q_1	0.5298	5.6016	-35.9075	387.2603	-6.62e+03	1.30e+05
q_2	0.0868	-0.7658	22.7645	-454.2934	9.99e+03	-2.39e+05
$1-x_1$	0.3917	-0.3409	-1.0703	2.8779	38.4908	158.1277
$1-x_2$	1.2958	-1.9386	1.7869	25.6580	62.2862	-124.9227
w_1	0.1197	0.6556	-1.8890	-4.5152	46.8271	18.8599
w_2	0.1383	0.0358	1.8823	1.6989	-29.6814	-78.2035
w_3	0.1826	-1.0283	0.8572	5.4231	-35.5716	57.4784

determined with high precision. Figure 4.3 shows the extremal polynomial

$$\frac{(p^T f_{(1)}(x))^2 + (q^T f_{(2)}(x))^2}{p^T p + q^T q}$$

in the equivalence theorem for various values of a (note that by Lemma 4.3.5, this function has to be less than or equal to the minimum eigenvalue of the information matrix corresponding to the E-optimal design with equality at the support points).

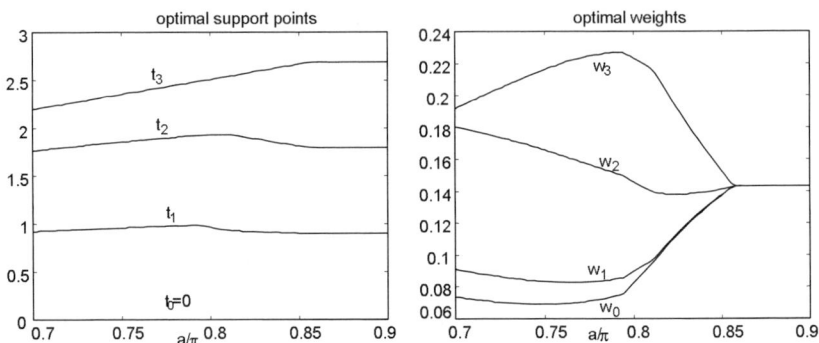

Figure 4.2: Support points and weights of the E-optimal design in a cubic trigonometric regression models on the interval $[-a, a]$ for various values of a

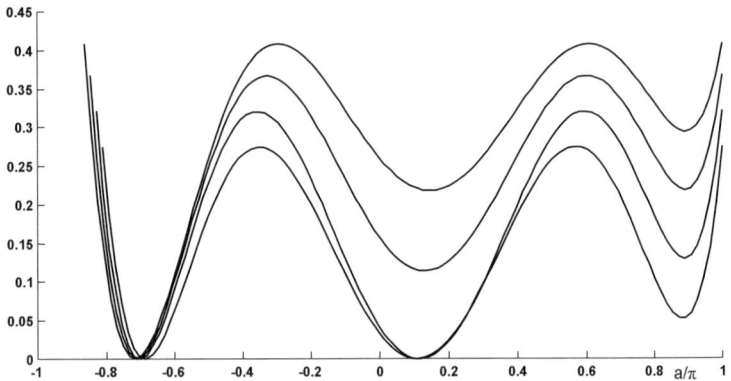

Figure 4.3: The extremal polynomials for various values of a

4.4   Numerical Comparison of D- and E-Optimal Designs

Let us now compare D- and E-optimal designs for trigonometrical regression models on intervals $[-a, a]$, $0 < a \leq \pi$. It is interesting to know how good E-optimal designs are in the sense of the D-criterion and how good D-optimal designs are from the E-optimality point of view.

Formally, we will calculate the value

$$\operatorname{eff} D(\xi) = \left(\frac{\det M(\xi)}{\det(M(\xi_D))} \right)^{1/m}$$

for $\xi = \xi_E$ and the value

$$\operatorname{eff} E(\xi) = \frac{\lambda_{\min}(M(\xi))}{\lambda_{\min}(M(\xi_E))}$$

for $\xi = \xi_D$, where ξ_E denotes an E-optimal design and ξ_D denotes a D-optimal design.

We will include in the comparison some equidistant designs usually used in practice. For brevity, a design will be called a uniform design on $[-a, a]$ if it is concentrates in $2k + 1$ equidistant points with equal weights and these points are $-a \left(1 - \frac{2i}{2k+1}\right)$, $i = 1, 2, \ldots, 2k + 1$ for $a \leq \hat{a}$, $\hat{a} = (2k + 1)\pi/(2k + 2)$. For $a > \hat{a}$, the points of uniform designs are $-a + \alpha + \frac{2i\pi}{2k+2}$, $i = 1, \ldots, 2k + 1$, where α is arbitrary number in $[0, a - \hat{a}]$. The uniform designs will be denoted ξ_u.

D- and E-optimal designs constructed by the methods described are listed in Table 4.9 for $k = 2, 3, 4, 5$ and $a = 0.2, 0.3, \ldots, 1$. In all cases when the designs are not presented, they coincide with the uniform designs.

In Table 4.10, we show the efficiency of the uniform, D- and E-optimal designs.

It can be noted that for small α, the uniform designs perform rather poorly, especially, in the sense of E-criteria. Also, one can see from Table 4.10 that the D-efficiency of E-optimal designs are higher than the E-efficiency of D-optimal design. It allows one to recommend E-optimal designs for small design intervals. One more advantage of E-optimal designs for small design intervals in comparison with D-optimal designs is that the first can be calculated explicitly by the formulas given in Section 4.3.

Table 4.9: Optimal designs for trigonometrical model on $[-a, a]$

$k = 2$								
a/π	0.2	0.3	0.4	0.5	0.6	0.7	0.8	0.9
$t_1 D/a$	0.646	0.634	0.618	0.596	0.569	0.536		
$t_1 E/a$	0.701	0.694	0.680	0.667	0.646	0.620		
$w_0 E$	0.237	0.220	0.197	0.162	0.122	0.095		
$w_1 E$	0.250	0.249	0.245	0.238	0.226	0.212		
$w_2 E$	0.132	0.142	0.156	0.181	0.213	0.241		
$k = 3$								
$t_1 D/a$	0.461	0.452	0.438	0.421	0.400	0.376	0.349	
$t_2 D/a$	0.825	0.817	0.806	0.789	0.768	0.735	0.694	
$t_1 E/a$	0.494	0.486	0.475	0.460	0.442	0.420	0.354	
$t_2 E/a$	0.862	0.858	0.850	0.839	0.824	0.802	0.751	
$w_0 E$	0.158	0.147	0.133	0.116	0.098	0.085	0.121	
$w_1 E$	0.162	0.156	0.146	0.131	0.110	0.081	0.081	
$w_2 E$	0.171	0.176	0.183	0.190	0.194	0.189	0.156	
$w_3 E$	0.088	0.094	0.105	0.121	0.147	0.188	0.203	
$k = 4$								
$t_1 D/a$	0.358	0.349	0.339	0.325	0.309	0.290	0.269	
$t_2 D/a$	0.671	0.660	0.646	0.628	0.603	0.573	0.536	
$t_3 D/a$	0.896	0.892	0.884	0.873	0.856	0.831	0.794	
$t_1 E/a$	0.378	0.371	0.361	0.349	0.334	0.316	0.296	
$t_2 E/a$	0.700	0.694	0.682	0.667	0.646	0.620	0.587	
$t_3 E/a$	0.922	0.918	0.913	0.906	0.896	0.879	0.854	
$w_0 E$	0.118	0.111	0.100	0.087	0.071	0.051	0.034	
$w_1 E$	0.121	0.114	0.106	0.094	0.080	0.065	0.061	
$w_2 E$	0.125	0.124	0.122	0.117	0.107	0.088	0.061	
$w_3 E$	0.129	0.136	0.144	0.156	0.168	0.178	0.165	
$w_4 E$	0.066	0.070	0.078	0.090	0.110	0.144	0.196	
$k = 5$								
$t_1 D/a$	0.287	0.288	0.276	0.264	0.251	0.234	0.219	0.202
$t_2 D/a$	0.561	0.556	0.535	0.517	0.495	0.469	0.437	0.403
$t_3 D/a$	0.778	0.779	0.759	0.741	0.720	0.690	0.651	0.605
$t_4 D/a$	0.936	0.936	0.924	0.917	0.904	0.885	0.853	0.805
$t_1 E/a$	0.297	0.299	0.290	0.281	0.268	0.254	0.237	0.201
$t_2 E/a$	0.596	0.571	0.561	0.546	0.526	0.501	0.472	0.402
$t_3 E/a$	0.794	0.798	0.789	0.776	0.757	0.732	0.699	0.603
$t_4 E/a$	0.938	0.934	0.934	0.940	0.932	0.919	0.900	0.810
$w_0 E$	0.097	0.091	0.082	0.070	0.058	0.045	0.037	0.094
$w_1 E$	0.094	0.093	0.085	0.073	0.061	0.047	0.029	0.082
$w_2 E$	0.104	0.099	0.094	0.084	0.075	0.062	0.050	0.094
$w_3 E$	0.113	0.112	0.112	0.105	0.102	0.092	0.068	0.082
$w_4 E$	0.097	0.105	0.114	0.131	0.146	0.163	0.168	0.092
$w_5 E$	0.043	0.045	0.053	0.072	0.087	0.114	0.167	0.104

Table 4.10: Comparison of optimal designs for trigonometrical model on $[-a, a]$

$k = 2$									
a/π	0.2	0.3	0.4	0.5	0.6	0.7	0.8	0.9	1
$\text{eff}_D(\xi_u)$	0.910	0.923	0.939	0.958	0.977	0.993	1	1	1
$\text{eff}_D(\xi_E)$	0.942	0.951	0.960	0.960	0.946	0.915	1	1	1
$\text{eff}_E(\xi_u)$	0.439	0.477	0.539	0.630	0.743	0.829	1	1	1
$\text{eff}_E(\xi_D)$	0.847	0.862	0.882	0.905	0.914	0.895	1	1	1
$k = 3$									
a/π	0.2	0.3	0.4	0.5	0.6	0.7	0.8	0.9	1
$\text{eff}_D(\xi_u)$	0.782	0.804	0.835	0.873	0.917	0.961	0.993	1	1
$\text{eff}_D(\xi_E)$	0.944	0.949	0.953	0.951	0.933	0.884	0.915	1	1
$\text{eff}_E(\xi_u)$	0.167	0.185	0.217	0.273	0.373	0.545	0.745	1	1
$\text{eff}_E(\xi_D)$	0.825	0.833	0.841	0.850	0.866	0.840	0.819	1	1
$k = 4$									
a/π	0.2	0.3	0.4	0.5	0.6	0.7	0.8	0.9	1
$\text{eff}_D(\xi_u)$	0.655	0.683	0.724	0.776	0.840	0.910	0.974	1	1
$\text{eff}_D(\xi_E)$	0.945	0.951	0.953	0.946	0.930	0.870	0.786	1	1
$\text{eff}_E(\xi_u)$	0.059	0.066	0.080	0.104	0.155	0.270	0.515	1	1
$\text{eff}_E(\xi_D)$	0.825	0.819	0.823	0.827	0.818	0.795	0.740	1	1
$k = 5$									
a/π	0.2	0.3	0.4	0.5	0.6	0.7	0.8	0.9	1
$\text{eff}_D(\xi_u)$	0.543	0.574	0.619	0.681	0.760	0.890	0.945	1.000	1
$\text{eff}_D(\xi_E)$	0.951	0.954	0.950	0.889	0.920	0.901	0.752	0.993	1
$\text{eff}_E(\xi_u)$	0.019	0.023	0.028	0.044	0.060	0.117	0.303	0.947	1
$\text{eff}_E(\xi_D)$	0.809	0.811	0.815	0.943	0.796	0.583	0.701	0.972	1

Chapter 5

D-Optimal Designs for Rational Models

Rational functions are often used for approximating an arbitrary continuous function. Rational approximations are usually better than polynomial ones since they include sufficiently fewer sets of parameters.

The present chapter is intended to construct and analyze locally D-optimal designs for the rational model with the help of the functional approach developed in the previous chapters. Note that maximin efficient designs can be studied in a way similar to that demonstrated in Section 2.2 for exponential models.

5.1 Introduction

The problem considered in this chapter was studied in a number of papers. He, Studden, and Sun (1996) proved that the problem of constructing locally D-optimal designs for rational models is equivalent to that of finding D-optimal designs for polynomial regression models with variance functions that are also polynomials. The last problem was considered in Karlin, Studden (1966), Fedorov (1972), Huang, Chang, and Wong (1995), Chang and Lin (1997), Ortiz and Rodziquez (1998), Imhof and Studden (1998), Cheng, Kleijnen, and Melas (2000). The connection between the problems is discussed in Dette, Haines, and Imhof (1999). Two basic approaches to constructing locally D-optimal designs were used: the implementation of numerical procedures and construction of designs in a closed analytical form.

The numerical construction proves to be easy enough since the determinant of the information matrix can be represented explicitly. As for analytical characterizations, they are based on the idea suggested by Stieltjes (Karlin and Studden (1966)). Stieltjes considered the problem of maximizing Vandermonde determinant multiplied by a weight function. He introduced a differential equation of the second order for the polynomial having roots in the design points. This equation proves to have a unique solution to be one of the classical orthogonal polynomials.

This approach allows one to obtain an analytical characterization of locally D-optimal designs in some special cases. (A review of previous results on this matter and some novel ones can be found in Dette, Haines, and Imhof (1999)). In this chapter, we develop this approach into an algebraic approach.

However, the main purpose of the present chapter is to apply the theory developed in Chapter 2. It proves that an optimal design function under a mild restriction on the parameters is uniquely determined and it is a real analytic vector function. Thus, it can be presented by a Taylor series to be constructed through Theorem 2.4.4.

The chapter is organized in the following way. Section 5.2 is devoted to the description of rational models. Two basic representations are introduced: rational models in the form of ratio of two polynomials and such models as algebraic sums of simplest fractions. Section 5.3 contains the study of the number of points in a locally D-optimal design. It proves that under our basic assumptions, described in Section 5.2, this number coincides with the number of parameters. This result is a slight modification of Theorem 5 in He, Studden, and Sun (1996). Section 5.4 considers the application of the functional approach to rational models. To find the zero term in the Taylor expansion of optimal design functions, an algebraic approach mentioned earlier is developed in Section 5.5. Here, we also give the full analytical solution of the problem for models presented as simplest fraction or the sum of two such functions. In the last section, an example is introduced. The Taylor expansion is built for the model in the form of an algebraic sum of three simplest fractions. The influence of the number of Taylor coefficients taken into account on the efficiency of the designs is studied numerically. Some results of this chapter were published in Russian [Melas (1999)] and were announced without proof [Melas (2001)].

5.2 Description of the Model

Let $\Theta_1 = (\theta_1, \ldots, \theta_{m-k})^T \in \mathbf{R}^{m-k}$, $\Theta_2 = (\theta_{m-k+1}, \ldots, \theta_m)^T \in \mathbf{R}^k$, $P(x) = P(x, \Theta_1)$, and $Q(x) = Q(x, \Theta_2)$ be polynomials of one variable $x \in \mathbf{R}$:

$$P(x) = \sum_{j=1}^{m-k} \theta_j x^{j-1}, \quad Q(x) = \sum_{j=1}^{k} \theta_{j+m-k} x^{j-1} + x^k. \tag{5.1}$$

Let $\Theta = (\theta_1, \ldots, \theta_m)^T$. Consider the function

$$\eta_1(x, \Theta) = P(x, \Theta_1)/Q(x, \Theta_2). \tag{5.2}$$

Let $\mathfrak{X}$ be some interval, Ω be some given bounded subset in $\mathbf{R}^m$.

Assume that the following conditions are satisfied:

(a) The fraction in the right-hand side of (5.2) is irreducible for $\Theta \in \Omega$.

(b) Polynomial $Q(x, \Theta_2)$ does not vanish at $x \in \mathfrak{X}$, $\Theta \in \Omega$.

(c) $m \geq 2k$.

Consider the case of $\mathfrak{X} = [0, d]$, $d < \infty$. The case of arbitrary intervals can be analyzed in a similar way.

In this chapter, we will analyze the locally D-optimal designs at points $\Theta^{(0)} \in \Omega$ for regression function (5.2) as conditions (a), (b), and (c) hold.

In the following section, we will prove that a locally D-optimal design under these conditions exists, it is unique, and it consists of m points with equal weights. We will also derive a representation for the determinant of the information matrix of a design concentrated at m points.

As some additional condition holds, regression function (5.2) can be represented as an algebraic sum of the simplest fractions.

Consider the condition

(d) Multiplicity of all roots (both real and complex) of the polynomial $Q(x) = Q(x, \Theta_2)$ is equal to 1 for $\Theta \in \Omega$.

Let $\gamma_1 < \gamma_2 < \cdots < \gamma_t$ be real roots of the polynomial $Q(x)$ and $c_1 \pm ib_1, \ldots, c_p \pm ib_p$ be the complex ones $(b_1, \ldots, b_p \neq 0)$, $t + 2p = k$.

It is easy to verify that for $m = 2k + l$ the right-hand side of equality (5.2) can be represented as

$$
\begin{aligned}
\eta_2(x, \tilde{\Theta}) = \sum_{j=1}^{l} \tilde{\theta}_j x^{j-1} + \sum_{j=1}^{t} \frac{\tilde{\theta}_{j+l}}{x - \gamma_j} \\
+ \sum_{j=1}^{p} \left(\tilde{\theta}_{j+l+t} + \tilde{\theta}_{j+l+t+p} x \right) / \left[(x - c_j)^2 + b_j^2 \right],
\end{aligned}
\tag{5.3}
$$

where

$$
\tilde{\Theta} = (\tilde{\theta}_1, \ldots, \tilde{\theta}_{m-k}, \gamma_1, \ldots, \gamma_t, c_1, \ldots, c_p, b_1, \ldots, b_p)^T,
$$

and we also have

$$
\tilde{\theta}_j \neq 0, \, j = l + 1, \ldots, m - k \text{ for } \Theta \in \Omega.
$$

Regression functions (5.2) and (5.3) are equivalent in the following sense.

Lemma 5.2.1 *Let conditions (a), (b), (c), and (d) be satisfied. Then the determinant of the information matrix for regression function (5.3) and any experimental design coincides with such a determinant for regression function (5.2) with a constant precision that does not depend on the design.*

This lemma will be proved in the following section. The case of the roots of more than multiplicity 1 can be considered in a similar way.

Let us also consider the following condition:

(e) $\gamma_j < 0, j = 1, \ldots, t, c_j < 0, j = 1, \ldots, p.$

It validates some additional results.

5.3 The Number of Points

Let ξ be an arbitrary experimental design, concentrated at m points,

$$\xi = \left(\begin{array}{ccc} x_1 & \cdots & x_m \\ \omega_1 & \cdots & \omega_m \end{array} \right). \tag{5.4}$$

Without loss of generality, assume that

$$0 \le x_1 < \cdots < x_m \le d.$$

Consider regression function (5.2).

Lemma 5.3.1 *Let conditions (a) and (b) hold. Then the determinant of the information matrix for designs of form (5.3) and regression function (5.2) assumes the form*

$$\det M(\xi, \Theta) = \prod_{i=1}^{m} \mu_i \left(C \prod_{1 \le i < j \le m} (x_j - x_i) \Big/ \prod_{i=1}^{m} Q^2(x_i) \right)^2, \tag{5.5}$$

where $Q(x) = Q(x, \Theta_2)$, $C = C(\Theta)$ is independent on the design, and $C(\Theta) \ne 0$.

Proof. Note that in this case, the basic functions assume the form

$$f_i(x) = f_i(x, \Theta) = \frac{\partial}{\partial \theta_i} \eta_1(x, \Theta), \quad i = 1, \ldots, m,$$

$$f_1(x) = \frac{1}{Q(x)}, f_2(x) = \frac{x}{Q(x)}, \ldots, f_{m-k}(x) = \frac{x^{m-k-1}}{Q(x)},$$

$$f_{m-k+1}(x) = -\frac{P(x)}{Q^2(x)}, \ldots, f_m(x) = -\frac{x^{k-1}P(x)}{Q^2(x)}.$$

Consider the determinant of the matrix

$$F = (f_j(x_i))_{i,j=1}^{m}.$$

Now, multiply the elements of the first line of matrix F by $Q^2(x_1)$ and represent the determinant of the matrix by Laplace's rule through elements of this line. Since the elements of other lines does not depend on x_1, derive

$$Q^2(x_1) \det F = Q(x_1)R(x_1) + P(x_1)S(x_1), \qquad (5.6)$$

where $R(x)$ is a polynomial of $\leq m-k-1$-st order and $S(x)$ is a polynomial of $\leq k-1$-st order.

Consider function (5.6) as the function of $x = x_1$ for fixed values of $x_2, \ldots, x_m$ and denote it by $L(x)$. Then $L(x_i) = 0, i = 2, \ldots, m$, since for $x_1 = x_i$ $(i = 2, \ldots, m)$, two lines of matrix F coincides with one another.

Let us demonstrate that $\det F \neq 0$. Indeed, if $\det F = 0$, then $L(x_1) = 0$, so

$$Q(x)R(x) + P(x)S(x) \equiv 0.$$

Since $P(x)$ and $Q(x)$ do not have a common divisor, $R(x) \equiv 0$ and $S(x) \equiv 0$. By induction, it is easy to check that this is impossible. So $L(x) \not\equiv 0$, which means that

$$L(x) = \text{const} \prod_{m \geq i > 1} (x - x_i),$$

where const $\neq 0$. Similarly, representing the determinant through elements of other lines, we obtain that $\det F$ has the form

$$C \prod_{1 \leq i < j \leq m} (x_j - x_i) \Big/ \prod_{i=1}^{m} Q^2(x_i),$$

where $C = C(\Theta) \neq 0$ and is independent of the design. Therefore

$$\det M(\xi, \Theta) = \omega_1 \cdots \omega_m \det{}^2 F,$$

implies formula (5.5). ∎

The following statement holds for regression function (5.3).

Lemma 5.3.2 *Let $\gamma_1 < \cdots < \gamma_t$, $(c_i, b_i) \neq (c_j, b_j)$ for $i \neq j$, and $\tilde{\theta}_i \neq 0, i = l+1, \ldots, m-k$. Then the determinant of the information matrix of design (5.3) for regression function (5.2) has form (5.5), where*

$$Q(x) = \prod_{j=1}^{t} (x - \gamma_j) \prod_{j=1}^{p} \left((x - c_j)^2 + b_j^2 \right)$$

and $C = C(\tilde{\Theta}) \neq 0$.

This lemma can be proved similarly to the previous one.

Note that for designs of form (5.4), the statement of Lemma 5.2.1 follows from Lemmas 5.2.1 and 5.3.1. For arbitrary designs, this statement can be proved using the Binet–Cauchy formulas and the above lemmas.

Now, analyze the problem of the number of points of a locally D-optimal design. The following result holds.

Theorem 5.3.1 *If conditions (a), (b), and (c) are satisfied, then a locally D-optimal design for regression function (5.2) at any point $\Theta^{(0)} \in \Omega$ exists, it is unique, and it is concentrated at m points with equal weights. If $m > 2k$, two points of the design coincide with the endpoints of interval $\mathfrak{X}$. If $m = 2k$ and conditions (d) and (e) are also satisfied, then the left endpoint of the interval is one of the design points.*

Proof. The existence of a locally D-optimal design follows from that functions (5.2) and (5.3) are continuously differentiable and the interval $\mathfrak{X}$ is compact.

Let

$$\xi = \left(\begin{array}{ccc} x_1 & \cdots & x_n \\ \omega_1 & \cdots & \omega_n \end{array} \right)$$

be a locally D-optimal design. Without loss of generality, assume that

$$0 \le x_1^* < x_2^* < \cdots < x_n^* \le d.$$

Note that $n \ge m$, since otherwise $\det M(\xi^*, \Theta) = 0$. Set

$$g(x) = f^T(x) M^{-1}(\xi^*, \Theta) f(x),$$

where $f(x) = f(x, \Theta)$ is defined above and Θ is fixed. By the Kiefer–Wolfowitz equivalence theorem (see Section 1.5),

$$g(x) \le m, x \in \mathfrak{X}, \ g(x_i^*) = m, \ i = 1, \dots, m.$$

Since function $g(x)$ is differentiable, that implies

$$g'(x_i^*) = 0, i = 2, \dots, m-1,$$

where if $x_1^* \ne 0$, then $g'(x_1^*) = 0$, and if $x_n^* \ne d$, then $g'(x_n^*) = 0$. Function $\tilde{g}(x) = g(x) - m$ is represented with the sum of polynomials and fractions. Bringing these fractions to a common denominator reveals that the function $\tilde{g}(x)$ has the form

$$\tilde{g}(x) = \tilde{P}(x)/Q^4(x),$$

where $Q(x) = \prod_{i=1}^{k}(x + \theta_{i+l+k})$ and $\tilde{P}(x)$ is a polynomial of degree $\le 2m-2$ for $l > 0$ and degree $2m$ for $l = 0$. Since the function $\tilde{g}(x)$ has at least $2n - 2$ zeros (taking into account the multiplicity), then we have $n = m$, $x_1^* = 0$, and $x_m^* = d$ for $l > 0$. Therefore, for $l > 0$, the function $\tilde{g}(x)$ has the form

$$\text{const } x(x - d) \prod_{i=2}^{m-1} (x - x_i^*)^2 / Q^4(x).$$

Now, let $l = 0$. Since by formula (5.5), $\det M(\xi, \Theta)$ decreases while increasing all design points by the same value, $x_1^* = 0$ and x_1^* is a zero of odd multiplicity of the function $\tilde{g}(x)$. Moreover, for $x \to \infty$ as well as for

$x \to -\infty$ we have $\tilde{P}(x) \sim -(mx^m)^2$. Therefore, the function $\tilde{g}(x)$ has a zero at $c_1 < 0$, zeros of at least second multiplicity at points x_i^*, $i = 2, \ldots, n-1$, and either zero of multiplicity ≥ 2 at $x_n^* < d$ or one zero at $c_2 = d$ and another at $c_3 > d$. Since $\tilde{P}(x)$ is a polynomial of degree $2m$, $n = m$ and the function $\tilde{g}(x)$ has the form

$$\text{const } x(x - c_1)(x - c_2)(x - c_3) \prod_{i=2}^{m-1} (x - x_i^*)^2 / Q^4(x),$$

where $c_1 < 0$ and either $x_m^* = c_2 = d, c_3 > d$, or $c_2 = c_3 = x_m^* < d$. Since the number of design points is equal to the number of parameters, all of the points of a locally D-optimal design are of the same weight. Assume that there exist two different optimal designs ξ_1 and ξ_2. Then by the Equivalence Theorem, design $(\xi_1 + \xi_2)/2$ is optimal. However, this design includes at least $m + 1$ distinct points, which contradicts the above layout. ∎

Lemma 5.3.3 *In the hypothesis of Theorem 5.3.1 for $m = 2k$ and sufficiently large d, inequality $x_m^* < d$ is valid.*

Proof. Let $x_1 = 0 < x_2 < \cdots < x_m$, $\xi = \{x_1, \ldots, x_m, 1/m, \ldots, 1/m\}$, and $\delta = \min_{1 \leq i \leq k} \theta_{i+k}$. Then, by formula (5.5), for $\bar{x}_i = x_i + \delta$,

$$\det M(\xi, \Theta) \leq \text{const} \frac{1}{\delta^m} \frac{1}{m^m x_m \bar{x}_{m-1}^2 \cdots \bar{x}_2^{m-1}}$$

$$= \frac{1}{x_m} w(x_2, \ldots, x_{m-1}),$$

where $w(x_2, \ldots, x_{m-1})$ is a bounded function (i. e., $w(x_2, \ldots, x_{m-1}) \leq C_1$, where C_1 is a constant).

Therefore, $\det M(\xi, \Theta)$ is small at large x_m. Hence, $x_m^* < d$ for sufficiently large d. ∎

5.4 Optimal Design Function

Consider regression function (5.2). Let $\Theta_1 \neq 0$ be fixed and

$$Z_{(1)} = \left\{ z \in \mathbf{R}^k : \text{fraction } \frac{P(x, \Theta_1)}{Q(x, z)} \text{ is irreducible} \right\}.$$

Consider a locally D-optimal design for $\Theta^T = (\Theta_1^T, z^T)$, $z \in Z_{(1)}$. This design exists, it is unique, and it concentrates at m points, as have been demonstrated in Theorem 5.3.1. Let u be the number of support points of this design on the left border of $\mathfrak{X}$, $u = 0$ or 1, $m - s$ be the number of points on the right border of $\mathfrak{X}$, and $m - s = 0$ or 1. Let the values u and $m - s$ be constant at $z \in Z \subset Z_{(1)}$, $x_u = 0$ at $u = 1$, and $x_m = d$ at $m - s = 1$,

$$\tilde{\tau} = (\tilde{\tau}_1, \ldots, \tilde{\tau}_{s-u}) = (x_{1+u}, \ldots, x_s),$$

where $0 < x_{1+u} < cdots < x_s < d$ are arbitrary points. The above results permits one to prove the following theorem.

Theorem 5.4.1 *Let conditions (a), (b), and (c) from Section 5.2 be satisfied. Then an optimal design function for regression function (5.2) at $z \in Z$ is uniquely determined, formed by m coordinates, and it is a real analytic vector function.*

Proof. Consider the function

$$\tilde{\varphi}(\tilde{\tau}, z) = \left(\prod_{1 \le i < j \le m} (x_j - x_i) \Big/ \prod_{i=1}^{m} Q^2(x_i, z) \right)^{2/m}.$$

By Theorem 5.3.1,

$$C(\Theta)\tilde{\varphi}(\tilde{\tau}, z) = \varphi(\tau, z) = (\det M(\xi, z))^{1/m},$$

where $\xi = \{x_1, \ldots, x_m; 1/m, \ldots, 1/m\}$, $\tau = (x_1, \ldots, x_m)$, and $C(\Theta) \ne 0$ for $\Theta_2 \in Z$. Since for any points $0 \le x_1 < x_2 < \cdots < x_m \le d$ the equality $\det M(\xi, z) > 0$, $z \in Z_{(1)}$ obviously holds, the functions $f_i(x, \Theta) = \frac{\partial}{\partial \theta_i} \eta(x, \Theta)$, $i = 1, \ldots, m$, for $\Theta_2 \in Z, \Theta_1 \ne 0$ form a Chebyshev system on $\mathfrak{X}$. Using the explicit formula for the system determinant, it is easy to verify that they form an ET-system of order m.

Let $\tau^*(z)$ be an optimal design function and $\tilde{\tau}^*(z)$ be a function, formed by all unfixed points of the locally D-optimal design (i. e., $\tilde{\tau}_i^* = \tau_{i+u}^*, i = 1, \ldots, s - u$). Since a locally D-optimal design is unique, $\tau^*(z)$ includes m coordinates and both functions are uniquely determined. Thus, assumptions A1-A4 from Section 2.3 are satisfied. By Theorem 2.3.1 for

$$J(\tilde{\tau}, z) = \left(\frac{\partial^2}{\partial \tilde{\tau}_i \partial \tilde{\tau}_j} \tilde{\varphi}(\tilde{\tau}, z) \right)_{i,j=1}^{m},$$
$$J(z) = J(\tilde{\tau}^*(z), z),$$

the inequality $\det J(z) \ne 0$ holds for $z \in Z$ and all unfixed points of a locally D-optimal design are real analytic functions at $z \in Z$. Thus, $\tau^*(z)$ is a real analytic function at $z \in Z$. ∎

Let us state a similar result for regression function (5.2) with $p = 0$.

Theorem 5.4.2 *Let $\tilde{\theta}_i \ne 0, i = 1, \ldots, m - k, \varepsilon > 0, p = 0$, and*

$$Z = \{z \in \mathbf{R}^k : z_i \ne z_j \, (i \ne j), z_i > \varepsilon, i = 1, \ldots, k\},$$
$$\overline{Z} = \{z \in \mathbf{R}^k : z_i > \varepsilon, i = 1, \ldots, k\}.$$

Then, for any $\varepsilon > 0$, the optimal design function for regression function (5.2) at any $z = \Theta_2 \in Z$ exists, is uniquely determined, and consists of m coordinates, one of which vanishes.

For any $m > 2k$ and for $m = 2k$ with the additional condition that d is sufficiently large, it is a real analytic vector function at $z \in Z$. It can be analytically extended to $\overline{Z}$.

This theorem can be proved in a similar way as the previous one.

The following monotony theorem is also valid for regression function (5.2).

Theorem 5.4.3 *Under hypothesis of Theorem 5.4.2 and with $p = 0$, all the unfixed coordinates of an optimal design function decrease in a strictly monotonous manner with respect to any of the arguments $z_1, \ldots, z_k$.*

Proof. By Theorem 2.4.5, it is sufficient to verify that

$$\frac{\partial^2}{\partial \tilde{\tau}_i \partial z_j} \tilde{\varphi}(\tilde{\tau}, z) > 0,$$

$i = 1, \ldots, s - u$, $j = 1, \ldots, k$ and $z \in Z$. A direct calculation gives that the left-hand side of this condition is equal to

$$\frac{1}{(z_j + \tilde{\tau}_i)^2} > 0.$$

■

5.5 Algebraic Approach and Limiting Designs

By Lemmas 5.3.1 and 5.3.2, the problem of finding a locally D-optimal design for regression function (5.2) for $\mathfrak{X} = [0, d]$ can be reduced to finding the maximum of the following function:

$$T(x_1, \ldots, x_m) = \prod_{i=1}^{h} x_i \prod_{1 \le i < j \le m} (x_j - x_i)^2,$$

for $0 \le x_1 < \cdots < x_m \le d$,

$$h(x) = Q^{-4}(x, \Theta_2^{(0)}).$$

This problem for some other functions $h(x)$ and, in particular, for $h(x) = x(d - x)$ was stated and solved by Stieltjes (see, e.g., Karlin and Studden (1966, Chap. X). Further, let us develop a technique for finding locally D-optimal designs based on his idea, described in Section 5.1.

Example 5.5.1 Let $k = 1$ and $l = 0$; that is, regression function takes the form

$$\eta(x) = \frac{\theta_1}{x + \theta_2}, \quad \theta_2 > 0, \theta_1 \neq 0.$$

A direct calculation demonstrates that locally D-optimal design has the form

$$\xi = \begin{pmatrix} 0 & \theta_2 \\ 1/2 & 1/2 \end{pmatrix}, \quad d > \theta_2$$

$$\xi = \begin{pmatrix} 0 & d \\ 1/2 & 1/2 \end{pmatrix}, \quad d \leq \theta_2.$$

Now, let $k \geq 2$. Consider the case $l = 0$ and d sufficiently large. Set $r = k - 1$. Let $\psi = (\psi_0, \dots, \psi_s)^T$, where $\psi_0 = 1$. Consider the polynomial $\psi(x) = \sum_{i=0}^{m-1} \psi_i x^{m-1-i}$. Let $x_2, \dots, x_m$ be the roots of this polynomial. Then

$$\psi(x) = \prod_{i=2}^{m} (x - x_i).$$

It is easy to verify (see also Fedorov (1972)) that the following equality holds:

$$\frac{1}{2} \frac{\psi''(x_i)}{\psi'(x_i)} = \sum_{j \neq i} \frac{1}{x_i - x_j}, \quad i = 2, \dots, m. \tag{5.7}$$

Let $\psi^*(x)$ stand for a polynomial that vanishes at points $x_2^*, \dots, x_m^*$, $0 < x_2^* < \cdots < x_m^*$, which are the points of a locally D-optimal design for some fixed $\Theta_2^{(0)}$. It has been demonstrated above (see Theorem 5.3.1) that $x_1^* = 0$ is the minimal point of such a design and function $T(0, x_2, \dots, x_m) = \text{const} \det M(\xi, \Theta)$,

$$\xi = \begin{pmatrix} 0 & x_1 & \cdots & x_m \\ 1/m & 1/m & \cdots & 1/m \end{pmatrix}$$

has a unique stationary point. Therefore,

$$\frac{\partial}{\partial x_i} T(0, x_2, \dots, x_m) = 0, \quad i = 2, \dots, m$$

for $x_j = x_j^*$, $j = 2, \dots, m$. Using the explicit form of function T, write this equality in the following form:

$$\sum_{j \neq i} \frac{1}{x_i^* - x_j^*} + \frac{1}{x_i^*} - 2 \frac{Q'(x_i^*)}{Q(x_i^*)} = 0, \quad i = 2, \dots, m.$$

Using relation (5.7) for $\psi(x) = \psi^*(x)$, derive

$$\frac{1}{2} \frac{\psi''(x)}{\psi'(x)} + \frac{1}{x} - 2 \frac{Q'(x)}{Q(x)} = 0$$

for $x = x_i^*, i = 2, \ldots, m$. Bringing the left-hand side of the equality to the common denominator, we have

$$\psi''(x)xQ(x) + 2\psi'(x)Q(x) - 4x\psi'(x)Q'(x) = 0 \qquad (5.8)$$

for $x = x_i^*, i = 2, \ldots, m$, and $\psi(x) = \psi^*(x)$. Denote $z_1 = \theta_{m-k+1}^{(0)}, \ldots, z_k = \theta_m^{(0)}$, $z_{k+1} = 1$. Then

$$Q(x) = Q(x, \Theta_2) = \sum_{i=1}^{k+1} z_i x^{i-1}.$$

Denote the left-hand side of equality (5.8) by $R(x, z)$. Differentiating $R(x, z)$ with respect to z_l, $l = 1, \ldots, k+1$, and collecting the similar terms, derive

$$(R(x, z))'_{z_l} = \mathcal{F}^T(x)D_{(l)}\psi,$$

where $\mathcal{F}(x) = (x^{m+r-1}, x^{m+r-2}, \ldots, 1)^T$ and $D_{(l)}$ is the $(m+r) \times m$ matrix, whose elements are defined by the following formula:

$$D_{(l)\nu\mu} = \begin{cases} 0, & \nu - \mu \neq l - 1 \\ (m - \nu)(m + 1 - \nu - 4l), & \nu - \mu = l - 1, \end{cases}$$

$\mu = 1, \ldots, m, \nu = 1, \ldots, m + r, l = 1, \ldots, k + 1$. From this, it follows that

$$R(x, z) = \mathcal{F}^T(x)A(z)\psi,$$

where

$$A(z) = \sum_{l=1}^{k} z_l D_{(l)} + D_{(k+1)}. \qquad (5.9)$$

At the same time, since $R(x, z) = 0$ for $x = x_i^*, i = 2, \ldots, m$, and $\psi = \psi^*$,

$$R(x, z) = \left(\sum_{i=0}^{k-1} \lambda_i x^{k-1-i} \right) \psi^*(x), \qquad (5.10)$$

where $\lambda_0 = \lambda_G^* = (m - 1)(m - 4k)$, $\lambda_1, \ldots, \lambda_{k-1}$ are some real numbers. Represent the left-hand side of equality (5.10) as

$$\mathcal{F}^T(x) \sum_{l=0}^{k-1} \lambda_l E_{(l)} \psi^*,$$

where $E_{(l)}$ is an $(m + r) \times m$ matrix and

$$E_{(l)\nu\mu} = \begin{cases} 0, & \nu - \mu \neq l - 1, \\ 1, & \nu - \mu = l - 1, \end{cases} \quad \mu = 1, \ldots, m, \nu = 1, \ldots, m + r.$$

Thus, vector ψ^*, corresponding to polynomial $\psi^*(x)$, solves

$$\left(A(z) - \sum_{l=0}^{k-1} \lambda_l E_{(l)} \right) \psi = 0, \tag{5.11}$$

where $\lambda_0 = \lambda_0^*$, for some real values of $\lambda_1, \ldots, \lambda_{k-1}$.

Let, vice versa, the vector ψ solve (5.11) for $\lambda_0 = \lambda_0^*$ and some real values of $\lambda_1, \ldots, \lambda_{k-1}$ and let polynomial $\psi(x)$ have the positive roots $0 < \tilde{x}_2 < \cdots < \tilde{x}_m$. Set $v_i = (2\tilde{x}_i Q(\tilde{x}_i)\psi'(\tilde{x}_i))^{-1}, i = 2, \ldots, m$. It is evident that $v_i \neq 0$. Then

$$\frac{\partial}{\partial x_i} T(0, \tilde{x}_2, \ldots, \tilde{x}_m) = \mathcal{F}^T(\tilde{x}_i) A(z) \psi v_i$$

$$= \mathcal{F}^T(\tilde{x}_i) \left(\sum_{l=0}^{k-1} \lambda_l E_{(l)} \right) \psi v_i$$

$$= \left(\sum_{l=0}^{k-1} \lambda_l \tilde{x}_i^{k-1-l} \right) \psi(\tilde{x}_i) v_i = 0,$$

$i = 2, \ldots, m$. Since the stationary point of function $T(0, x_2, \ldots, x_m)$ is unique, we have $\tilde{x}_2 = x_2^*, \ldots, \tilde{x}_m = x_m^*$.

Consider the matrix

$$A(z) - \lambda_0^* E_{(0)} - \sum_{l=1}^{k-1} \lambda_l E_{(l)}.$$

Set $\lambda = (\lambda_1, \ldots, \lambda_{k-1})$. Derive from formula (5.9) that the first line of this matrix is all zeros. Reject this line and denote the matrix obtained with $B(\lambda)$ under any fixed λ. Equation (5.11) takes the form

$$B(\lambda)\psi = 0. \tag{5.12}$$

Let λ be fixed. Denote the elements of matrix $B(\lambda)$ by $b_{ij} = (B(\lambda))_{ij}$, $i = 1, \ldots, m + r - 1$, $j = 1, \ldots, m$. Formula (5.9) implies that

$$b_{ij} = 0, \ j - i > 1,$$
$$b_{\nu\nu+1} = \nu(\nu + 1), \ \nu = 1, \ldots, m - 1,$$

from which it follows that (5.12) has no more than one solution ψ under fixed λ. Let $\psi = \psi_{(\lambda)}$ be a solution of the equation. Then, by recursion, we can derive

$$\psi_0 = 1, \ \psi_\nu = -\sum_{j=1}^{\nu} b_{\nu j}\psi_{j-1}/(\nu + \nu^2), \tag{5.13}$$

$\nu = 1, \ldots, m - 1$. Moreover,

$$b^T_{(m+i-1)} \psi(\lambda) = 0, \ i = 1, \ldots, k - 1,$$

where $b_{(m+i-1)}$ is the $(m + i - 1)$-st line of matrix $B(\lambda)$. Set

$$Q_i(\lambda) = b^T_{(m+i-1)} \psi(\lambda), \ i = 1, \ldots, k - 1.$$

Here, $Q_i(\lambda)$ is a polynomial of no more than the $m - 1$-st degree of $k - 1$ variables $\lambda_1, \ldots, \lambda_{k-1}$.

Now, consider the system of equations

$$Q_1(\lambda) = 0, \ldots, Q_{k-1}(\lambda) = 0 \tag{5.14}$$

for any real vectors $\lambda = (\lambda_1, \ldots, \lambda_{k-1})$. Since (5.11) at some fixed λ (denote it with λ^*) has the solution $\psi = \psi^*$, the system of equations (5.14) has at least one solution. Moreover, vector ψ^* can be determined by formulas (5.13) and $\psi^* = \psi_{(\lambda^*)}$. If $\tilde{\lambda}$ is a solution of the system of equations (5.14), then, evidently, the pair $(\tilde{\lambda}, \psi_{(\tilde{\lambda})})$ solves (5.11).

Let $\tilde{x}^*_m$ be the point of the locally optimal design for $l = 0$, $\mathfrak{X} = [0, \infty]$. Consider now the case $l > 0$ or $l = 0$ and $d < \tilde{x}^*_m$.

Consider the equation

$$\mathcal{F}^T(x) A(z) \psi = \psi''(x) Q(x) x(x - d)$$
$$+ 2(2x - d) \psi'(x) Q(x)$$
$$- 4x(x - d) \psi'(x) Q'(x).$$

The explicit form of matrix A can be found in a similar way, as in the previous case. The layout remains the same.

Denote $\kappa = m - 1$ for $l = 0$ and $d > \tilde{x}^*_m$, $\kappa = m - 2$ for $l = 0$, $d < \tilde{x}^*_m$ and for $l > 0$.

Because of the above arguments, the following theorem has been proven.

Theorem 5.5.1 *Let conditions* (a), (b), *and* (c) *be satisfied and conditions* (d) *and* (e) *be satisfied for* $m = 2k$. *Then* (5.11) *with respect to vectors* λ *and* ψ *has the unique solution* $(\tilde{\lambda}, \tilde{\psi})$, *such that the polynomial* $\tilde{\psi}(x)$ *has exactly* κ *roots on* $[0, d]$. *Moreover, the system of equations* (5.14) *has the unique solution* $\overline{\lambda}$, *such that polynomial* $\psi_{(\overline{\lambda})}(x)$ *has* κ *roots on* $[0, d]$. *Here,* $\overline{\lambda} = \tilde{\lambda}$, $\psi_{(\overline{\lambda})}(x) = \tilde{\psi}(x) = \psi^*(x)$, *and the mentioned roots are support points of the locally D-optimal design for regression function* (5.1).

Calculating locally D-optimal design by solving (5.11) or the system of equations (5.14) will be called *the algebraic approach*.

In some cases with $r = 1$, the algebraic approach provides finding locally D-optimal design in an explicit form. Consider the following example.

Example 5.5.2 Let $k = 2$ and $l = 0$. Consider the regression function of the form

$$\eta(x, \Theta) = \frac{\theta_1}{x + \theta_3} + \frac{\theta_2}{x + \theta_4},$$

$\theta_1, \theta_2 \neq 0$, $\theta_3, \theta_4 > 0$, $x \in \mathfrak{X} = [0, d]$, and d is sufficiently large. In this case $m - 1 = 3$ and $r = k - 1 = 1$. Set $z_1 = \theta_3$, $z_2 = \theta_4$, and $\Delta = z_1 + z_2$. At first, consider $z_1 z_2 = 1$. Matrix $B_{(1)}$ has the form

$$\begin{pmatrix} -\lambda & 2 & 0 & 0 \\ 12 & -2\Delta - \lambda & 6 & 0 \\ 0 & 6 & -2\Delta - \lambda & 12 \\ 0 & 0 & 2 & -\lambda \end{pmatrix};$$

$\det B_{(1)} = (\lambda^2 - 2\Delta\lambda - 24)^2 - 36\lambda^2$, $\lambda_0 = -12$. From this, it follows that $\psi_1 = \lambda/2$, $\psi_2 = ((\lambda + 2\Delta)\lambda/2 - 12)/6$, and $\psi_3 = (-(\lambda + 2\Delta)\psi_2 - 6\lambda/2)/12$, $\psi_3 = 2\psi_2/\lambda$. Derive from these equalities that $\psi_2 = \pm\lambda/2$ and $\psi_3 = \pm 1$. Since $\psi(x)$ has only the positive zeros, $\psi_1 < 0$ and $\psi_2 > 0$; hence, $\lambda < 0$, $\psi_2 = -\lambda/2$, and $\lambda^2 + (2\Delta + 6)\lambda - 24 = 0$. Therefore,

$$\lambda^* = -\Delta - 3 - \sqrt{(\Delta + 3)^2 + 24}$$

is the unique solution of $\det B_{(1)} = 0$ at which all the zeros of the corresponding polynomial $\psi(x)$ are positive numbers, and $\psi(x)$ assumes the form

$$x^3 + \frac{\lambda^*}{2} x^2 - \frac{\lambda^*}{2} x - 1 = (x - 1)\left(x^2 + \left(1 + \frac{\lambda^*}{2}\right)x + 1\right).$$

Therefore,

$$x_3^* = 1, \quad x_{2,4}^* = \frac{1}{2}\left(-\frac{\lambda^*}{2} - 1 \mp \sqrt{\left(\frac{\lambda^*}{2} + 1\right)^2 - 4}\right).$$

For arbitrary z_1 and z_2, derive by the linear transformation of the model

$$x_3^* = \sqrt{z_1 z_2}, \quad x_{2,4}^* = \frac{\sqrt{z_1 z_2}}{2}\left(-\frac{\lambda^*}{2} - 1 \mp \sqrt{\left(\frac{\lambda^*}{2} + 1\right)^2 - 4}\right).$$

Locally D-optimal design has the form

$$\xi = \begin{pmatrix} 0 & x_2^* & x_3^* & x_4^* \\ 1/4 & 1/4 & 1/4 & 1/4 \end{pmatrix}$$

For arbitrary $k \geq 1$ and m, the solution of (5.11) with the needed properties can be found by means of Maple software.

Deriving explicit analytic expressions for the points of a locally D-optimal design by means of the above algebraic approach seems to be impossible for $k \geq 3$. Nevertheless, in this case the functional approach provides expanding the design points into the Taylor series with respect to the powers of $z_1, \ldots, z_k$. In this view, an expression for the points of the design at some specially selected point $z_{(0)}$ is needed.

Note that for $z \to z_\alpha = (\alpha, \ldots, \alpha)^T$ we have $\det M(\xi, z) \to 0$. However, the vector $\tilde{\tau}$, composed by the design's points that are distinct from the ends of the interval $[0, d]$, can be evaluated for $z_\alpha = (\alpha, \ldots, \alpha)^T$ and its coordinates are the limits of the locally D-optimal design for $z \to z_\alpha$. Set $\alpha = 1$. At first, consider $l = 0$ and sufficiently large d. In this case, we have the following equation for function $\psi(x)$, which has been defined earlier:

$$\psi''(x)x(x+1) + 2\psi'(x)(x(1-2k)+1) = \lambda_0 \psi(x), \qquad (5.15)$$
$$\lambda_0 = (m-1)(m-2) + 2(m-1)(1-2k).$$

Equating the coefficients to one another under the same degrees of x in the left- and right-hand sides, derive

$$\psi_1 = \frac{m(m-1)}{2(m-2k-1)},$$
$$\psi_{\nu+1} = \psi_\nu \frac{(m-\nu)(m-\nu-1)}{\nu(2m-\nu-4k-1)}, \qquad (5.16)$$
$$\nu = 1, \ldots, m-2.$$

Thus, the following theorem is valid.

Theorem 5.5.2 *For $m = 2k$ and $\mathfrak{X} = [0, d]$, where d is sufficiently large, nonzero points of the locally D-optimal design for regression function (5.3) with $z \to (1, \ldots, 1)^T$ converge to the zeros of the polynomial $\psi(x)$, whose coefficients can be calculated by formulas (5.16) for any $k \geq 1$.*

Demonstrate the result of Theorem 5.5.2 by applying it to the following example.

Example 5.5.3 Let $k = 3$, $m = 6$, the regression function be of the form

$$\eta(x, \Theta) = \frac{\theta_1}{x + \theta_4} + \frac{\theta_2}{x + \theta_5} + \frac{\theta_3}{x + \theta_6},$$

$\theta_1, \theta_2, \theta_3 \neq 0$, $\theta_4, \theta_5, \theta_6 > 0$, $x \in \mathfrak{X} = [0, d]$, and d sufficiently large. Let $z_i = \theta_{i+3} \to 1$, $i = 1, 2, 3$. Applying Theorem 5.5.2, one obtains

$$\psi(x) = x^5 - 15x^4 + 50x^3 - 50x^2 + 15x - 1 =$$
$$= (x-1)(x^4 - 14x^3 + 36x^2 - 14x + 1)$$
$$= (x-1)(x^2 + (-7 + \sqrt{15})x + 1)(x^2 + (-7 - \sqrt{15})x + 1),$$

from which it follows that

$$x_2^* = (7 + \sqrt{15} - \sqrt{60 + 14\sqrt{15}})/2 \approx 0.0927,$$

$$x_3^* = (7 - \sqrt{15} - \sqrt{60 - 14\sqrt{15}})/2 \approx 0.3616,$$

$$x_4^* = 1,$$

$$x_5^* = (7 - \sqrt{15} + \sqrt{60 - 14\sqrt{15}})/2 \approx 2.765,$$

$$x_6^* = (7 + \sqrt{15} + \sqrt{60 + 14\sqrt{15}})/2 \approx 10.78.$$

In the following section we will use this design to find the Taylor expansion of the optimal design's points.

Now, consider the case that either $d > 0$ is arbitrary and $l > 0$ or that $l = 0$ and d is sufficiently small. By Theorem 5.3.1, in both cases points x_1^* and x_m^* have the form $x_1^* = 0, x_m^* = d$. Let $\psi(x) = \prod_{i=2}^{m-1}(x - x_i^*) = \sum_{i=0}^{m-2} \psi_i x^{m-2-i}$, $\psi_0 = 1$. Note, that for some $\lambda \in \mathbb{R}$ this function satisfies

$$\psi''(x)x(x - d)(x + 1) + 2\psi'(x)((2x - d) - x(x - d)2k)) =$$
$$= (\lambda_0 x + \lambda)\psi(x), \qquad (5.17)$$
$$\lambda_0 = (m - 2)(m - 3) + 2(m - 2)(2 - 2k).$$

Let A be a $m \times (m - 1)$ matrix such that the left-hand side of the equation is equal to

$$f^T(x)A\psi, \quad f(x) = (x^{m-1}, \dots, x, 1)^T.$$

Let

$$B = B(\lambda) = A - \lambda_0 E_0 - \lambda E_1, \quad B_{(1)} = B_-,$$

where "$-$"means rejecting the first line of a matrix; matrices E_0 and E_1 were introduced earlier. The following result can be verified in a manner similar to the proof of Theorem 5.4.2.

Theorem 5.5.3 *For $m = 2k$ and sufficiently small $d > 0$ as well as for $m > 2k$ and arbitrary $d > 0$, there exists the unique solution of*

$$\det B_{(1)}(\lambda) = 0$$

for $\lambda = \lambda^* \in \mathbb{R}$ such that the points of locally D-optimal design that are neither 0 nor d, converge at $z \to (1, \dots, 1)^T$ to the zeros of the polynomial $\psi(x)$, which solves (5.15) for $\lambda = \lambda^*$. Coefficients of this polynomial can be evaluated by recursive formulas (5.16).

Example 5.5.4 Let us consider the regression function

$$\eta(x, \Theta) = \frac{\theta_1}{x + \theta_3} + \frac{\theta_2}{x + \theta_4}$$

for $x \in \mathfrak{X} = [0, d]$, $d < (5 + \sqrt{21})/2$.

In this case,

$$\psi(x) = (x - x_2^*)(x - x_3^*) = x^2 + \psi_1 x + \psi_2.$$

Equation (5.15) takes the form

$$2x(x + 1)(x - d) + 2(2x + \psi_1)(-2x^2 + (1 + 4d)x - d)$$
$$= (\lambda_0 x + \lambda)(x^2 + \psi_1 x + \psi_2), \quad \lambda_0 = -6.$$

Matrix $B_{(1)}$ has the form

$$\begin{pmatrix} 14d + 6 - \lambda & 2d & 0 \\ -6 & 2(1 + 4d) - \lambda & 6 \\ 0 & -2d & -\lambda \end{pmatrix},$$

and

$$\psi_1 = (\lambda - 6 - 14d)/2d, \quad \psi_2 = (6 + 14d - \lambda)/\lambda. \tag{5.18}$$

Note that with $d = 1$,

$$\det B_{(1)} = (\lambda - 10)(\lambda^2 - 20\lambda + 24)$$

and $\lambda^* = 10 + 2\sqrt{19}$ is a unique solution of $\det B_{(1)}(\lambda) = 0$ such that the corresponding polynomial $\psi(x)$ has two roots inside the interval $[0, d]$. Due to the continuity argument from this, it follows that for an arbitrary d, the value λ^* is the maximal positive root of $\det B_{(1)}(\lambda) = 0$, which can be write in the form

$$\lambda^3 - (22d + 8)\lambda^2 + (112d^2 + 100d + 12)\lambda - 12d(14d + 16) = 0.$$

Thus, support points of the optimal design are $x_1^* = 0$, $x_4^* = d$,

$$x_{2,3}^* = -\frac{\psi_1}{2} \mp \sqrt{\frac{\psi_1^2}{4} - \psi_2},$$

where ψ_1 and ψ_2 can be found by formulas (5.18) with $\lambda = \lambda^*$.

5.6 The Taylor Expansion

As it have been proven earlier, the matrix $J = J(z)$ is nonsingular for any $z \in Z$ as well as for $z = z_{(0)} = (1, \ldots, 1)^T$. Moreover, the previous section presents a technique for solving

$$\frac{\partial}{\partial \tilde{\tau}} \tilde{\varphi}(\tilde{\tau}, z) = 0 \tag{5.19}$$

at point $z = z_{(0)}$. Therefore, Theorem 2.4.3 can be used to expand the points of a locally D-optimal design into the Taylor series.

Consider the case $k = 3$,

$$\eta(x, \Theta) = \frac{\theta_1}{x + \theta_4} + \frac{\theta_2}{x + \theta_5} + \frac{\theta_3}{x + \theta_6}, \quad \theta_1, \theta_2, \theta_3 \neq 0,$$

$\theta_4 > \theta_5 > \theta_6 > 0$, $\mathfrak{X} = [0, \infty)$, $(\theta_4 + \theta_5 + \theta_6)/3 = 1$.

In the previous section, we found $\tau_{(0)} \approx (0.09, 0.36, 1, 2.76, 10.78)$.

Let $z_1 = \theta_4 = 1 + \Delta_1 + \Delta_2$, $z_2 = \theta_5 = 1 - \Delta_1$, and $z_3 = \theta_6 = 1 - \Delta_2$. For arbitrary z_1, z_2, and z_3, the design can be obtained by multiplying of the points by $(z_1 + z_2 + z_3)/3$.

Equation (5.19) in the present case assumes the form

$$\frac{1}{x_j} + \sum_{i \neq j} \frac{1}{x_j - x_i} - 2 \left(\frac{1}{x_j + 1 + \Delta_1 + \Delta_2} + \frac{1}{x_j + 1 - \Delta_1} + \frac{1}{x_j + 1 - \Delta_2} \right) = 0,$$

$j = 2, \ldots, 6$. Set $u = \Delta_1 \Delta_2$ and $v = \Delta_1 + \Delta_2$. Rewrite the equation in the form

$$\frac{1}{x_j} + \sum_{i \neq j} \frac{1}{x_j - x_i} - 2 \left(\frac{1}{x_j + 1 + v} + \frac{2 x_j + 2 - v}{x_j^2 + (2 - v) x_j + 1 + u - v} \right) = 0,$$

$j = 2, \ldots, 6$. It follows from this equation that the points of the optimal design are functions of arguments u and v and they can be expanded into the Taylor series with respect to the powers of u and v in a vicinity of point $(0, 0)$:

$$\tau_i(u, v) = \sum_{s_1=0}^{\infty} \sum_{s_2=0}^{\infty} \tau_{i(s_1, s_2)} u^{s_1} v^{s_2}, \tag{5.20}$$

$i = 1, \ldots, 5$. Applying Theorem 2.4.3, calculate the coefficients $\{\tau_{i(s_1, s_2)}\}$.

Results of the seven steps of this algorithm are presented in Table 5.1.

Let $\tau_{<i>} = \tau_{<i>}(u, v)$ be the segment of the series (5.20) containing coefficients with $s_1 + s_2 \leq i$, $i = 1, 2, \ldots$. The efficiency of designs obtained from $\tau_{(0)}$, $\tau_{<i>}$, $i = 1, \ldots, 6$, by adding the point $x_1^* = 0$ is shown in Table 5.2. Under the efficiency, we understand the magnitude

$$[\det M(\xi_\tau, z) / \det M(\xi^*, z)]^{1/m},$$

where ξ^* is a locally optimal design,

$$\xi_\tau = \begin{pmatrix} 0 & \tau_1 & \cdots & \tau_{m-1} \\ 1/m & 1/m & \cdots & 1/m \end{pmatrix}.$$

We see from Table 5.2 that the efficiency of design $\xi_{\tau_{(0)}}$ proves to be very high. This table shows how many terms in the Taylor representation should be used to obtain optimal designs with a desirable precision.

The approach can be applied for arbitrary rational models, described in Section 5.2.

Table 5.1: Coefficients for the rational model, $k = 3$

$j\backslash i$	0	1	2	3	4	5	6	7
	0.0928	0.0449	-0.0155	0.0080	-0.0049	0.0033	-0.0024	0.0018
	0.3616	0.1514	-0.0481	0.0254	-0.0163	0.0116	-0.0088	0.0069
0	1.0000	0.3333	-0.0955	0.0493	-0.0312	0.0220	-0.0166	0.0130
	2.7654	0.6861	-0.1803	0.0900	-0.0558	0.0387	-0.0288	0.0224
	10.7802	1.9661	-0.5087	0.2534	-0.1571	0.1090	-0.0811	0.0632
	0	0.0540	-0.0310	0.0232	-0.0189	0.0160	-0.0139	
	0	0.1577	-0.0961	0.0758	-0.0648	0.0576	-0.0523	
1	0	0.2864	-0.1798	0.1422	-0.1212	0.1073	-0.0972	
	0	0.4991	-0.3166	0.2482	-0.2096	0.1839	-0.1654	
	0	1.3892	-0.8869	0.6968	-0.5893	0.5180	-0.4663	
	-0.0449	0.0310	-0.0386	0.0417	-0.0437	0.0451		
	-0.1514	0.0963	-0.1239	0.1404	-0.1544	0.1670		
2	-0.3333	0.1909	-0.2334	0.2622	-0.2868	0.3088		
	-0.6861	0.3606	-0.4130	0.4542	-0.4907	0.5238		
	-1.9661	1.0174	-1.1592	1.2758	-1.3804	1.4756		
	0	0.0310	-0.0464	0.0635	-0.0811			
	0	0.0961	-0.1517	0.2192	-0.2940			
3	0	0.1798	-0.2844	0.4079	-0.5439			
	0	0.3166	-0.4964	0.7010	-0.9241			
	0	0.8869	-1.3936	1.9704	-2.6011			
	-0.0155	0.0240	-0.0515	0.0873				
	-0.0481	0.0763	-0.1731	0.3088				
4	-0.0955	0.1479	-0.3246	0.5736				
	-0.1803	0.2700	-0.5658	0.9814				
	-0.5087	0.7601	-1.5900	2.7607				
	0	0.0232	-0.0566					
	0	0.0758	-0.1944					
5	0	0.1422	-0.3637					
	0	0.2482	-0.6287					
	0	0.6968	-1.7680					
	-0.0080	0.0197						
	-0.0254	0.0652						
6	-0.0493	0.1250						
	-0.0900	0.2232						
	-0.2534	0.6284						

Table 5.2: The efficiency of designs $\tau_{<i>}$, $i = 0, \ldots, 6$

Δ_1	Δ_2	0	1	2	3	4	5	6
0.1	0.2	0.99	0.99	0.99	0.99	0.99	0.99	1.00
0.0	0.3	1.00	1.00	1.00	1.00	1.00	1.00	1.00
-0.1	0.4	1.00	1.00	1.00	1.00	1.00	1.00	1.00
-0.2	0.5	0.99	1.00	1.00	1.00	1.00	1.00	1.00
0.2	0.3	1.00	0.99	1.00	1.00	1.00	1.00	1.00
0.0	0.5	0.99	0.99	1.00	1.00	1.00	1.00	1.00
-0.2	0.7	0.96	0.98	1.00	1.00	1.00	1.00	1.00
-0.3	0.8	0.92	0.96	1.00	1.00	1.00	1.00	1.00
-0.4	0.9	0.80	0.88	0.97	0.99	0.99	1.00	1.00
0.3	0.4	0.99	0.98	1.00	1.00	1.00	1.00	1.00
0.2	0.5	0.99	0.98	1.00	1.00	1.00	1.00	1.00
0.0	0.7	0.96	0.96	1.00	1.00	1.00	1.00	1.00
-0.1	0.8	0.91	0.92	0.99	0.99	1.00	1.00	1.00
-0.2	0.9	0.79	0.83	0.96	0.98	0.99	0.99	1.00
0.4	0.5	0.97	0.94	1.00	1.00	1.00	1.00	1.00
0.2	0.7	0.94	0.92	1.00	0.99	1.00	1.00	1.00
0.1	0.8	0.90	0.88	0.99	0.99	1.00	1.00	1.00
0.0	0.9	0.78	0.78	0.96	0.96	0.99	0.99	1.00
0.5	0.6	0.93	0.88	1.00	0.98	1.00	1.00	1.00
0.4	0.7	0.91	0.86	0.99	0.98	1.00	1.00	1.00
0.3	0.8	0.87	0.82	0.99	0.97	1.00	0.99	1.00
0.2	0.9	0.75	0.70	0.95	0.92	0.98	0.97	0.99
0.6	0.7	0.85	0.76	0.99	0.94	1.00	0.97	1.00
0.5	0.8	0.81	0.72	0.98	0.92	1.00	0.96	1.00
0.4	0.9	0.70	0.61	0.93	0.85	0.98	0.92	1.00
0.7	0.8	0.71	0.57	0.95	0.81	1.00	0.87	0.98
0.6	0.9	0.61	0.49	0.90	0.74	0.98	0.80	0.99
0.7	0.9	0.55	0.41	0.86	0.64	0.98	0.69	0.94
0.8	0.9	0.45	0.31	0.80	0.51	0.98	0.53	0.42

Chapter 6

D-Optimal Designs for Exponential Models

In the present chapter, we will analyze the behavior of the locally D-optimal designs for the regression functions of the following kind:

$$\eta(x, \Theta) = \sum_{j=1}^{j_0} \theta_{0j} x^{j-1} + \sum_{i=1}^{k} \sum_{j=1}^{j_i} \theta_{ij} x^{j-1} e^{-\theta_{s+i} x}, \qquad (6.1)$$

where $\Theta = (\theta_1, \ldots, \theta_m)^T$ is the vector of parameters to be estimated, $\theta_1 = \theta_{01}, \ldots, \theta_s = \theta_{kj_k}, s = \sum_{i=0}^{k} j_i, m = s + k, x \in \mathfrak{X}_1 = [b, d]$, and b and d are arbitrary real numbers such that $d > b$.

Let us denote the nonlinear parameters of model (6.1) by $\lambda_1 = \theta_{s+1}, \ldots, \lambda_k = \theta_{s+k}$.

If the first sum in the right-hand side of (6.1) vanishes, set $j_0 = 0$. If, moreover, $\lambda_i > 0$, $i = 1, \ldots, k$, consider also the set $\mathfrak{X} = \mathfrak{X}_2 = [b, \infty)$. Without loss of generality, we can assume $\lambda_1 > \lambda_2 > \cdots > \lambda_k$. Let us also consider the particular form of function (6.1) at $j_0 = 0, j_i = 1$:

$$\eta(x, \Theta) = \sum_{i=1}^{k} \theta_i e^{-\theta_{k+i} x}. \qquad (6.2)$$

To simplify notation, let us analyze mainly the regression function (6.2), since the layout for model (6.1) is very similar.

Functions of types (6.1) and (6.2) form an important class of solutions of linear differential equations, so they are widespread in practice.

These regression functions were analyzed in Chapter 2. It has already been noted that a locally D-optimal design for such a function depends only on nonlinear parameters $\lambda_1, \ldots, \lambda_k$. Now, we will present a more careful formulation and a proof of some results announced in Section 2.2.

In Section 6.1, the problem of the number of support points of a locally D-optimal design is considered. It is proved that in many cases, this number coincides with the number of the parameters. For this reason, the following investigations are performed for designs with such number of support points.

Section 6.2 is devoted to studying optimal design functions, defined in Section 2.3. It will be shown that under mild restrictions, these functions are uniquely determined and are real analytical vector functions monotonously depending on each parameter. Results of this section were previously obtained in Melas (1978) and presented in more detail in Ermakov and Melas (1995, Chap. 5). However, here they are derived from the general theory of Chapter 2 and in a slightly more general form.

In Section 6.3, the Taylor expansions are constructed for optimal design functions for model (6.2) with $k = 3$. Tables of coefficients of these series are given and the figures illustrating the behavior of the design functions are built. The influence of the number of coefficients used for constructing designs on their efficiency is numerically studied.

Remember that in Section 2.2, maximin efficient D-optimal designs for the regression function (6.2) were already constructed and studied.

6.1   The Number of Support Points

The existence of a locally D-optimal design for model (6.1) at $x \in \mathfrak{X}_1 = [\omega, d]$ is implied by the continuity of functions $\frac{\partial}{\partial \theta_i} \eta(x, \theta), i = 1, \ldots, m$, and the compactness of the interval. For $l_0 = 0$ and $\lambda_k > 0$, we can verify (see below) that there exists a value d' such that a locally D-optimal design for $\mathfrak{X} = [b, d]$ is independent of d at $d > d'$. So, the case $\mathfrak{X} = \mathfrak{X} = [b, \infty)$ can be reduced to $\mathfrak{X} = \mathfrak{X}_1 = [b, d]$.

Let

$$\xi = \left(\begin{array}{ccc} x_1 & \cdots & x_{n^*} \\ \omega_1 & \cdots & \omega_{n^*} \end{array} \right)$$

be an arbitrary locally D-optimal design. For $k > 2$, let $\hat{\lambda} = (\hat{\lambda}_1, \ldots, \hat{\lambda}_k)^T$ be any vector such that $\hat{\lambda}_1 > \cdots > \hat{\lambda}_k$, $\hat{\lambda}_{i+1} = (\hat{\lambda}_i + \hat{\lambda}_{i+2})/2$, $i = 1, \ldots, k-2$.

Theorem 6.1.1 *For regression function (6.1), the number of points n^* of any locally D-optimal design satisfies the inequalities*

$$m \le n^* \le m - 2k + k(k+1)/2 + 1.$$

Moreover, if $k > 2$, then $n^* = m$ for the vectors λ, lying inside a vicinity of a point of $\hat{\lambda}$ type.

Proof.   To simplify notation, consider the case $j_0 = 0$, $j_1 = \cdots = j_k = 1$, and also $m = 2k$. The general case can be analyzed in a similar way.

We begin the proof with a study of the corresponding D-optimal design problem in the linear regression model

$$\sum_{i=1}^{2k} \beta_i e^{-\tilde{\lambda}_i x}, \tag{6.3}$$

where $0 < \tilde{\lambda}_1 < \cdots < \tilde{\lambda}_{2k}$ are fixed known values and $\beta_1, \ldots, \beta_{2k}$ are the unknown parameters to be estimated. It is easy to see that for a design with masses $\mu_1, \ldots, \mu_n$ at the points $x_1, \ldots, x_n$ $(n \geq 2k)$, the information matrix in this model is of the form

$$A(\xi, \tilde{\lambda}) = \left(\sum_{s=1}^{n} e^{-\tilde{\lambda}_i x_s} e^{-\tilde{\lambda}_j x_s} \omega_s \right)_{i,j=1}^{2k}. \tag{6.4}$$

In the following, we investigate the maximum of $\det A(\xi, \tilde{\lambda})$, where the components of the vector $\tilde{\lambda} = (\tilde{\lambda}_1, \ldots, \tilde{\lambda}_{2k})^T$ are defined by

$$\tilde{\lambda}_{2i-1} = \lambda_i, \tilde{\lambda}_{2i} = \lambda_i + \Delta, 0 < \Delta < \min_{i=1,\ldots,k-1}(\lambda_{i+1} - \lambda_i), i = 1, \ldots, k, \tag{6.5}$$

where $0 < \lambda_1 < \cdots < \lambda_k$ (in the case $k = 1$, the value $\Delta > 0$ can be chosen arbitrarily). In the following, let

$$\xi^* = \operatorname{argmax} \det A(\xi, \tilde{\lambda})$$

denote a design maximizing the determinant, where the maximum is taken over the set of all approximate designs on X. Note that designs maximizing this determinant exist, because the induced design space

$$\left\{ (e^{-\tilde{\lambda}_1 x}, \ldots, e^{-\tilde{\lambda}_{2k} x})^T \mid x \in X \right\}$$

is compact [see Pukelsheim (1993)]. By the well-known Kiefer–Wolfowitz equivalence theorem, we have

$$\max_{x \in \mathfrak{X}} f^T(x) A^{-1}(\xi^*, \tilde{\lambda}) f(x) = 2k,$$

where $f^T(x) = (e^{-\tilde{\lambda}_1 x}, \ldots, e^{-\tilde{\lambda}_{2k} x})$ denotes the vector of regression functions in the model (6.3). It follows from Gantmacher (1998) that any minor of the matrix $(e^{-\tilde{\lambda}_i x_j})_{i,j=1}^{2k}$ with $x_1 > x_2 > \cdots > x_{2k}$ and $\tilde{\lambda}_1 < \tilde{\lambda}_2 < \cdots < \tilde{\lambda}_{2k}$ is positive. Therefore, the Cauchy-Binet formula implies that

$$\operatorname{sign}(A^{-1})_{ij} = (-1)^{i+j}, \tag{6.6}$$

where we use the notation $A = A(\xi^*, \tilde{\lambda})$ for the sake of brevity.

We will need the following auxiliary result.

Lemma 6.1.1 *Consider the functions*

$$\varphi_i(x) = \sum_{j=1}^{t_i} \alpha_{i,j} e^{-\mu_{i,j} x},$$

where t_i are arbitrary integers, $i = 0, \ldots, s$ and $\{\alpha_{ij}, \mu_{ij}\}$ are some real numbers.

Let the following conditions hold:

(i) $\min_{1 \le j \le t_{i+1}} \mu_{i+i,j} > \max_{1 \le i \le t_i} \mu_{ij}, \ i = 0, 1, \ldots, s - 1.$

(ii) $\text{sign} \, \alpha_{i,j} = +1, \ j = 1, \ldots, t_i, \ i = 0, \ldots, s.$

Then the function $\sum_{i=0}^{s} b_i \varphi_i(x)$, where $b_0, \ldots, b_s$ are arbitrary real numbers, has at most s roots counted with their multiplicity.

Proof of Lemma 6.1.1. Denote $\tau = (x_0, \ldots, x_s)$,

$$J(\tau) = \det \left(\varphi_i(x_j) \right)_{i,j=0}^{x}.$$

Using the expansion of the determinant by a line several times, we obtain

$$J(\tau) = \sum_{l_0=1}^{t_0} \cdots \sum_{l_s=1}^{t_s} \left[\left(\prod_{i=0}^{s} \alpha_{il_i} \right) \det \left(e^{-\mu_{jl_j} x_\nu} \right)_{j,\nu=0}^{s} \right].$$

Due to the Chebyshev property of exponential functions (see Karlin and Studden, 1966, Chap. 1), each term on the right-hand side is positive whenever $x_0 > x_1 > \cdots > x_s$. For fact, $(\prod_{i=0}^{s} \alpha_{il_i}) > 0$ due to condition (ii) and $\det \left(e^{-\mu_{jl_j} x_\nu} \right)_{j,\nu=0}^{s} > 0$ due to condition (i) and the result from Gantmacher (1998).

Thus, $J(\tau) > 0$ for arbitrary $x_0 > x_1 > \cdots > x_s$. Moreover, we have for any $\bar{\tau} = (\bar{x}_0, \ldots, \bar{x}_t)$ with $\bar{x}_0 \ge \bar{x}_1 \ge \cdots \ge \bar{x}_t$,

$$\lim_{\tau \to \bar{\tau}} J(\tau) / \prod_{i<j} (x_i - x_j) > 0, \tag{6.7}$$

since

$$\lim_{\tau \to \bar{\tau}} \det \left(e^{-\theta_{i_s} x_j} \right)_{s,j=0}^{t} / \prod_{i<j} (x_i - x_j) > 0$$

if $\theta_{i_0} < \cdots < \theta_{i_s}$.

This property can easily be verified considering the number of the same coordinates in the vector $\bar{x}$.

It is known (see Karlin abd Studden (1966, Chap. 1)), that under conditions $J > 0$ and (6.7), any generalized polynomial of the form $\sum_{i=0}^{t} b_i \varphi_i(t)$ has at most t roots counted with their multiplicity.   ∎

Let us represent the function

$$m - f^T(x)A^{-1}f(x)$$

in the form

$$\sum_{i=0}^{s} b_i\varphi_i(x).$$

Let us define

$$\varphi_0(x) \equiv m,$$

$$\varphi_{l-1}(x) = (-1)^l \sum_{i=1}^{l-1} A_{l-i,i}e^{-(\tilde{\lambda}_i+\tilde{\lambda}_{l-i})x}, \ l = 2, \ldots, 2k,$$

$$\varphi_{l-1}(x) = (-1)^l \sum_{j=1}^{4k-l+1} A_{2k+1-j,l-2k+j-1}e^{-(\tilde{\lambda}_{2k+1-j}+\tilde{\lambda}_{l-2k+j-1})x},$$

$$l = 2k+1, \ldots, 4k,$$

where $A_{i,j} = (A^{-1})_{ij}$.

Consider the cases $k = 1, 2$. Note that the coefficients in the functions are positive since sign $A_{i,j} = (-1)^{i+j}$ (i.e., condition (ii) holds). Moreover, observing the definition of $\tilde{\lambda}$ in (6.5), condition (i) can also easily be verified for $k = 1, 2$.

Now, we have

$$m - f^T(x)A^{-1}(\xi^*, \tilde{\lambda})f(x) = \varphi_0(x) + \sum_{i=1}^{4k-1}(-1)^i\varphi_i(x) := g(x),$$

and from the Equivalence Theorem for the D-optimality criterion, it follows that $g(x) \leq 0$ for all x. This implies for the support points, say $x_1^*, \ldots, x_n^*$, of a design ξ^* maximizing det $A(\xi, \tilde{\lambda})$,

$$g(x_i^*) = 0, \quad i = 1, 2, \ldots, n,$$
$$g'(x_i^*) = 0, \quad i = 2, 3, \ldots, n-1.$$

A careful counting of the multiplicities and an application of Lemma 6.1.1 now show $2n - 2 \leq 4k - 1$, which implies $n = 2k$ in the case $k = 1$ or 2.

In the case $k \geq 3$, the same arguments are applicable for any vector $\tilde{\lambda}$ satisfying (6.5). In fact, for such $\tilde{\lambda}$ and the functions φ_i, $i = 0, \ldots, 4k$ defined above, both conditions of Lemma 6.1.1 can be easily verified. An argument of continuity therefore shows $n^*(\lambda) = 2k$ for the number of supports of a D-optimal design for the model (6.3) with respect to any λ in a neighborhood of the point $\hat{\lambda}$.

For a proof of the second bound in the case $k \geq 3$, we consider an arbitrary point of the form (6.5), say $\tilde{\lambda} = (\tilde{\lambda}_1, \ldots, \tilde{\lambda}_{2k})$, and define $s \leq k(k+1)/2$ as the number of distinct values in the set

$$\{2\lambda_1, \ldots, 2\lambda_k, \lambda_1 + \lambda_2, \ldots, \lambda_1 + \lambda_k, \lambda_2 + \lambda_3, \ldots, \lambda_{k-1} + \lambda_k\}.$$

We denote with $u_1 < \cdots < u_s$ the distinct values from this set arranged by increasing and introduce the functions

$$\tilde{\varphi}_0(x) \equiv m,$$
$$\tilde{\varphi}_1(x) = A_{11}e^{-u_1 x} = A_{11}e^{-2\lambda_1 x},$$
$$\tilde{\varphi}_2(x) = -2A_{12}e^{-(u_1+\Delta)x},$$
$$\tilde{\varphi}_{2l-1}(x) = a_l e^{-(u_l+2\Delta)x} + c_l e^{-u_{l+1} x}, \quad l = 2, \ldots, s,$$
$$\tilde{\varphi}_{2l}(x) = -b_l e^{-(u_l+\Delta)x}, \quad l = 2, \ldots, s,$$
$$\tilde{\varphi}_{2s+1}(x) = a_{s+1}e^{-(u_s+2\Delta)x}.$$

It can be easily checked that that the coefficients a_l, b_l and c_l can be chosen such that the representation

$$f^T(x)A^{-1}(\xi^*, \tilde{\lambda})f(x) = \sum_{i=1}^{2s+1} \tilde{\varphi}_i(x) \tag{6.8}$$

is satisfied. Due to equalities $\text{sign}A_{i,j} = (-1)^{i+j}$, we have $a_l, b_l, c_l > 0$, $l = 1, \ldots, s$. By the same arguments as in the previous paragraph, we obtain for the determinant

$$\tilde{J}(\tau) = \det\left(\tilde{\varphi}_i(x_j)\right)_{i,j=0}^{2s+1},$$

with $x_0 > x_1 > \cdots > x_{2s+1}$, the inequality $\tilde{J}(\tau) > 0$. Moreover, for any vector $\bar{\tau} = (\bar{x}_0, \ldots, \bar{x}_{2s+1})^T$ with components satisfying $\bar{x}_0 \geq \bar{x}_1 \geq \cdots \geq \bar{x}_{2s+1}$, it follows,

$$\lim_{\tau \to \bar{\tau}} \tilde{J}(\tau) / \prod_{j>i}(x_i - x_j) > 0.$$

Due to (6.8) and the Equivalence Theorem for the D-optimality criterion we have for the generalized polynomial,

$$g(x) = \tilde{\varphi}_0(x) - \sum_{i=1}^{2s+1} \tilde{\varphi}_i(x),$$

$$g(x_i^*) = 0, \quad i = 1, 2, \ldots, n,$$

$$g'(x_i^*) = 0, \quad i = 2, 3, \ldots, n-1.$$

Moreover, g has at most $2s+1$ roots counted with corresponding multiplicity. Consequently, $2n - 2 \leq 2s + 1 \leq k(k+1) + 1$, which yields

$$n \leq \frac{k(k+1)}{2} + 1 + 1/2.$$

This proves the assertion of the theorem for the regression model of the form (6.3).

To prove the assertion of the theorem for the regression model (6.2), we consider for an arbitrary approximate design ξ the polynomial

$$q(x) = m - f^T(x)A^{-1}(\xi, \tilde{\lambda})f(x)$$
$$= m - f^T(x)L^T(LA(\xi)L^T)^{-1}Lf(x),$$

(6.9)

where $\tilde{\lambda} = (\tilde{\lambda}_1, \ldots, \tilde{\lambda}_{2k})$ is defined by (6.5) and the $2k \times 2k$ matrix L is given by

$$\begin{pmatrix} Q & 0 & 0 & \cdots & 0 \\ 0 & Q & 0 & \cdots & 0 \\ \vdots & \vdots & \vdots & & \\ 0 & 0 & 0 & \cdots & Q \end{pmatrix},$$

with

$$Q = \begin{pmatrix} 1 & 0 \\ 1/\Delta & -1/\Delta \end{pmatrix}.$$

Note that $\det L = (-1/\Delta)^k \neq 0$ and that $\lim_{\Delta \to 0} f^T(x)L^T$ is equal to

$$\lim_{\Delta \to 0} \left(e^{-\lambda_1 x}, \frac{e^{-(\lambda_1 + \Delta)x} - e^{-\lambda_1 x}}{\Delta}, \ldots, e^{-\lambda_k x}, \frac{e^{-(\lambda_k + \Delta)x} - e^{-\lambda_k x}}{\Delta} \right)$$
$$= \left(e^{-\lambda_1 x}, -xe^{-\lambda_1 x}, \ldots, e^{-\lambda_k x}, -xe^{-\lambda_k x} \right),$$

Consequently, we have for any design ξ,

$$\lim_{\Delta \to 0} LA(\xi, \tilde{\lambda})L^T = M(\xi, \lambda),$$

(6.10)

where the matrix $M(\xi, \lambda)$ is defined in Section 6.1.

If ξ^* denotes a locally D-optimal design for the regression model (6.2) with support points by $x_1^* < \cdots < x_{n^*}^*$, then it follows from (6.9) and (6.10) that

$$m - \tilde{f}^T(x)\tilde{M}^{-1}(\xi^*, \lambda)\tilde{f}(x) = \lim_{\Delta \to 0} m - f^T(x)A^{-1}(\xi^*, \tilde{\lambda})f(x),$$

(6.11)

where the vector $\tilde{f}^T(x)$ corresponds to the gradient in model (6.2) and is defined by

$$\tilde{f}^T(x) = \left(e^{-\lambda_1 x}, -xe^{-\lambda_1 x}, \ldots, e^{-\lambda_k x}, -xe^{-\lambda_k x} \right).$$

By the Equivalence Theorem, the polynomial on the left-hand side has roots $x_1^*, \ldots, x_{n^*}^*$, where $x_2^*, \ldots, x_{n^*-1}^*$ are roots of multiplicity. Consequently, we obtain $2n^* - 1 \leq h$, where h is the number of roots of the polynomial on the right-hand side of (6.11). By the arguments of the first part of the proof, we have $h \leq 4k - 1$ for $k = 1, 2$ and for $k \geq 3$ in a neighborhood of points λ satisfying (6.5). Moreover, we have $h \leq k(k+1)/2$ in general, which completes the proof of the theorem. ∎

By Theorem 6.1.1, we have $n^* = m$ at $k = 1, 2$.

The same is true for $k > 2$, but for a subset of possible values of parameters λ_i. In addition, we performed a numerical study of locally optimal designs for the case $k = 3$ and many distinct values of $\{\lambda_i\}$. It was not found for any case when $n^* > m$. All of this gives the ground to restrict attention for designs with $n = m$. Therefore, in the remainder of this chapter, we will study designs that are locally optimal among the designs with $n = m$. These designs are called locally D-optimal designs with minimal support. It is easy to check that in this case, all of the design points have the $1/m$ weight. The behavior of the design points at changing the parameters will be studied by means of the basic differential equation that has been introduced in Chapter 2.

6.2   Optimal Design Function

First, consider a regression function of form (6.2).

Let $\theta_i \neq 0$, $i = 1, \ldots, k$, $\mathfrak{X} = [b, \infty)$, and

$$Z = \{z = (z_1, \ldots, z_k)^T; z_i \neq z_j \ (i \neq j), z_i > \varepsilon, i = 1, \ldots, k\},$$

$$\bar{Z} = \{z = (z_1, \ldots, z_k)^T; z_i > \varepsilon, i = 1, \ldots, k\},$$

where ε is an arbitrarily small positive number.

Let $\zeta = \{x_1, \ldots, x_m\}$, $x_i \in \mathfrak{X}$, $i = 1, \ldots, m$, be an arbitrary m-points experimental design. Without loss of generality, assume that

$$b \leq x_1 < \cdots < x_m.$$

Determinant of the information matrix of such a design for regression function (6.2) has the form

$$\theta_1^2 \cdots \theta_k^2 \det{}^2 F(\zeta, z),$$

where $z = (\theta_{k+1}, \ldots, \theta_{2k})^T$,

$$F(\zeta, z) = \left(e^{-z_1 x_s}, -x_s e^{-z_1 x_s}, \ldots, e^{z_k x_s}, -x_s e^{-z_k x_s}\right)_{s=1}^{m}.$$

Thus the problem of finding a locally D-optimal design with minimal support transforms to

$$\det{}^2 F(\zeta, z) \to \sup_{\zeta \in \Xi}, \tag{6.12}$$

where

$$\Xi = \{\zeta = \{x_1, \ldots, x_m\}; b \leq x_1 < x_2 < \cdots < x_m < \infty\}.$$

Let $\zeta_l = \{x_{1,l}, x_{2,l}, \ldots, x_{m,l}\}$, $l = 1, 2, \ldots$, be an arbitrary sequence of designs such that $x_{m,l} \to \infty$ with $l \to \infty$. Decomposing the determinant of matrix F by its lowest line (only its elements depend on x_m), we have that the determinant is the sum of values, each of which tends to zero at $l \to \infty$. Therefore,

$$\det F(\zeta_l, z) \to 0$$

for $l \to \infty$ and any fixed $z \in Z$. Thus, the upper bound in problem (6.12) is attained at some design. Since the exponential system of functions is a Chebyshev one, we have

$$\det F(\zeta, z) > 0, \ z \in Z.$$

Moreover, increasing all of the design points by the same value $\Delta > 0$ leads to decreasing the matrix F determinant by $\exp(-2\Delta \sum_i^k z_i)$. Therefore, for any locally D-optimal design with minimal support (LDMS design) on $\mathfrak{X} = [b, \infty)$, we have $x_1 = b$, and all the points of such a design have the form

$$x_1 = b + \tilde{x}_1, \ldots, x_m = b + \tilde{x}_m,$$

where $\{\tilde{x}_1, \ldots, \tilde{x}_m\}$ is the vector of support points of the LDMS design on $\mathfrak{X} = [0, \infty)$. Thus, without loss of generality, we can assume that $\mathfrak{X} = [0, d]$, where d is sufficiently large, and consider only the case $x_1 = 0$.

Thus, problem (6.12) can be reduced to

$$\det F(\zeta, z) \to \max_{\zeta},$$

where the maximum is taken over all of the designs of the form

$$\{\zeta; \zeta = \{0, x_2, \ldots, x_m\}, \ 0 < x_2 < \cdots < x_m < d\}.$$

Introduce the design $\zeta_\tau = \{0, x_2, \ldots, x_m\}$, corresponding to vector $\tau = (\tau_1, \ldots \tau_{m-1}) = (x_2, \ldots, x_m)$. Consider the function

$$\bar{\varphi}(\tau, z) = \det F(\zeta_\tau, z) / \prod_{i<j} (z_i - z_j)^4.$$

Evidently, it is a real analytic function for $z \in Z$.

Expanding the elements of the matrix F in series with respect to the powers of $x_j z_i$ $(j = z, \ldots, m; \ i = 1, \ldots, k)$ we can verify (see also Ermakov and Melas (1995, Chap. 5)) that the function $\bar{\varphi}(\tau, z)$ has the form

$$\bar{\varphi}(\tau, z) = \mathrm{const} \prod_{i<j} (x_j - x_i) \exp\left(-\bar{z}\sum_2^m x_i\right)\left(1 + O\left(\sum_{i=1}^k \delta_i\right)^2\right), \quad (6.13)$$

where $\delta_i = z_i - \bar{z}$, $\bar{z} = \sum_{i=1}^k z_i/k$, and it can be extended to points $z \in \bar{Z} \backslash Z$ in a way to remain analytical at $z \in \bar{Z}$.

Let $\varphi(\tau, z)$ stand for the function $\bar{\varphi}(\tau, z)$, extended as mentioned earler. Since the exponential functions form an ET-system, $\varphi(\tau, z) > 0$ for $z \in \bar{Z}$.

Let $\tau^* = \tau^*(z) = (\tau_1^*, \ldots, \tau_{m-1}^*) = (x_2^*, \ldots, x_m^*)$ be the vector of nonzero points of some LDMS design

$$V = \{\tau = (\tau_1, \ldots, \tau_{m-1}), \, 0 < \tau_1 < \cdots < \tau_{m-1} < d\}.$$

By the necessary extreme conditions, we have

$$\frac{\partial}{\partial \tau} \varphi(\tau, z) = 0$$

for $z \in Z$, and $\tau = \tau^* = \tau^*(z)$.

Let $J = J(z) = J(\tau^*(z), z)$, where

$$J(\tau, z) = \left(\frac{\partial^2}{\partial \tau_i \partial \tau_j} \varphi(\tau, z) \right)_{i,j=1}^{m-1}.$$

Now, we can state the following theorem.

Theorem 6.2.1 *Let* $\mathfrak{X} = [b, \infty)$, $\theta_i \neq 0$, $i = 1, \ldots, k$, $z \in \bar{Z}$. *Then an optimal design function for regression function (6.2) exists, is uniquely determined, and is a real analytic vector function. One of its coordinates coincides with b as the other ones strictly decrease along each of the arguments* $z_i = \theta_{i+1}$, $i = 1, \ldots, k$.

Proof. Consider the system of functions

$$e^{-z_1 x}, -xe^{-z_1 x}, \ldots, e^{-z_k x}, -xe^{-z_k x} \tag{6.14}$$

for $z_i \neq z_j$ ($i \neq j$). At the points of coordinates of the form

$$z_1 = \cdots = z_{j_1} = \bar{z}_1,$$

$$z_{j_1+1} = \cdots = z_{j_2} = \bar{z}_2, \cdots, z_{j_{t-1}+1} = \ldots = z_{j_t} = \bar{z}_t,$$

redefine this system in the following way:

$$e^{-\bar{z}_1 x}, -xe^{-\bar{z}_1 x}, \ldots, (-x)^{2i_1-1} e^{-\bar{z}_1 x}, \ldots,$$
$$e^{-\bar{z}_t x}, -xe^{-\bar{z}_t x}, \ldots, (-x)^{2i_t-1} e^{-\bar{z}_t x}, \tag{6.15}$$

$i_1 = j_1$, $i_2 = j_2 - j_1, \ldots, i_t = j_t - j_{t-1}$. Let $\zeta^* = \zeta^*(z) = \{x_1^*, \ldots, x_m^*\}$ be the maximum point of the determinant of system of functions (6.14) or (6.15). Then $x_1^*(z) = 0$ and

$$\frac{\partial}{\partial \tau} \varphi(\tau, z) = 0 \tag{6.16}$$

for $\tau = (x_2^*, \ldots, x_m^*)$. It has been already demonstrated for the case of $z \in Z$ and can be verified in a similar way for $z \in \bar{Z}$. Since systems of functions (6.14) and (6.15) are ET-systems,

$$\det J(\tau^*(z), z) \neq 0.$$

Let us demonstrate that (6.16) has the only solution at point $z_{(0)} = (\alpha, \ldots, \alpha)$, where $\alpha > 0$.

Lemma 6.2.1 *At the points of form $z_{(0)} = (\alpha, \dots, \alpha)$, $\alpha > 0$ (6.16) has the unique solution, which has the form $\tau = (\tau_1^*, \dots, \tau_{m-1}^*)$, $\tau_i^* = \gamma_i/2\alpha$, where $\gamma_1, \dots, \gamma_{m-1}$ are the zeros of the Laguerre polynomial of $m - 1$-st order with parameter 0.*

Proof. It is known (Karlin and Studden, 1966, Chap. 10.2), that point $(\gamma_1, \dots, \gamma_{m-1})$ is the only stationary point of the function

$$e^{-\bar{\alpha} \sum_{i=1}^{m-1} \tau_i} \prod_{1 \leq i < j \leq m-1} (\tau_j - \tau_i) \prod_{i=1}^{m-1} \tau_i \tag{6.17}$$

for $\bar{\alpha} = 1/2$, $\tau \in V$, and also $\sum \gamma_i = m(m-1)$. By equality (6.13) the function $\varphi(\tau, z)$ has form (6.17) for $z = z_{(0)}$, $\alpha = \bar{\alpha}$.

Thus, the lemma is complete for $\alpha = \bar{\alpha} = 1/2$. In the case of arbitrary α, it can be verified by the direct differentiation that (6.16) holds if and only if $\tau = (\gamma_1, \dots, \gamma_{m-1})/(2\alpha)$. ∎

Since all, the conditions of Theorem 2.3.1 are satisfied, then for any $z \in \hat{Z}$ $\tau^*(z)$ is uniquely determined. Moreover, by the same theorem, $\tau^*(z)$ is a real analytic vector function.

By Theorem 2.4.5, it is sufficient to verify that

$$\frac{\partial^2 \varphi(\tau, z)}{\partial \tau_i \partial z_j} < 0 \tag{6.18}$$

for $i = 1, \dots, m - 1$, $j = 1, \dots, k$ with $z = z_{(0)} = (\alpha, \dots, \alpha)$ to check the monotony of the coordinates of the design function $\tau^*(z)$.

The direct calculation shows that the expression at the left-hand side of (6.18) is equal to -1 for any $i = 1, \dots, m - 1$, $j = 1, \dots, k$. ∎

Now, consider more general regression function (6.1).
Let $\Theta_1 = (\theta_1, \dots, \theta_{m-k})^T$,

$$(\theta_{11}, \dots, \theta_{1j_1}) \neq (0, \dots, 0), \dots, (\theta_{k1}, \dots \theta_{kj_k}) \neq (0, \dots, 0),$$

$$\Theta_2 = (\theta_{m-k+1}, \dots, \theta_m)^T, \ z_1 = \theta_{m-k+1}, \dots, z_k = \theta_m,$$

$$\mathfrak{X} = [b, d],$$

where $b < d$ are arbitrary real numbers.

Let Z be an arbitrary simply connected open set in R^k, such that the number of such points of a saturated locally D-optimal design that coincide with ω or d is fixed.

Theorem 6.2.2 *Under the above conditions, an optimal design function for regression function (6.1) exists, it is uniquely determined, and it is a real analytic vector function. All of its coordinates that do not coincide with b or d strictly decrease along each of the arguments $z_1, \dots, z_k$.*

This theorem can be proved similarly to the previous one.

6.3   Taylor Expansions

Coordinates of an optimal design function can be expanded into the Taylor series with respect to the powers of parameters by means of Theorem 2.4.4. Let the regression function have form (6.2). The write it again for the convenience of the reader,

$$\eta(x, \Theta) = \sum_{i=1}^{k} \theta_i e^{-\theta_{i+k} x},$$

$\theta_i \neq 0$, $i = 1, \ldots, k$, $\theta_{i+k} \neq \theta_{j+k}$ $(i \neq j)$, $i, j = 1, \ldots, k$.

Consider the case $\theta_{i+k} > 0$, $i = 1, \ldots, k$, $\mathfrak{X} = [0, \infty)$.

As it has been already demonstrated, there exists a unique LDMS design (under fixed θ_{i+k}) and this design corresponds to the vector

$$\zeta^* = \{0, x_2^*, \ldots, x_m^*\}, \ 0 < x_2^* < \cdots < x_m^*.$$

Denote $\bar{\theta} = \sum_{i=1}^{k} \theta_{i+k}/k$. Note that multiplying all of the parameters θ_{i+k} by some value implies dividing the design points by the same value. Therefore, without loss of generality, we can assume $\sum_{i=1}^{k} \theta_{i+k}/k = 1$. Denote $z_1 = 1 - \theta_{k+1}/\bar{\theta}, \ldots, z_{k-1} = 1 - \theta_{2k-1}/\bar{\theta}$.

The direct calculation for $k = 1$ demonstrates that $\zeta^* = \{0, 1\}$.

Let $k \geq 2$, $Z = [-1, 1]^{k-1}$,

$$\Psi(z) = \prod_{1 \leq i < j \leq 1} (\theta_{i+k} - \theta_{j+k})^4,$$

$$\tau(z) = (x_2^*(z), \ldots, x_m^*(z)), \ z_{(0)} = (0, \ldots, 0).$$

Expand the vector function $\tau(z)$ into the Taylor series in a vicinity of $z_{(0)}$. Value $\tau_{(0)} = \tau(z_{(0)})$ can be determined by Lemma 6.1.1. The case $k = 2$ was already considered in Section 2.2. Consider the case $k = 3$.

Example 6.3.1 Let $k = 3$,

$$\eta(x, \Theta) = \theta_1 e^{-\theta_4 x} + \theta_2 e^{-\theta_5 x} + \theta_3 e^{-\theta_6 x},$$

$\theta_{i+3} \neq \theta_{j+3}$ $(i \neq j)$, $\theta_1, \theta_2, \theta_3 \neq 0$, $\theta_{i+3} > 0$, $i = 1, 2, 3$, $(\theta_4 + \theta_5 + \theta_6)/3 = 1$. Denote $z_1 = 1 - \theta_5$ and $z_2 = 1 - \theta_6$.

Consider optimal design function $\tau(z) = (x_2^*(z), \ldots, x_6^*(z))$. Expand this function into a Taylor series in a vicinity of point $z_{(0)} = (0, 0)$:

$$x_{i+1}^*(z) = \bar{\tau}_i(z) = \sum_{s_1=0}^{\infty} \sum_{s_2=0}^{\infty} \tau_{i(s_1, s_2)} z_1^{s_1} z_2^{s_2},$$

$i = 1, \ldots, 5$. Applying Theorem 2.4.4 from Chapter 2, tabulate the coefficients $\{\tau_{i(s_1, s_2)}\}$ (Table 6.1). In this table, the block numbered (s_1, s_2) contains coefficient vector $(\tau_{1(s_1, s_2)}, \ldots, \tau_{5(s_1, s_2)})^T$.

Table 6.1: Taylor coefficients for Example 6.3.1

$s_2 \backslash s_1$	0	1	2	3	4	5
	0.3085	0.0000	0.0054	0.0000	0.0002	0.0000
	1.0565	0.0000	0.0638	0.0000	0.0086	0.0000
0	2.3054	0.0000	0.3037	0.0000	0.0896	0.0000
	4.1995	0.0000	1.0078	0.0000	0.5103	0.0000
	7.1301	0.0000	2.9050	0.0000	2.2077	0.0000
		0.0054	0.0027	0.0003	0.0003	
		0.0638	0.0232	0.0172	0.0094	
1		0.3037	0.0396	0.1791	0.0147	
		1.0078	−0.2267	1.0206	−0.5050	
		2.9050	−2.2496	4.4153	−4.3407	
			0.0005	0.0006		
			0.0258	0.0188		
2			0.2687	0.0293		
			1.5310	−1.0101		
			6.6230	−8.6815		

The form of Table 6.1 is because $\tau_{(s_1,s_2)} = \tau_{(s_2,s_1)}$ and Theorem 2.4.4 gives coefficients with $s_1 + s_2 = 1, 2, \ldots$. Thus, in this table, results of the first five steps of the algorithm are given.

The behavior of the saturated locally optimal design points with $z_1 = 0$ is presented at Figure 6.1. This figure shows that the design points are increasing functions of the parameter.

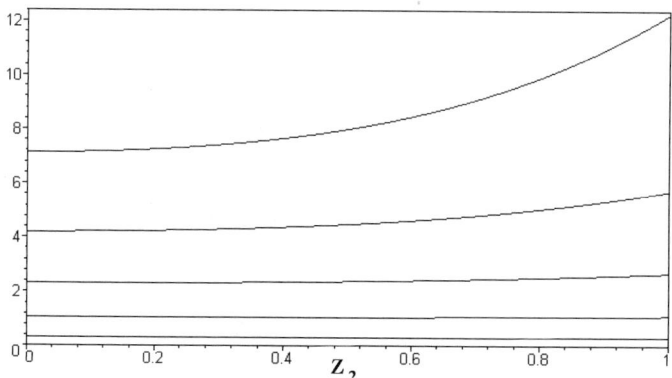

Figure 6.1: Optimal design function with $z_1 = 0$ for Example 6.3.1

Analyze the efficiency of designs that can be obtained from $\tau_{(0)}$, $\tau_{<1>}(z), \ldots, \tau_{<5>}(z)$ by appending them with zero, where $\tau_{<i>}(z)$ is a segment of the Taylor series for function $\tau(z)$, including $1 + i$ nonzero terms of the expansion.

Table 6.2: Efficiency of designs $\tau_{<n>}$

z_1	z_2	I_n					
		0	1	2	3	4	5
0.0	0.4	0.99	0.99	1.00	1.00	1.00	1.00
0.0	0.5	0.98	0.98	1.00	1.00	1.00	1.00
0.0	0.6	0.96	0.96	1.00	1.00	1.00	1.00
0.0	0.7	0.91	0.91	0.99	0.99	1.00	1.00
0.0	0.8	0.83	0.83	0.96	0.96	0.99	0.99
0.0	0.9	0.69	0.69	0.86	0.86	0.92	0.92
−0.2	0.5	0.99	0.99	1.00	1.00	1.00	1.00
−0.2	0.6	0.97	0.97	1.00	1.00	1.00	1.00
−0.2	0.7	0.93	0.93	0.99	0.99	1.00	1.00
−0.2	0.8	0.86	0.86	0.96	0.97	0.98	0.99
−0.2	0.9	0.71	0.71	0.86	0.87	0.92	0.93
−0.4	0.9	0.73	0.73	0.86	0.88	0.91	0.93
0.3	0.4	0.97	0.97	1.00	1.00	1.00	1.00
0.3	0.5	0.94	0.94	1.00	1.00	1.00	1.00
0.3	0.6	0.91	0.91	1.00	0.99	1.00	1.00
0.3	0.7	0.85	0.85	0.99	0.97	1.00	0.99
0.3	0.8	0.75	0.75	0.95	0.93	0.99	0.97
0.3	0.9	0.59	0.59	0.85	0.81	0.94	0.89

Denote

$$I_n = \left[\det M(\xi_{\tau_{<n>(z)}}, \Theta)/\det M(\xi^*, \Theta)\right]^{1/m},$$

$$\xi_{\tau_{<n>}} = \{0, \tau_{<n>}1, \ldots, \tau_{<n>m-1}; 1/m, \ldots, 1/m\}.$$

In Table 6.2, values of I_n, $n = 0, 1, 2, 3, 4, 5$, are given for different values of z_1 and z_2. Table 6.2 demonstrates the high efficiency of the designs. Similar results can be obtained for the case $k > 3$.

An extension of the functional approach for studying maximin efficient D-optimal designs was already described in Section 2.2.

Chapter 7

E- and c-Optimal Designs

In this chapter, we investigate locally E- and c-optimal designs for a wide class of nonlinear regression models. This class includes rational, logistic, and exponential models. A method of asymptotic analysis for such models is introduced. This method allows on to establish that if all nonlinear parameters of the model tend to same limit, then locally E-optimal and most of the locally c-optimal designs tend to the same limiting design, which is e_m-optimal for a heteroskedastic polynomial model. Based on this result, it is demonstrated that in many cases the locally E- and c-optimal designs are supported at the Chebyshev points, defined in Section 1.8. These points proved to be real analytic functions of the nonlinear parameters and these functions can be expanded into a Taylor series by the general technique of Section 2.4. For rational models, the optimal designs are found explicitly in many cases. It is also demonstrated that in the models under consideration, E-optimal designs are usually more efficient for estimating individual coefficients than D-optimal designs. Note that Sections 7.1–7.3, 7.5 and 7.7, are based on Dette, Melas, and Pepelyshev (2004a).

7.1 Introduction

It is the purpose of the present chapter to study locally E-optimal designs for a class of nonlinear regression models, which can be represented in the form

$$Y = \sum_{i=1}^{s} a_i h_i(t) + \sum_{i=1}^{k} a_{s+i}\varphi(t, b_i) + \varepsilon . \qquad (7.1)$$

Here, $h_1, \ldots, h_s$ and φ are given functions, the explanatory variable t varies in an interval $I \subset \mathbb{R}$, ε denotes a random error with mean zero and constant variance, and $a_1, \ldots, a_{s+k}, b_1, \ldots, b_k$ denote the unknown parameters of the model. The consideration of this type of model was motivated by the recent work of Imhof and Studden (2001), who considered a class of rational models of the form

$$Y = \sum_{i=1}^{s} a_i t^{i-1} + \sum_{i=1}^{k} \frac{a_{s+i}}{t - b_i} + \varepsilon, \tag{7.2}$$

where $t \in I, b_i \neq b_j (i \neq j)$, and the parameters $b_i \notin I$ are assumed to be known for all $i = 1, \ldots, k$. Note that model (7.2) is in fact linear, because Imhof and Studden (2001) assumed the b_i to be known. These models are very popular because they have appealing approximation properties (see Petrushev and Popov (1987) for some theoretical properties or Dudzinsky and Mykytowycz (1961), Ratkowsky (1983, p. 120) for an application of this model). In this chapter (in contrast to the work of Imhof and Studden (2001)), the nonlinear parameters in the model (7.1) are not assumed to be known, but also have to be estimated from the data. Moreover, model (7.1) considered here includes numerous other regression functions. For example, in environmental and ecological statistics, exponential models of the form

$$a_1 e^{b_1 t} + a_2 e^{b_2 t}$$

are frequently used in toxicokinetic experiments (see, e.g., Becka and Urfer (1996) or Becka, Bolt, and Urfer (1993)) and this corresponds to the choice $\varphi(t, x) = e^{tx}$ in (7.1). Another popular class of logarithmic models is obtained from (7.1) by the choice $\varphi(t, x) = \log(t - x)$.

Imhof and Studden (2001) studied E-optimal designs for model (7.2) with $s = 1$ under the assumption that the nonlinear parameters $b_1, \ldots, b_k$ are known by the experimenter and do not have to be estimated from the data. In particular they proved that the support of the E-optimal design for estimating a subset of the parameters $a_1, \ldots, a_{\ell+1}$ is given by the Chebyshev points corresponding to the regression functions in model (7.2). These points are the extremal points of the function

$$1 + \sum_{i=1}^{k} \frac{a_i^*}{x - b_i} = p^*(x),$$

in the interval I, which has the smallest deviation from zero; that is,

$$\operatorname*{supp}_{x \in I} |p^*(x)| = \min_{a_2, \ldots, a_{k+1}} \operatorname*{supp}_{x \in I} \left| 1 + \sum_{i=1}^{k} \frac{a_i}{x - b_i} \right|. \tag{7.3}$$

The universality of this solution is due to the fact that any subsystem of the regression functions in model (7.2), which is obtained by deleting one of the basis functions, forms a weak Chebyshev system on the interval I (see Karlin and Studden (1966) and the discussion in Section 7.2). However, in the case where the parameters $b_1, \ldots, b_k$ are unknown and also have to be estimated from the data, the locally optimal design problem for model (7.2) is equivalent to an optimal design problem in the linear regression model

$$Y = \sum_{i=1}^{s} \beta_i t^{i-1} + \sum_{i=1}^{2k} \left(\frac{\beta_{s+2i-1}}{t - b_i} + \frac{\beta_{s+2i}}{(t - b_i)^2} \right) + \varepsilon, \qquad (7.4)$$

for which the corresponding regression function does not satisfy the weak Chebyshev property mentioned above. Nevertheless, we will prove in this chapter that in cases with $k \geq 2$, where the quantity

$$\max_{i \neq j} |b_i - b_j|$$

is sufficiently small, locally E-optimal designs and many locally c-optimal designs for estimating linear combinations of the parameters are still supported on Chebyshev points. This substantially simplifies the construction of locally E-optimal designs. Moreover, we show that this result does not depend on the specific form of models (7.2) and (7.4) but can be established for the general model (7.1) (or its equivalent linearized model). Additionally, it can be shown numerically that in many cases, the E-optimal design is in fact supported on the Chebyshev points for all admissible values of the parameters $b_1, \ldots, b_k$ ($b_i \neq b_j; i \neq j$). Our approach is based on a study of the limiting behavior of the information matrix in model (7.1) in the case where all nonlinear parameters in model (7.1) tend to the same limit. We show that in this case, the locally E-optimal and many locally optimal designs for estimating linear combinations of the coefficients $a_{s+1}, b_{s+1}, \ldots, a_{s+k}, b_{s+k}$ in model (7.1) have the same limiting design. This indicates that E-optimal designs in models of the type (7.1) yield more precise estimates of the individual coefficients than the popular D-optimal designs and we will illustrate this fact in several examples.

The remaining part of the chapter is organized as follows. In Section 7.2*, we introduce the basic concepts and notation, and present some preliminary results. Section 7.3 is devoted to an asymptotic analysis of model (7.1), which is based on a linear transformation introduced in the appendix (see Section 7.7). In Section 7.4, we will establish some analytical properties of support points of locally E-optimal designs considered as functions of the nonlinear parameters for model (7.1). This allows one to apply the general technique of Section 2.4 for representing the support points by the Taylor series. Some applications of general results of Section 7.3 to the rational model (7.2) and its equivalent linear regression model (7.4) are presented in Section 7.5, which extends the results of Imhof and Studden (2001) to the case where the nonlinear parameters in model (7.2) are not known and

---

*Note that in Sections 7.2, 7.3, 7.5 and 7.7 a part of materials (theorems, tables, and figures) are taken from Dette, H., Melas, V.B., Pepelyshev, A. (2004a). Optimal designs for a class of nonlinear regression models. *Ann. Statist.*, **32**(3), 2142–2167. ©2004 Institute of Mathematical Statistics.

have to be estimated from the data. In Section 7.5, we also find locally c- and E-optimal designs for particular rational models (see Examples 7.5.1 - 7.5.4). Section 7.6 introduces similar results for exponential models. This section demonstrates also the usefulness of the functional approach in this case.

7.2 Preliminary Results

Consider the nonlinear regression model (7.1) and define

$$f(t, b) = (f_1(t, b), \ldots, f_m(t, b))^T$$

$$= (h_1(t), \ldots, h_s(t), \varphi(t, b_1), \varphi'(t, b_1), \ldots, \varphi(t, b_k), \varphi'(t, b_k))^T \tag{7.5}$$

as a vector of $m = s + 2k$ regression functions, where the derivatives of the function φ are taken with respect to the second argument. It is straightforward to show that the Fisher information for the parameter $(a_1, \ldots, a_s, a_{s+1}, b_{s+1}, \ldots, a_{s+k}, b_{s+k})^T = (\beta_1, \ldots, \beta_m)^T = \beta$ in the equivalent linear regression model

$$Y = \beta^T f(t, b) + \varepsilon = \sum_{i=1}^{s} \beta_i h_i(t)$$

$$+ \sum_{i=1}^{k} (\beta_{s+2i-1} \varphi(t, b_i) + \beta_{s+2i} \varphi'(t, b_i)) + \varepsilon \tag{7.6}$$

is given by

$$M(\xi, b) = \int f(t, b) f^T(t, b) \xi(dt), \tag{7.7}$$

where $\xi = \begin{pmatrix} x_1 & \cdots & x_n \\ \omega_1 & \cdots & \omega_n \end{pmatrix}$ is an approximate experimental design.

The dependence on the parameter b will be omitted whenever it is clear from the context. Among the numerous optimality criteria proposed in the literature, we consider the D-, E-, and c-optimality criteria in this chapter. A D-optimal design ξ_D^* for the regression model (7.6) maximizes the determinant

$$|M(\xi, b)| \tag{7.8}$$

over the set of all approximate designs on the interval I. Similarly, an E-optimal design ξ_E^* maximizes the minimum eigenvalue

$$\lambda_{\min}(M(\xi, b)), \tag{7.9}$$

whereas for a given vector $c \in \mathbb{R}^m$, a c-optimal design minimizes the expression

$$c^T M^-(\xi, b) c, \tag{7.10}$$

where the minimum is taken over the set of all designs for which the linear combination $c^T \beta$ is estimable (i.e., $c \in \text{range}(M(\xi, b)) \; \forall \; b$).

Note that a locally optimal design problem in a nonlinear model (7.1) corresponds to an optimal design problem in model (7.6) for the transformed vector of parameters $K_a^T b$, where the matrix $K_a \in \mathbb{R}^{m \times m}$ is given by

$$K_a = \text{diag}\Big(\underbrace{1, \ldots, 1}_{s}, \underbrace{1, \frac{1}{a_1}, 1, \ldots, 1, \frac{1}{a_k}}_{2k}\Big). \tag{7.11}$$

For example, a locally D-optimal design in model (7.1) maximizes the determinant

$$|K_a^{-1} M(\xi, b) K_a^{-1}| = |K_a^{-1}|^2 |M(\xi, b)|,$$

does not depend on the parameters $a_1, \ldots, a_k$, and coincides with the D-optimal design in model (7.6). Similarly, the c-optimal design for model (7.1) can be obtained from the $\bar{c}$-optimal design in the model (7.6), where the vector $\bar{c}$ is given by $\bar{c} = K_a c$. Finally, the locally E-optimal design in the nonlinear regression model (7.1) maximizes $\lambda_{\min}(K_a^{-1} M(\xi, b) K_a^{-1})$, where $M(\xi, b)$ is the information matrix in the equivalent linear regression model (7.6). For the sake of transparency, we will mainly concentrate on the linearized version (7.6). The corresponding results in the nonlinear regression model (7.1) will be briefly mentioned, whenever it is necessary.

It is well known (see Studden (1968), Pukelsheim and Studden (1993), Heiligers (1994), or Imhof and Studden (2001) among others and Sections 3.3 and 4.3 of this book) that for many linear regression models, the E- and c-optimal designs are supported at the Chebyshev points.

For a further discussion, assume that the functions $f_1, \ldots, f_m$ generate a Chebyshev system on the interval I with Chebyshev polynomial $c^{*T} f(t)$ and Chebyshev points $s_1, \ldots, s_m$, define the $m \times m$ matrix $F = (f_i(s_j))_{i,j=1}^m$, and consider a vector of weights given by

$$w = (w_1, \ldots, w_m)^T = \frac{JF^{-1}c^*}{\|c^*\|^2}, \tag{7.12}$$

where the matrix J is defined by $J = \text{diag}\{(-1), 1, \ldots, (-1)^m\}$. It is then easy to see that

$$\frac{c^*}{\|c^*\|^2} = FJw = \sum_{j=1}^m f(s_j)(-1)^j w_j \in \partial \mathcal{R}, \tag{7.13}$$

where

$$\mathcal{R} = \text{conv}(f(I) \cup f(-I))$$

denotes the Elfving set [see Elfving (1952)]. Consequently, if all weights in (7.12) are non-negative, it follows from Elfving's theorem that the design

$$\xi_{c^*}^* = \begin{pmatrix} s_1 & \cdots & s_m \\ w_1 & \cdots & w_m \end{pmatrix} \tag{7.14}$$

is c^*-optimal in the regression model (7.6) [see Elfving (1952)], where $c^* \in \mathbb{R}^m$ denotes the vector of coefficients of the Chebyshev polynomial defined in the previous paragraph. The following results relates this design to the E-optimal design.

Lemma 7.2.1 *Assume that $f_1, \ldots, f_m$ generate a Chebyshev system on the interval I such that the Chebyshev points are unique. If the minimum eigenvalue of the information matrix of an E-optimal design has multiplicity 1, then the design $\xi^*_{c^*}$ defined by (7.12) and (7.14) is E-optimal in the regression model (7.6). Moreover, in this case, the E-optimal design is unique.*

Lemma 7.2.2 *Assume that the functions $f_1, \ldots, f_m$ generate a Chebyshev system on the interval I with Chebyshev polynomial $c^{*T} f(t)$ and let $\xi^*_{c^*}$ denote the c^*-optimal design in the regression model (7.6) defined by (7.14). Then c^* is an eigenvector of the information matrix $M(\xi^*_{c^*}, b)$, and if the corresponding eigenvalue $\lambda = \frac{1}{\|c^*\|^2}$ is the minimal eigenvalue, then $\xi^*_{c^*}$ is also E-optimal in the regression model (7.6).*

These lemmas are reformulations of Theorem 3.3.5, parts (a) and (c).

We now discuss the c-optimal design problem in the regression model (7.6) for a general vector $c \in \mathbb{R}^m$ (not necessarily equal to the vector c^* of coefficients of the Chebyshev polynomial). Assume again that $f_1, \ldots, f_m$ generate a Chebyshev system on the interval I. As a candidate for the c-optimal design we consider the measure

$$\xi_c = \xi_c(b) = \begin{pmatrix} s_1 & \cdots & s_m \\ w_1 & \cdots & w_m \end{pmatrix}, \tag{7.15}$$

where the support points are the Chebyshev points and the weights are already chosen such that the expression $c^T M^{-1}(\xi_c, b)c$ becomes minimal; that is,

$$w_i = \frac{|e_i^T J F^{-1} c|}{\sum_{j=1}^m |e_j^T J F^{-1} c|}, \quad i = 1, \ldots, m \tag{7.16}$$

[see Pukelsheim (1993)]. The following result characterizes the optimal designs for estimating the individual coefficients.

Lemma 7.2.3 *Assume that the functions $f_1, \ldots, f_m$ generate a Chebyshev system on the interval I and let $e_j = (0, \ldots, 0, 1, 0, \ldots, 0)^T \in \mathbb{R}^m$ denote the j-th unit vector. The design ξ_{e_j} defined by (7.15) and (7.16) for the vector $c = e_j$ is e_j-optimal if the system*

$$\{f_i \mid i \in \{1, \ldots, m\} \setminus \{j\}\}$$

is a weak Chebyshev system on the interval I.

Proof. If $f_1, \ldots, f_m$ generate a weak Chebyshev system on the interval I, it follows from Theorem 2.1 in Studden (1968) that the design ξ_{e_j} defined in (7.15) and (7.16) is e_j-optimal if

$$\epsilon e_i^T J F^{-1} e_j \geq 0, \quad i = 1, \ldots, m,$$

for some $\epsilon \in \{-1, 1\}$. The assertion of the lemma now follows by Cramer's rule. ∎

Remark 7.2.1 It is worthwhile to mention that, in general, the sufficient condition of Lemma 7.2.3 is not satisfied. To see this, assume that $k \geq 3$, that the function φ is continuously differentiable with respect to the second argument, and that the functions $f_1(\cdot, b), \ldots, f_m(\cdot, b)$ defined by (7.5) generate a Chebyshev system for any b. Define an $(m-1) \times (m-1)$ matrix

$$F_j(x) := \Big(h_1(t_i), \ldots, h_s(t_i), \varphi(t_i, b_1), \varphi'(t_i, b_1), \ldots, \varphi(t_i, b_{j-1}), \varphi'(t_i, b_{j-1}),$$

$$\varphi(t_i, x), \varphi(t_i, b_{j+1}), \ldots, \varphi(t_i, b_k), \varphi'(t_i, b_k) \Big)_{i=1}^{m-1},$$

where $c < t_1 < \cdots < t_{m-1} < d, b_i \neq b_j$ whenever $i \neq j$ and $x \neq b_i$. We choose $t_1, \ldots, t_{m-1}$ such that

$$g(x) = \det F_j(x) \not\equiv 0$$

(note that the functions $f_1, \ldots, f_m$ form a Chebyshev system and, therefore, this is always possible) and observe that

$$g(b_i) = 0, \quad i = 1, \ldots, k; i \neq j.$$

Because $k \geq 3$ and g is continuously differentiable, it follows that there exist two points, say x^* and x^{**}, such that such that $g'(x^*) < 0$ and $g'(x^{**}) > 0$. Consequently, there exists an $\bar{x}$ such that

$$0 = g'(\bar{x}) = \det\Big(f_\nu(t_i, b_{\bar{x}}) \Big)_{i=1,\ldots,m-1}^{\nu=1,\ldots,m,\nu \neq s+2j-1},$$

where the vector $b_{\bar{x}}$ is defined by $b_{\bar{x}} = (b_1, \ldots, b_{j-1}, \bar{x}, b_{j+1}, \ldots, b_k)^T$. Note that the Chebyshev property of the functions $f_1, \ldots, f_{s+2j-2}, f_{s+2j}, \ldots, f_m$ would imply that all corresponding determinants were of the same sign. Therefore the conditions $g'(x^*) < 0, g'(x^{**}) > 0$ imply that there exists a $\tilde{x} \in (x^*, \bar{x})$ or $\tilde{x} \in (\bar{x}, x^{**})$, such that the system of regression functions

$$\Big\{ f_1(t, b_{\tilde{x}}), \ldots, f_{s+2j-2}(t, b_{\tilde{x}}), f_{s+2j}(t, b_{\tilde{x}}), \ldots, f_m(t, b_{\tilde{x}}) \Big\}$$

$$= \Big\{ h_1(t), \ldots, h_s(t), \varphi(t, b_1), \varphi'(t, b_1), \ldots, \varphi'(t, b_{j-1}),$$

$$\varphi'(t, \tilde{x}), \varphi(t, b_{j+1}), \varphi'(t, b_{j+1}), \ldots, \varphi'(t, b_k) \Big\}$$

is not a weak Chebyshev system on the interval I. Finally, in the case $k = 2$, if

$$\lim_{|b| \to \infty} \varphi(t, b) \to 0,$$

it can be shown by a similar argument that there exists an $\tilde{x}$ such that the system

$$\{h_1, (t), \ldots, h_s(t), \varphi(t, b_1)\varphi'(t, b_1)\varphi'(t, \tilde{x})\}$$

is not a Chebyshev system on the interval I.

7.3   Asymptotic Analysis of E- and c-Optimal Designs

Recall the definition of the information matrix in (7.7) for model (7.6) with, design space given by $I = [c_1, d_1]$ and assume that the nonlinear parameters vary in a compact interval, say

$$b_i \in [c_2, d_2]; \quad i = 1, \ldots, k.$$

We are interested in the asymptotic properties of E- and c-optimal designs if

$$b_i = x + \delta r_i, \quad i = 1, \ldots, k, \tag{7.17}$$

for some $x \in [c_2, d_2], \delta > 0, r_1 < r_2 < \cdots < r_k$, and $\delta \to 0$. For this purpose, we study for fixed $\varepsilon, \Delta > 0$ the set

$$\Omega_{\varepsilon, \Delta} = \{b \in \mathbb{R}^k \mid b_i - b_j = \delta(r_i - r_j); \, i, j = 1, \ldots, k;$$

$$\delta \leq \varepsilon; \, b_i \in [c_2, d_2], \, \min_{i \neq j} |r_i - r_j| \geq \Delta\}, \tag{7.18}$$

introduce the functions

$$\bar{f}_i(t, x) = \bar{f}_i(t) = h_i(t), \quad i = 1, \ldots, s,$$

$$\bar{f}_{s+i}(t, x) = \bar{f}_{s+i}(t) = \varphi^{(i-1)}(t, x), \quad i = 1, \ldots, 2k, \tag{7.19}$$

and the corresponding vector of regression functions

$$\bar{f}(t, x) = (\bar{f}_1(t, x), \ldots, \bar{f}_{s+2k}(t, x))^T, \tag{7.20}$$

where the derivatives are taken with respect to the second argument; that is,

$$\varphi^{(i)}(t, x) = \frac{\partial^i}{\partial^i u}\varphi(t, u)\Big|_{u=x}, \quad i = 0, \ldots, 2k - 1.$$

Again, the dependency of the functions $\bar{f}_i$ on the parameter x will be omitted whenever it is clear from the context. The linear model with vector of regression functions given by (7.20) will serve as an approximation for model (7.6) if the parameters b_i are sufficiently close to each other.

Lemma 7.3.1 *Assume that the function*

$$\varphi : [c_1, d_1] \times [c_2, d_2] \to \mathbb{R}$$

in model (7.1) satisfies

$$\varphi \in C^{0,2k-1}([c_1, d_1] \times [c_2, d_2])$$

and that for any fixed $x \in [c_2, d_2]$, the functions $\bar{f}_1, \ldots, \bar{f}_{s+2k}$ defined by (7.19) form a Chebyshev system on the interval $[c_1, d_1]$. For any $\Delta > 0$ and any design on the interval $[c_1, d_1]$ with at least $m = s + 2k$ support points, there exists an $\varepsilon > 0$ such that for all $b \in \Omega_{\varepsilon, \Delta}$, the maximum eigenvalue of the inverse information matrix $M^{-1}(\xi, b)$ defined in (7.20) is simple.

Proof. Recall the definition of the functions in (7.20) and let

$$\bar{M}(\xi, x) = \int_c^d \bar{f}(t, x) \bar{f}^T(t, x) \, d\xi(x) \tag{7.21}$$

denote the information matrix in the corresponding linear regression model. Because of the Chebyshev property of the functions $\bar{f}_1, \ldots, \bar{f}_{s+2k}$, it follows that $|\bar{M}(\xi, x)| \neq 0$ (note that the design ξ has at least $s+2k$ support points). It will be shown in the appendix (see Theorem 7.7.1) that under condition (7.17) with $\delta \to 0$, the asymptotic expansion

$$\delta^{4k-2} M^{-1}(\xi, b) = h \bar{\gamma} \bar{\gamma}^T + o(1) \tag{7.22}$$

is valid, where the vector $\bar{\gamma} = (\bar{\gamma}_1, \ldots, \bar{\gamma}_{s+2k})^T$ is defined by

$$\bar{\gamma}_{s+2i-1} = - \prod_{j \neq i} (r_i - r_j)^{-2} \cdot \sum_{j \neq i} \frac{2}{r_i - r_j}, \quad i = 1, \ldots, k,$$
$$\bar{\gamma}_1 = \cdots = \bar{\gamma}_s = 0; \ \bar{\gamma}_{s+2i} = 0, \ i = 1, \ldots, k, \tag{7.23}$$

and the constant h is given by

$$h = ((2k - 1)!)^2 (\bar{M}^{-1}(\xi, x))_{m,m}. \tag{7.24}$$

From (7.22) it follows that the maximal eigenvalue of the matrix $M^{-1}(\xi, b)$ is simple if δ is sufficiently small.

For a fixed value $r = (r_1, \ldots, r_k)$ and fixed $x \in \mathbb{R}$ in the representation (7.17), denote by $\varepsilon = \varepsilon(x, r)$ the maximal value (possibly ∞) such that the matrix $M^{-1}(\xi, b)$ has a simple maximal eigenvalue for all $\delta \leq \varepsilon$. Then the function $\varepsilon : (x, r) \to \varepsilon(x, r)$ is continuous and the infimum

$$\inf \left\{ \varepsilon(x, b) \Big| x \in [c_1, d_1], \min_{i \neq j} |r_i - r_j| \geq \Delta, \|r\|_2 = 1 \right\}$$

is attained for some $x^* \in [c_1, d_1]$ and r^*, which implies

$$\varepsilon^* = \varepsilon(x^*, r^*) > 0.$$

This means that for any $b \in \Omega_{\varepsilon^*, \Delta}$, the multiplicity of the maximal eigenvalue of the information matrix $M^{-1}(\xi, b)$ is equal to 1. ∎

Theorem 7.3.1 *Assume that the function $\varphi : [c_1, d_1] \times [c_2, d_2] \to \mathbb{R}$ satisfies*

$$\varphi \in C^{0,2k-1}([c_1, d_1] \times [c_2, d_2])$$

and that the systems of functions

$$\{f_1(t, b), \ldots, f_m(t, b)\},$$
$$\{\bar{f}_1(t, x), \ldots, \bar{f}_m(t, x)\}$$

defined by (7.5) and (7.19), respectively, are Chebyshev systems on the interval $[c_1, d_1]$ (for arbitrary but fixed $x, b_1, \ldots, b_k \in [c_2, d_2]$ with $b_i \neq b_j$ whenever $i \neq j$). If ε is sufficiently small, then for any $b \in \Omega_{\varepsilon,\Delta}$ the design $\xi_{c^*}^*$ defined by (7.12) and (7.14) is the unique E-optimal design in the regression model (7.6).

Proof.   The proof is a direct consequence of Lemma 7.2.2 and Lemma 7.3.1, which shows that the multiplicity of the maximum eigenvalue of the inverse information matrix of any design has multiplicity one, if $b \in \Omega_{\varepsilon,\Delta}$ and ε is sufficiently small.   ∎

From Remark 7.3.1 we may expect that, in general, c-optimal designs in the regression model (7.1) are not necessarily supported at the Chebyshev points. Nevertheless, an analog of Lemma 3.1 is available for specific vectors $c \in \mathbb{R}^m$. The proof is similar the proof of Lemma 3.1 and therefore omitted (see also the proof of Theorem 7.3.2, which uses similar arguments).

Lemma 7.3.2 *Let $e_i = (0, \ldots, 0, 1, 0, \ldots, 0)^T$ denote the i-th unit vector in $\mathbb{R}^m$. Under the assumptions of Lemma 3.1 define a vector $\tilde{\gamma} = (0, \ldots, 0, \gamma_1, \ldots, \gamma_{2k}) \in \mathbb{R}^m$ by*

$$\gamma_{2i} = \prod_{j \neq i} (r_i - r_j)^{-2} \quad i = 1, \ldots, k,$$

$$\gamma_{2i-1} = -\gamma_{2i} \sum_{j \neq i} \frac{2}{r_i - r_j} \quad i = 1, \ldots, k. \tag{7.25}$$

(i) *If $c \in \mathbb{R}^m$ satisfies $c^T \tilde{\gamma} \neq 0$, then for any $\Delta > 0$, sufficiently small ε, and any $b \in \Omega_{\varepsilon,\Delta}$, the design $\xi_c(b)$ defined in (7.15) and (7.16) is c-optimal in the regression model (7.6).*

(ii) *The assumption $c^T \tilde{\gamma} \neq 0$ is, in particular, satisfied for the vector $c = e_{s+2j-1}$ for any $j = 1, \ldots k$ and for the vector $c = e_{s+2j}$ for any $j = 1, \ldots, k$, which satisfies, condition*

$$\sum_{\ell \neq j} \frac{1}{r_j - r_\ell} \neq 0. \tag{7.26}$$

Remark 7.3.1 Note that it follows from the proof of Theorem 7.3.1 that the assumption of compactness of the intervals is only required for the existence of the set $\Omega_{\varepsilon,\Delta}$. In other words, if condition (7.17) is satisfied and δ is sufficiently small, the maximum eigenvalue of the matrix $M^{-1}(\xi, b)$ will have multiplicity 1 (independently of the domain of the function φ). The same remark applies to the statement of Theorem 7.3.1 and Lemma 7.3.2.

Our final result of this section shows that under assumption (7.17) with small δ, the locally E- and locally c-optimal designs for the vectors c considered in Lemma 7.3.2 are very close. To be precise, we assume that the assumptions of Theorem 7.3.1 are valid and consider the design

$$\bar{\xi}_c = \bar{\xi}_c(x) = \begin{pmatrix} \bar{s}_1 & \cdots & \bar{s}_m \\ \bar{w}_1 & \cdots & \bar{w}_m \end{pmatrix}, \tag{7.27}$$

where $\bar{s}_1, \ldots, \bar{s}_m$ are the Chebyshev points corresponding to the system $\{\bar{f}_i \mid i = 1, \ldots, m\}$ defined in (7.19),

$$\bar{w}_i = \frac{|e_i^T J \bar{F}^{-1} c|}{\sum_{j=1}^m |e_j^T J \bar{F}^{-1} c|}, \quad i = 1, \ldots, m, \tag{7.28}$$

with $\bar{F} = (f_i(\bar{s}_j))_{i,j=1}^m$ and $c \in \mathbb{R}^m$ is a fixed vector.

Theorem 7.3.2 *Assume that the assumptions of Theorem 7.3.1 are satisfied and that for the system $\{\bar{f}_1, \ldots, \bar{f}_m\}$ the Chebyshev points are unique.*

(i) *If $\delta \to 0$, the design $\xi_{c^*}^*(b)$ defined by (7.14) and (7.12) converges weakly to the design $\bar{\xi}_{e_m}(x)$ defined by (3.10) and (3.11) for $c = e_m$.*

(ii) *If $c \in \mathbb{R}^m$ satisfies $c^T \tilde{\gamma} \neq 0$ for the vector $\tilde{\gamma}$ defined in (7.25) and $\delta \to 0$, then the design $\xi_c^*(b)$ defined by (7.15) and (7.16) converges weakly to the design $\bar{\xi}_{e_m}(x)$.*

(iii) *The assumption $c^T \tilde{\gamma} \neq 0$ is, in particular, satisfied for the vector $c = e_{s+2j-1}$ for any $j = 1, \ldots k$ and for the vector $c = e_{s+2j}$ for any $j = 1, \ldots, k$, which satisfies condition (7.26).*

Proof. It follows from Theorem 7.3.1 that the design $\xi_{c^*}^* = \xi_{c^*}^*(b)$ is locally E-optimal for sufficiently small $\delta > 0$. In other words, if δ is sufficiently small, the design $\xi_{c^*}^*$ minimizes

$$\max_{\|c\|_2=1} c^T M^{-1}(\xi, b) c$$

in the class of all designs. Note that the components of the vector $r = (r_1, \ldots, r_k)$ are ordered, which implies

$$e_{s+2i-1}^T \tilde{\gamma} \neq 0, \quad i = 1, k.$$

Multiplying (7.32) in the appendix with δ^{4k-2}, it then follows from Theorem 7.7.1 in the appendix that for some subsequence $\delta_k \to 0$,

$$\xi_{c^*}^* \to \hat{\xi}(x),$$

where the design $\hat{\xi}(x)$ minimizes the function

$$\max_{\|c\|_2=1} (c^T\tilde{\gamma})^2 e_m^T \bar{M}^{-1}(\xi, x) e_m$$

and the vector $\tilde{\gamma}$ is defined by (7.23). The maximum is attained for $c = \tilde{\gamma}/\|\tilde{\gamma}\|_2$ (independently of the design ξ) and, consequently, $\hat{\xi}(x)$ is e_m-optimal in the linear regression model defined by the regression function in (7.20). Now, the functions $\bar{f}_1, \ldots, \bar{f}_m$ generate a Chebyshev system and the corresponding Chebyshev points are unique, which implies that the e_m-optimal design $\bar{\xi}_{e_m}(x)$ is unique. Consequently, every subsequence of designs $\xi_{c^*}^*(b)$ contains a weakly convergent subsequence with limit $\bar{\xi}_{e_m}(x)$ and this proves the first part of the assertion. For a proof of the second part, we note that a c-optimal design minimizes

$$c^T M^{-1}(\xi, b) c$$

in the class of all designs on the interval I. Now, if $c^T\tilde{\gamma} \neq 0$ and

$$e_{s+2i-1}^T \tilde{\gamma} = -\prod_{j\neq i}(r_i - r_j)^{-2} \sum_{j\neq i} \frac{2}{r_i - r_j} \neq 0$$

for some $i = 1, \ldots, k$, the same argument as in the previous paragraph shows that $\xi_c^*(b)$ converges weakly to the design, which maximizes the function

$$(\tilde{\gamma}^T c)^2 e_m^T \bar{M}^{-1}(\xi, x) e_m.$$

If $e_{s+2i-1}^T\tilde{\gamma} = 0$ for all $i = 1, \ldots, k$, the condition $c^T\tilde{\gamma} \neq 0$ implies $e_{s+2i}^T\tilde{\gamma} \neq 0$ for some $i = 1, \ldots, k$ and the assertion follows by multiplying (7.32) in the appendix with δ^{4k-4} and similar arguments. Finally, the third assertion follows directly from the definition of the vector $\tilde{\gamma}$ in (7.23). ∎

Remark 7.3.2 Note that Theorem 7.3.1, Lemma 7.3.2, and Theorem 7.3.2 remain valid for the locally optimal designs in the nonlinear regression model (7.1). This follows by a careful inspection of the proofs of the previous results. For example, Theorem 7.7.1 in the appendix shows that

$$\delta^{4k-2} K_a M^{-1}(\xi, b) K_a = h(K_a\tilde{\gamma})(K_a\tilde{\gamma})^T + o(1),$$

where the vector $\tilde{\gamma}$ is defined in Lemma 3.3 and, consequently, there exists a set $\Omega_{\varepsilon,\Delta}$ such that for all $b \in \Omega_{\varepsilon,\Delta}$, the maximum eigenvalue of the inverse information matrix in model (7.1) is simple. Similarly, if $\delta \to 0$ and (7.17) is satisfied, c-optimal designs in the nonlinear regression model are given by the design $\xi_{\bar{c}}(b)$ in (7.15) and (7.16) with $\bar{c} = K_a c$ whenever $\tilde{\gamma}^T\bar{c} \neq 0$ and all these designs converge weakly to the e_m-optimal design in the linear regression model defined by the functions (7.20).

We finally remark that Theorem 7.3.2 and Remark 7.3.2 indicate that E-optimal designs are very efficient for estimating the parameters $a_{s+1}, b_1, \ldots, a_{s+k}, b_k$ in the nonlinear regression model (7.1) and the linear model (7.6), because for small differences $|b_i - b_j|$, the E-optimal design and the optimal design for estimating the individual coefficients are close to the optimal design for estimating the coefficient b_k. Therefore, we expect E-optimal designs to be more efficient for estimating these parameters than D-optimal designs. We will illustrate this fact in Section 7.5, which discusses the rational model in more detail.

7.4 Analytical Properties of Optimal Designs

As it was shown in the previous section with $b \in \Omega$, where Ω is an open set in R^k locally E- and c-optimal designs for model (7.7) are located in the Chebyshev points of the function system $f_1(t, b), \ldots, f_m(t, b)$. Also, the corresponding weight coefficients can be found by (7.12) and (7.16), respectively. Thus, it will do to study the Chebyshev points $x_1^*, \ldots, x_m^*$ and coefficients of vector c^* as functions of vector b.

We will show that under some not very restrictive additional conditions, these functions are real analytic and can be expanded into a Taylor series by formulas of Section 2.4. These conditions are certainly satisfied for rational and exponential models.

We will need the following result.

Lemma 7.4.1 *Let the following conditions be satisfied:*

(a) *Function,* $f_1(t) = f_1(t, b), \ldots, f_m(t) = f_m(t, b)$ *are functions of a general form generating a Chebyshev system on the interval I with $b \in B$, where B is an open set in R^k, and such that the Chebyshev points are uniquely determined.*

(b) *Functions $f_i(t, b)$, $i = 1, \ldots, m$, are real analytic in t with $t \in I$ for any $b \in B$.*

(c) *The number (say u_1) of Chebyshev points coinciding with the left bound of I and the number (say u_2) of such points for the right bound of I remains the same with $b \in B$.*

(d) *For any $b \in B$ and for any nonzero vector c, the equation $c^T f'(t) = 0$, where $f(t) = f(t, b) = (f_1(t, b), \ldots, f_m(t, b))^T$, has at most $m-1$ roots and multiple roots are counted twice.*

Then coefficients of the Chebyshev polynomial $c^{*T} f(t, b)$ and the Chebyshev point are real analytic functions of b with $b \in B$.

Proof. Let, for certainly, $I = [d_1, d_2]$, $u_1 = 1$, $u_2 = 0$; that is,

$$d_1 = t_1^*(b) < \cdots < t_m^*(b) < d_2$$

where $t_i^*(b)$, $i = 1, \ldots, m$, are Chebyshev points of the function system $\{f_1(t, b), \ldots, f_n(t, b)\}$. All other cases can be considered in a similar way.

Denote

$$t = (t_2, \ldots, t_m)^T, \quad t^*(b) = (t_2^*(b), \ldots, t_m^*(b))^T,$$

$$c = (c_1, \ldots, c_m)^T,$$

$$\tau = (c^T, t^T, \quad \tau^*(b) = (c^{*T}(b), t^{*T}(b))^T.$$

Due to the definition of the Chebyshev polynomial (see Section 1.8), the equation system

$$c^T f(t_i, b) = (-1)^i, \quad i = 1, \ldots, m, \quad t_1 = d_1,$$

$$c^T f'(t_i, b) = 0, \quad i = 2, \ldots, m,$$

has the solution $c = c^*(b)$, $t = t^*(b)$ and determine, the vector $\tau^*(b)$ as an implicit function of b, $b \in B$.

The Jacobi matrix of this system has the form

$$Q = Q(\tau, b) = \left(\begin{array}{cc} F & 0 \\ \tilde{F} & D \end{array} \right),$$

where

$$F = (f_i(t_j, b))_{i,j=1}^m,$$

$$\tilde{F} = (f_i'(t_j, b))_{i=1, j=2}^m,$$

$$D = \mathrm{diag} \left\{ c^T f''(t_2), \ldots, c^T f''(t_m) \right\}.$$

Assume in the following that b is any fixed vector in B.

Denote

$$J(b) = Q\left(\tau^*(b), b\right).$$

Note that

$$\det Q = \det F \det D$$

and

$$\det F \neq 0$$

for any points $t_1 < t_2 < \cdots < t_m$ due to the definition of the Chebyshev system. Since

$$c^{*T} f'(t_i^m) = 0, \quad i = 2, \ldots, m,$$

due to condition (d) it follows that $c^{*T} f''(t_i^*) \neq 0$, $i = 2, \ldots, m$, and

$$\det D = \prod_{i=2}^m c^{*T} f''(t_i^*) \neq 0.$$

Therefore, $\det J(b) \neq 0$ for $b \in B$.

Now, the proposition of the lemma follows from the Implicit Function Theorem (see Section 1.8). ∎

Taylor expansions for the vector function $\tau^*(b)$ will be constructed for exponential models in Section 7.6. In the next section we will obtain a characterization of the Chebyshev points as roots of some polynomials for rational models.

7.5 Rational Models

In this section, we discuss the rational model (7.2) in more detail, where the design space is a compact or semi-infinite interval I. In contrast to the work of Imhof and Studden (2001), we assume that the nonlinear parameters $b_1, \ldots, b_k \notin I$ are not known by the experimenter but have to be estimated from the data. A typical application of this model can be found in the work of Dudzinski and Mykytowycz (1961), where this model was used to describe the relation between the weight of the dried eye lens of the European rabbit and the age of the animal. In the notation of Sections 7.2 and 7.3 we have $f(t) = f(t, b) = (f_1(t), \ldots, f_m(t))^T$ with

$$f_i(t) = f_i(t, b) = t^{i-1}, \quad i = 1, \ldots, s,$$

$$f_{s+2i-1}(t) = f_{s+2i-1}(t, b) = \frac{1}{t - b_i}, \quad i = 1, \ldots, k, \tag{7.29}$$

$$f_{s+2i}(t) = f_{s+2i}(t, b) = \frac{1}{(t - b_i)^2}, \quad i = 1, \ldots, k,$$

and the equivalent linear regression model is given by (7.4). The corresponding limiting model is determined by the regression functions $\bar{f}(t) = \bar{f}(t, x) = (\bar{f}_1(t, x), \ldots, \bar{f}_m(t, x))^T$, with

$$\bar{f}_i(t) = t^{i-1}, \quad i = 1, \ldots, s,$$

$$\bar{f}_{i+s}(t) = \bar{f}_{s+i}(t, x) = \frac{1}{(t - x)^i}, \quad i = 1, \ldots, 2k. \tag{7.30}$$

Some properties of the functions defined by (7.29) and (7.30) are discussed in the following lemma.

Lemma 7.5.1 *Define*

$$\mathcal{B} = \{b = (b_1, \ldots, b_k)^T \in \mathbb{R}^k \mid b_i \notin I; b_i \neq b_j\};$$

then the following assertions are true;

(i) If I is a finite interval or $I \subset [0, \infty)$ and $b \in \mathcal{B}$, then the system

$$\{f_1(t_1, b), \ldots, f_m(t, b)\}$$

defined in (7.29) is a Chebyshev system on the interval I. If $x \notin I$, then the system

$$\{\bar{f}_1(t, x), \ldots, \bar{f}_m(t, x)\}$$

defined by (7.30) is a Chebyshev system on the interval I.

(ii) Assume that $b \in \mathcal{B}$ and that one of the following conditions is satisfied:

(a) $I \subset [0, \infty)$

(b) $s = 1$ or $s = 0$.

For any $j \in \{1, \ldots, k\}$, the system of regression functions

$$\{f_i(t, b) \mid i = 1, \ldots, m, i \neq s + 2j\}$$

is a Chebyshev system on the interval I.

(iii) If I is a finite interval or $I \subset [0, \infty)$, $k \geq 2$, and $j \in \{1, \ldots, k\}$, then there exists a nonempty set $W_j \subset \mathcal{B}$ such that for all $b \in W_j$, the system of functions

$$\{f_i(t, b) \mid i = 1, \ldots, m; i \neq s + 2j - 1\}$$

is not a Chebyshev system on the interval I.

Proof. Part (iii) follows from Remark 7.2.1. Parts (i) and (ii) are proved similarly and we restrict ourselves to the first case. For this purpose, we introduce the functions $\psi(t, b) = (\psi_1(t, \tilde{b}), \ldots, \psi_m(t, \tilde{b}))^T$, with

$$
\begin{aligned}
\psi_i(t, \tilde{b}) &= t^{i-1}, \quad i = 1, \ldots, s, \\
\psi_{s+i}(t, \tilde{b}) &= \frac{1}{t - \tilde{b}_i}, \quad i = 1, \ldots, 2k,
\end{aligned}
\tag{7.31}
$$

where $\tilde{b} = (\tilde{b}_1, \ldots, \tilde{b}_{2k})^T$ is a fixed vector with pairwise different components. With the notation

$$L(\Delta) = \begin{pmatrix} I_s & 0 \\ 0 & G_k(\Delta) \end{pmatrix} \in \mathbb{R}^{m \times m},$$

$$G_k(\Delta) = \begin{pmatrix} G(\Delta) & & \\ & \ddots & \\ & & G(\Delta) \end{pmatrix} \in \mathbb{R}^{2k \times 2k},$$

$$G(\Delta) = \begin{pmatrix} 1 & 0 \\ -\frac{1}{\Delta} & \frac{1}{\Delta} \end{pmatrix} \in \mathbb{R}^{2 \times 2}$$

(here I_s is the $s \times s$ identity matrix), it is easy to verify that

$$f(t, b) = L(\Delta)\psi(t, \tilde{b}_\Delta) + o(1) , \tag{7.32}$$

where $\tilde{b}_\Delta = (b_1, b_1 + \Delta, \ldots, b_k, b_k + \Delta)^T$. For a fixed vector $T = (t_1, \ldots, t_m)^T \in \mathbb{R}^m$ with ordered components $t_1 < \cdots < t_m$ such that $t_i \in I$ $(i = 1, \ldots, m)$, define the matrices

$$F(T, b) = (f_i(t_j, b))_{i,j=1}^m ,$$

$$\psi(T, \tilde{b}) = (\psi_i(t_j, \tilde{b}))_{i,j=1}^m$$

then we obtain from (7.32),

$$\det F(T, b) = \lim_{\Delta \to 0} \tfrac{1}{\Delta^k} \psi(T, \tilde{b}_\Delta)$$

$$= \frac{\prod_{1 \leq i < j \leq m}(t_j - t_i) \prod_{1 \leq i < j \leq k}(b_i - b_j)^4}{\prod_{i=1}^k \prod_{j=1}^m (t_j - b_i)^2}, \tag{7.33}$$

where the last identity follows from the fact that $\psi(T, \tilde{b})$ is a Cauchy–Vandermonde matrix, which implies

$$\det \psi(T, \tilde{b}) = \frac{\prod_{1 \leq i < j \leq m}(t_j - t_i) \prod_{1 \leq i < j \leq 2k}(\tilde{b}_i - \tilde{b}_j)}{\prod_{i=1}^{2k} \prod_{j=1}^m (t_j - \tilde{b}_i)}.$$

Now for any $b \in \mathcal{B}$ the right hand side does not vanish and is of one sign independently of T. Consequently $\{f_i(t, b) \mid i = 1, \ldots, m\}$ is a Chebyshev system on the interval I. The assertion regarding the system $\{\bar{f}_i(t, x) \mid i = 1, \ldots, m\}$ is proved similarly and therefore left to the reader. ∎

The case $k = 1$ will be studied more explicitly in Examples 7.5.1 and 7.5.2. Note that the third part of Lemma 7.5.1 shows that for $k \geq 2$, the main condition in Theorem 7.5 in the paper of Imhof and Studden (2001) is *not* satisfied in general for the linear regression model with the functions given by (7.29). These authors assumed that every subsystem of $\{f_1, \ldots, f_m\}$ that consists of $m - 1$ of these functions is a weak Chebyshev system on the interval I. Because the design problem for this model is equivalent to the design problem for model (7.2) (where the nonlinear parameters are not known and have to be estimated), it follows that, in general, we cannot expect locally E-optimal designs for the rational model to be supported at the Chebyshev points. However, the linearized regression model (7.4) is a special case of the general model (7.6) with $\varphi(t, b) = (t - b)^{-1}$ and all results of Section 7.3 are applicable here. In particular, we obtain that the E-optimal designs and the optimal designs for estimating the individual coefficients $a_{s+1}, b_1, \ldots, a_{s+k}, b_k$ are supported at the Chebyshev points if the nonlinear parameters $b_1, \ldots, b_k$ are sufficiently close (see Theorem 7.3.1, Lemma 7.3.2, and Remark 7.3.2).

Theorem 7.5.1 *Consider the rational model (7.29) on the interval $[-1, 1]$ with $s \geq 1$ and unknown parameters $a_1, \ldots, a_{s-1}, a_s, b_1, \ldots, a_{s+k}, b_k$.*

(i) *If $s = 1$, then the Chebyshev points $s_1 = s_1(b), \ldots, s_m = s_m(b)$ for the system of regression functions in (7.29) on the interval $[-1, 1]$ are given the roots of the polynomial*

$$(1 - t^2) \sum_{i=0}^{4k} d_i U_{-2k+s+i-2}(t), \tag{7.34}$$

where $U_j(x)$ denotes the j-th Chebyshev polynomial of the second kind (see Szegö (1975)), $U_{-1}(x) = 0, U_{-n}(x) = -U_{n-2}(x)$, and the factors $d_0, \ldots, d_{4k}$ are defined as the coefficients of the polynomial

$$\sum_{i=0}^{4k} d_i t^i = \prod_{i=1}^{k} (t - \tau_i)^4, \tag{7.35}$$

where

$$2b_i = \tau_i + \frac{1}{\tau_i}, \quad i = 1, \ldots, k.$$

(ii) *Let $\Omega_E \subset \mathcal{B}$ denote the set of all b such that an E-optimal design for the model (7.4) is given by (7.14) and (7.12); then $\Omega_E \neq \emptyset$.*

Proof. The second part of the theorem is a direct consequence of Lemma 7.5.1 and Theorem 7.3.1, and the first part of the proposition follows by Theorem A.2 in Imhof and Studden (2001). ∎

Remark 7.5.1 The following notes can be useful as an addition to Theorem 7.5.1:

(a) The Chebyshev points for the system (7.29) on an arbitrary finite interval $I \subset \mathbb{R}$ can be obtained by rescaling the points onto the interval $[-1, 1]$. The case $s = 0$ and $I = [0, \infty)$ will be discussed in more detail in Examples 7.5.1 and 7.5.3.

(b) It follows from Theorem 7.3.1 that the set Ω_E defined in the second part of Theorem 7.5.1 contains the set $\Omega_{\varepsilon, \Delta}$ defined in (7.18) for sufficiently small ε. In other words, if the nonlinear parameters $b_1, \ldots, b_k$ are sufficiently close, the locally E-optimal design will be supported at the Chebyshev points with weights given by (7.12). Moreover, we will demonstrate in the subsequent examples that in many cases, the set Ω_E coincides with the full set $\mathcal{B}$.

(c) In applications, the Chebyshev points can be calculated numerically with the Remez algorithm (see Studden and Tsay (1976) or De Vore and Lorentz (1993)). In some cases, these points can be obtained explicitly (see Examples 7.5.1 and 7.5.2).

Remark 7.5.2 We note that a similar result is valid for c-optimal designs in the rational regression model (7.4). For example, assume that one of the assertions of Lemma 7.5.1 is valid and that we are interested in estimating a linear combination $c^T \beta$ of the parameters in the rational model (7.4). We obtain from Lemma 7.3.2 that if $c \in \mathbb{R}^m$ satisfies $c^T \tilde{\gamma} \neq 0$, then for sufficiently small ε and any $b \in \Omega_{\varepsilon, \Delta}$, the design $\xi_c(b)$ defined in (7.15) and (7.16) is c-optimal. In particular, this is true for $c = e_{s+2j-1}$ (for all $j = 1, \ldots, k$) and the vector $c = e_{s+2j}$ if the index j satisfies the condition (7.26). Note that due to the third part of Lemma 7.5.1, in the case $k \geq 2$ there exists $b \in \mathcal{B}$ such that the e_{s+2j}-optimal design is not necessarily supported at the Chebyshev points. However, from Theorem 7.3.2, it follows that for a vector $b \in \mathcal{B}$ satisfying (7.17), with $\delta \to 0$ and any vector c with $c^T \tilde{\gamma} \neq 0$ we have for the designs $\xi_{c^*}^*(b)$ and $\xi_c^*(b)$ defined by (7.14) and (7.15),

$$\xi_{c^*}^*(b) \quad \to \quad \bar{\xi}_{e_m}(x),$$

$$\xi_c^*(b) \quad \to \quad \bar{\xi}_{e_m}(x),$$

where the design $\bar{\xi}_{e_m}(x)$ is defined in (7.27) and (7.28), respectively, and e_m-optimal in the limiting model with the regression functions (7.30).

Example 7.5.1 Consider the rational model

$$Y = \frac{a}{t - b} + \varepsilon, \quad t \in [0, \infty), \tag{7.36}$$

with $b < 0$ (here we have $k = 1$, $s = 0$, and $I = [0, \infty)$). The corresponding equivalent linear regression model is given by

$$Y = \beta^T f(t, b) = \frac{\beta_1}{t - b} + \frac{\beta_2}{(t - b)^2}. \tag{7.37}$$

In this case, it follows from the first part of Lemma 7.5.1 that the system of regression functions

$$\left\{ \frac{1}{t - b}, \frac{1}{(t - b)^2} \right\} = \{ f_1(t), f_2(t) \}$$

is a Chebyshev system on the interval $[0, \infty)$ whenever $b < 0$. Moreover, any subsystem (consisting of one function) is obviously a Chebyshev system on the interval $[0, \infty)$. The Chebyshev points are the (local) extremal of the function

$$g(t) = \rho \left(\frac{1}{t - b} + \frac{\kappa}{(t - b)^2} \right),$$

where ρ and κ are determined by the conditions

$$g(t) \leq 1 \quad \forall\, t \in [0, \infty),$$

$$g(s_j) = (-1)^j, \quad j = 1, 2.$$

It is easy to see that $s_1 = 0$ and that s_2 is the positive solution of the equation $g'(t) = 0$, which implies

$$\kappa = \frac{b - s_2}{2}.$$

Observing the relation $g(s_1) = -g(s_2)$, by a straightforward calculation we obtain

$$s_2 = \sqrt{2}|b| = -\sqrt{2}b$$

and the condition $g(s_1) = g(0) = -1$ implies

$$\rho = \frac{-2}{\sqrt{2} - 1}b,$$

which determines the Chebyshev polynomial explicitly. Now, we consider the design $\xi_c^*(b)$ defined in (7.15) as a candidate for the c-optimal design in model (7.37). The weights (for any $c \in \mathbb{R}^2$) are obtained from (7.16), where the matrix F is given by

$$F = (f_i(s_j))^2_{i,j=1} = \begin{pmatrix} \dfrac{1}{|b|} & \dfrac{1}{(\sqrt{2}+1)|b|} \\ \dfrac{1}{b^2} & \dfrac{1}{(\sqrt{2}+1)^2 b^2} \end{pmatrix}.$$

A straightforward calculation shows that

$$F^{-1}c = \frac{1}{2} \begin{pmatrix} |b|(-\sqrt{2}c_1 + (2 + \sqrt{2})c_2 b) \\ -|b|(4 + 3\sqrt{2})(-c_1 + c_2 b) \end{pmatrix},$$

which gives

$$\xi_c^*(b) = \begin{pmatrix} 0 & \sqrt{2}|b| \\ w_1 & w_2 \end{pmatrix}, \tag{7.38}$$

where the weights are given by

$$w_1 = 1 - w_2 = \frac{|b(-\sqrt{2}c_1 + (2 + \sqrt{2})c_2 b)|}{|b|\{-\sqrt{2}c_1 + (2 + \sqrt{2})c_2 b \mid +(4 + 3\sqrt{2})| - c_1 + c_2 b|\}}.$$

It can easily be checked by Elfving's theorem [see Elfving (1952)] or by the equivalence theorem for c-optimality [see Pukelsheim (1993)] that this design is in fact c-optimal in the regression model (7.37) whenever

$$\frac{c_2}{c_1} \notin \left[\frac{1}{b}, \frac{1}{(1 + \sqrt{2})b}\right].$$

In the remaining cases, the c-optimal design is a one-point design supported at $t = b - \frac{c_1}{c_2}$. In particular, by Lemma 7.2.3, the e_1- and e_2-optimal design for estimating the coefficients β_1 and β_2 in model (7.37) are given by

$$\xi^*_{e_1}(b) = \begin{pmatrix} 0 & \sqrt{2}|b| \\ \frac{1}{4}(2 - \sqrt{2}) & \frac{1}{4}(2 + \sqrt{2}) \end{pmatrix},$$

$$\xi^*_{e_2}(b) = \begin{pmatrix} 0 & \sqrt{2}|b| \\ 1 - \frac{1}{\sqrt{2}} & \frac{1}{\sqrt{2}} \end{pmatrix},$$

(7.39)

respectively. It follows from the results of Imhof and Studden (2001) that an E-optimal design in the regression model (7.37) is given by the c^*-optimal design for the Chebyshev vector

$$c^* = (1 + \sqrt{2})|b|(-2, |b|(1 + \sqrt{2}))^T;$$

that is,

$$\xi^*_E = \begin{pmatrix} 0 & \sqrt{2}|b| \\ w_1 & w_2 \end{pmatrix},$$

(7.40)

where

$$w_1 = \frac{1}{2} \frac{(2 - \sqrt{2})(6 - 4\sqrt{2} + b^2)}{b^2 + 12 - 8\sqrt{2}} = 1 - \frac{1}{2} \frac{\sqrt{2}(2\sqrt{2} - 2 + b^2)}{b^2 + 12 - 8\sqrt{2}} = 1 - w_2.$$

The corresponding information matrix is obtained by a tedious calculation,

$$M(\xi^*_E(b), b) = \begin{pmatrix} \dfrac{(\sqrt{2} - 1)(b^2 + 6\sqrt{2} - 8)}{b^2(b^2 + 12 - 8\sqrt{2})} & \dfrac{2(3 - \sqrt{2})(b^2 + \sqrt{2} - 1)}{b^3(b^2 + 12 - 8\sqrt{2})} \\ \dfrac{2(3 - \sqrt{2})(b^2 + \sqrt{2} - 1)}{b^3(b^2 + 12 - 8\sqrt{2})} & \dfrac{(8\sqrt{2} - 11)(7b^2 + 16\sqrt{2} - 20)}{7b^4(b^2 + 12 - 8\sqrt{2})} \end{pmatrix},$$

(7.41)

and has a minimum eigenvalue

$$\lambda_{\min}(M(\xi^*_E(b), b) = \frac{17 - 2\sqrt{2}}{b^2(b^2 + 12 - 8\sqrt{2})} = \frac{1}{\|c^*\|^2}$$

of multiplicity 1 with corresponding eigenvector c^*. Note that for $b \to -\infty$, this design approximates the optimal design $\xi^*_{e_2}(b)$ for estimating the individual coefficient β_2 in the rational model (7.37).

It is of some interest to compare these designs with the locally D-optimal design. It follows from the results in He, Studden, and Sun (1996) and a straightforward calculation that this design is given by

$$\xi^*_D = \begin{pmatrix} 0 & |b| \\ \frac{1}{2} & \frac{1}{2} \end{pmatrix}.$$

(7.42)

The designs are now compared by their efficiencies for estimating the coefficients β_1 and β_2; that is,

$$\text{eff}_i(\xi) = \left(\frac{e_i^T M^{-1}(\xi, b) e_i}{e_i^T M^{-1}(\xi^*_{e_i}, b) e_i} \right)^{-1}, \quad i = 1, 2.$$

(7.43)

The values $e_i^T M^{-1}(\xi_{e_i}^*, b)e_i$ can be directly obtained from the Chebyshev vector, which gives

$$e_i^T M^{-1}(\xi_{e_i}^*, b)e_i = \begin{cases} 4(1+\sqrt{2})^2 b^2 & \text{if } i = 1 \\ (1+\sqrt{2})^4 b^4 & \text{if } i = 2. \end{cases}$$

Now, a straightforward calculation yields

$$\text{eff}_i(\xi_D^*) = \begin{cases} \dfrac{4(\sqrt{2}+1)^2}{34} \approx 0.6857 & \text{if } i = 1 \\[3mm] \dfrac{(\sqrt{2}+1)^4}{40} \approx 0.8493 & \text{if } i = 2. \end{cases}$$

for the efficiencies of the D-optimal design defined by (7.42). The corresponding efficiencies of the E-optimal design in the regression model (7.37) depend on the parameter b and are obtained by a straightforward but tedious inversion of the matrix $M(\xi_E^*(b), b)$ defined in (7.41); that is,

$$\text{eff}_i(\xi_E^*(b)) = \begin{cases} \dfrac{28(b^4(5\sqrt{2}-7) + b^2(34\sqrt{2}-48) + 396 - 280\sqrt{2})}{(9\sqrt{2}-11)(b^2 - 8\sqrt{2}+12)(7b^2 + 16\sqrt{2}-20)} & \text{if } i = 1 \\[4mm] \dfrac{b^4(\sqrt{2}-1) + (6\sqrt{2}-8)b^2 + 68 - 48\sqrt{2}}{(\sqrt{2}-1)(b^2 - 8\sqrt{2}+12)(b^2 - 6\sqrt{2}+8)} & \text{if } i = 2. \end{cases}$$

$$(7.44)$$

The corresponding efficiencies are depicted in Figure 7.1 for the range $b \in [-2.5, -1]$. We observe for the e_1-efficiency for all $b \leq -1$,

$$0.9061 \approx \frac{4(5\sqrt{2}-7)}{(8\sqrt{2}-11)} = \lim_{b\to-\infty} \text{eff}_1(\xi_E^*(b))$$

$$\leq \text{eff}_1(\xi_E^*(b)) \leq \text{eff}_1(\xi_E^*(-1)) \approx 0.9595,$$

and similarly for the e_2-efficiency,

$$0.9805 \approx \text{eff}_2(\xi_E^*(-1)) \leq \text{eff}_2(\xi_E^*(b)) \leq \lim_{b\to-\infty} \text{eff}_2(\xi_E^*(b)) = 1.$$

This demonstrates that the E-optimal designs yield substantially more accurate estimates for the individual parameters in the regression model (7.37) than the D-optimal design.

We finally mention the results for the locally optimal design in the rational model (7.36), which maximize or minimize the corresponding functional for the matrix $K_a^{-1} M(\xi, b) K_a^{-1}$, where $K_a = \text{diag}(1, -\frac{1}{a})$. Obviously, the locally e_1-, e_2-, and D-optimal designs are given by (7.39) and (7.42), respectively, and coincide with the corresponding designs in the equivalent linear regression model (7.37). On the other hand, the c-optimal design for the rational model (7.36) is obtained from the $\bar{c}$-optimal design $\xi_{\bar{c}}^*(b)$ in

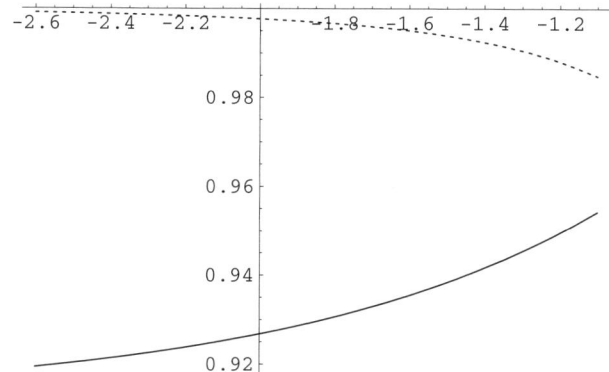

Figure 7.1: Efficiencies of the E-optimal design $\xi^*(b)$ for estimating the individual coefficients in the regression model (7.37) for various values dotted line: $\mathrm{eff}_2(\xi^*(b))$.

(7.38) for model (7.37), with $\bar{c} = K_a c = (c_1, -c_2/a)^T$. Similarly, the locally E-optimal design for the rational model (7.36) is given by

$$\xi_E^* = \begin{pmatrix} 0 & \sqrt{2}|b| \\ w_1^* & w_2^* \end{pmatrix},$$

where the weights are given by

$$w_1^* = \frac{2\sqrt{2}a^2 + (4 + 3\sqrt{2})b^2}{2\{4(1 + \sqrt{2})a^2 + (7 + 5\sqrt{2})b^2\}}$$

$$= 1 - \frac{(4 + 3\sqrt{2})(2a^2 + (1 + \sqrt{2})b^2)}{2\{4(1 + \sqrt{2})a^2 + (7 + 5\sqrt{2})b^2\}} = 1 - w_2^*.$$

A comparison of the efficiencies for the D- and E-optimal designs, in the rational model (7.36) yields similar results as in the corresponding equivalent linear regression model (7.37). For a broad range of parameter values (a, b), the locally E-optimal designs in the rational model (7.36) are substantially more efficient for estimating the individual parameters than the locally D-optimal designs.

Example 7.5.2 We now consider the rational model

$$Y = a_1 + \frac{a_2}{t - b} + \varepsilon, \quad t \in [-1, 1], \tag{7.45}$$

where $|b| > 1$. The corresponding equivalent linear regression model is given by

$$Y = \beta_1 + \frac{\beta_2}{t - b} + \frac{\beta_3}{(t - b)^2} + \varepsilon, \quad t \in [-1, 1], \tag{7.46}$$

and the first part of Lemma 7.5.1 shows that this system is a Chebyshev system on the interval $[-1, 1]$. Moreover, the three subsystems obtained by deleting one of the regression functions form also weak Chebyshev systems (this follows partially from Lemma 7.5.1(ii), and the remaining case has to be checked directly). Therefore, the optimal designs for estimating the individual coefficients and the E-optimal design are supported at the Chebyshev points, which are given by $s_1 = -1$, $s_2 = 1/b$, and $s_3 = 1$. A similar calculation in Example 7.5.1 shows that the E-optimal design in the equivalent linear regression model (7.46) is given by

$$\xi_E^* = \left(\begin{array}{ccc} -1 & \frac{1}{b} & 1 \\ w_1 & w_2 & w_3 \end{array} \right),$$

where

$$w_1 = \frac{b+1}{2} \cdot \frac{2b^7 - 2b^6 + 2b^5 + 2b^4 - 4b^3 - 2b^2 + b + 2}{4b^8 - 4b^4 - 4b^2 + 5},$$

$$w_2 = \frac{(b^2 - 1)(2b^6 + 2b^4 - 3)}{4b^8 - 4b^4 - 4b^2 + 5},$$

$$w_3 = \frac{b-1}{2} \cdot \frac{2b^7 + 2b^6 + 2b^5 - 2b^4 - 4b^3 + 2b^2 + b - 2}{4b^8 - 4b^4 - 4b^2 + 5},$$

Here, we have used Lemma 7.2.2 and the fact that the vector of the coefficients of the Chebyshev polynomial is given by

$$c^* = (2b^2 - 1, 4b(b^2 - 1), 2(b^2 - 1)^2)^T.$$

The optimal designs for estimating the individual coefficients β_1, β_2, and β_3 are given by

$$\xi_{e_1}^* = \left(\begin{array}{ccc} -1 & \frac{1}{b} & 1 \\ \frac{b(1+b)}{2(2b^2-1)} & \frac{b^2-1}{2b^2-1} & \frac{b(b-1)}{2(2b^2-1)} \end{array} \right),$$

$$\xi_{e_2}^* = \left(\begin{array}{ccc} -1 & \frac{1}{b} & 1 \\ \frac{1}{8}(2 + \frac{1}{b}) & \frac{1}{2} & \frac{1}{8}(2 - \frac{1}{b}) \end{array} \right),$$

$$\xi_{e_3}^* = \left(\begin{array}{ccc} -1 & \frac{1}{b} & 1 \\ -\frac{1}{4} & \frac{1}{2} & \frac{1}{4} \end{array} \right),$$

respectively. We note again that for $|b| \to \infty$, all designs are approximated by the optimal design $\xi_{e_3}^*$ for estimating the individual coefficient β_3. The corresponding efficiencies $\text{eff}_i(\xi_E^*(b))$, $i = 1, 2, 3$, are depicted in Figure 7.2 for the interval $[2, 4]$ and demonstrate again that the locally E-optimal design is highly efficient for estimating the coefficients β_1, β_2, and β_3 in model (7.46).

The locally D-optimal design can be obtained by similar arguments as given in Example 7.5.1; that is,

$$\xi_D^*(b) = \left(\begin{array}{ccc} -1 & \frac{1}{b} & 1 \\ \frac{1}{3} & \frac{1}{3} & \frac{1}{3} \end{array} \right),$$

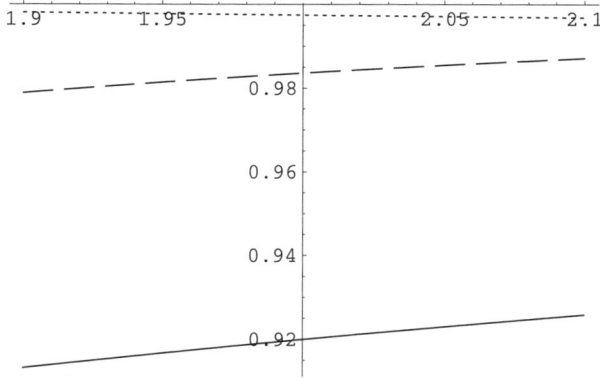

Figure 7.2: Efficiencies of the E-optimal design $\xi^*(b)$ for estimating the individual coefficients in the regression model (7.46) for various values of $b \in [2, 4]$. Solid line: $\mathrm{eff}_1(\xi^*(b))$; dotted line: $\mathrm{eff}_2(\xi^*(b))$, dashed line: $\mathrm{eff}_3(\xi^*(b))$

and the corresponding efficiencies can be calculated explicitly and are given by

$$
\mathrm{eff}_i(\xi_D^*(b)) = \begin{cases} \dfrac{2(2b^2 - 1)^2}{3(3b^4 - 3b^2 + 2)} & \text{if } i = 1 \\[3mm] \dfrac{32b^2}{3 + 36b^2} & \text{if } i = 2 \\[3mm] \dfrac{2}{9} & \text{if } i = 3. \end{cases}
$$

Again, we observe that, the locally E-optimal design yield substantially more accurate estimates of the individual parameters than D-optimal designs. Finally, the locally optimal designs for the rational model (7.45) are obtained as follows. The optimal designs for estimating the individual coefficients and the locally D-optimal design coincide with the corresponding designs in the linear regression model (7.46) whereas the locally E-optimal design puts masses

$$
w_1^* = \frac{2(b^2 - 1)^4 + a_2^2 b(8b^5 + 4b^4 - 14b^3 - 6b^2 + 7b + 3)}{2\{4(b^2 - 1)^4 + a_2^2(16b^6 - 28b^4 + 12b^2 + 1\}},
$$

$$
w_2^* = \frac{(b^2 - 1)\{2(b^2 - 1)^3 + a_2^2(8b^4 - 6b^2 - 1)\}}{4(b^2 - 1) + a_2^2(16b^6 - 28b^4 + 12b^2 + 1)},
$$

$$
w_3^* = \frac{2(b^2 - 1)^4 + a_2^2 b_2(8b^5 - 4b^4 - 14b^2 + 6b^2 + 7b - 3)}{2\{4(b^2 - 1)^4 + a_2^2(16b^6 - 28b^4 + 12b^2 + 1)\}}
$$

at the points $-1, 1/b$, and 1, respectively.

Example 7.5.3 We now discuss optimal designs for the rational model

$$
Y = \frac{a_1}{t - b_1} + \frac{a_2}{t - b_2} + \varepsilon, \quad t \in [0, \infty), \tag{7.47}
$$

where $b_1, b_2 < 0$ and $|b_2 - b_2| > 0$ ($k = 2, s = 0$). The corresponding equivalent linear regression model is given by

$$Y = \frac{\beta_1}{t - b_1} + \frac{\beta_2}{(t - b_1)^2} + \frac{\beta_3}{t - b_2} + \frac{\beta_4}{(t - b_2)^2} + \varepsilon. \tag{7.48}$$

Locally D-optimal designs for model (7.47) [or, equivalently, (7.48)] have been determined in Chapter 5, and the optimal designs for estimating the individual coefficients can be obtained numerically from the results of this chapter. We now compare these designs by looking at D-, E-, and e_i-efficiencies. For the sake of brevity, we restrict ourselves to model (7.48), which corresponds to the locally optimal design problem for model (7.47), with $(a_1, a_2) = (1, 1)$. In our comparison, we will also include the E-optimal design in the limiting model under assumption (7.17); that is,

$$Y = \frac{\beta_1}{t - x} + \frac{\beta_2}{(t - x)^2} + \frac{\beta_3}{(t - x)^3} + \frac{\beta_4}{(t - x)^4} + \varepsilon, \tag{7.49}$$

where the parameter x is chosen as $x = (b_1 + b_2)/2$. Without loss of generality, we assume that $x = -1$, because, in the general case, the optimal designs can be obtained by a simple scaling argument. The limiting optimal design was obtained numerically and is given by

$$\bar{\xi}_E(-1) = \begin{pmatrix} 0 & 0.18 & 1.08 & 7.9 \\ 0.13 & 0.26 & 0.27 & 0.34 \end{pmatrix}. \tag{7.50}$$

¿From Theorem 7.3.1, we obtain that for sufficiently small

$$\Delta = \left| \frac{b_1 - b_2}{2} \right|,$$

the E-optimal designs for model (7.47) is given the design $\xi_{c^*}^*(b)$ defined in (7.12) and (7.14). From Lemma 7.2.2, it follows that the design $\xi_{c^*}^*(b)$ is E-optimal whenever

$$\lambda_{c^*} := \frac{c^{*T} M(\xi_E^*(b), b) c^*}{c^{*T} c^*} \le \lambda_{(2)}(M(\xi_E^*(b), b)) = \lambda_{(2)},$$

where $\lambda_{\min}(M(\xi_E^*(b), b)) \le \lambda_{(2)} \le \cdots \le \lambda_{(m)}$ denote the ordered eigenvalues of the matrix $M(\xi_E^*(b), b)$. The ratio $\lambda_{(2)}/\lambda_{c^*}$ is exemplarily depicted in Figure 7.3 for $b_1 = 1$ and a broad range of b_2 values, which shows that it is always larger than 1. Other cases yield a similar picture and practically the locally E-optimal design for the rational model (7.47), and the equivalent linear regression model (7.48) is always supported at the Chebyshev points and given by (7.12) and (7.14). In Tables 7.1 and 7.2, we give the main characteristics and efficiencies for the locally E- and D-optimal design $\xi_E^*(b), \xi_D^*(b)$ and for the E-optimal design $\bar{\xi}_E^*(\frac{b_1 + b_2}{2})$ in the limiting regression model (7.49). The efficiencies are calculated with respect to the

Table 7.1: D- and E-optimal designs for linear regression model (7.48) on the interval $[0, \infty)$, where $b_1 = -1 - z$ and $b_2 = -1 + z$. These designs are locally D- and E-optimal in the rational model (7.47) for the initial parameter $a_1 = a_2 = 1$. Note that the smallest support point, of the D-optimal design (t_{1D}^*) and E-optimal design (t_{1E}^*) are equal to 0 and that the masses of the D-optimal design are equal to each other

z	0.1	0.2	0.3	0.4	0.5	0.6	0.7	0.8	0.9	0.95
t_{2D}^*	0.21	0.20	0.20	0.19	0.17	0.15	0.13	0.10	0.06	0.04
t_{3D}^*	1.00	0.98	0.95	0.92	0.87	0.80	0.71	0.60	0.44	0.31
t_{4D}^*	4.78	4.73	4.65	4.54	4.39	4.19	3.94	3.60	3.13	2.78
t_{2E}^*	0.18	0.17	0.17	0.16	0.15	0.13	0.11	0.09	0.05	0.03
t_{3E}^*	1.08	1.06	1.03	0.99	0.94	0.87	0.77	0.65	0.47	0.34
t_{4E}^*	7.85	7.77	7.65	7.46	7.21	6.88	6.45	5.88	5.05	4.43
w_{1E}^*	0.13	0.13	0.13	0.13	0.12	0.10	0.08	0.07	0.05	0.03
w_{2E}^*	0.26	0.26	0.27	0.26	0.25	0.22	0.20	0.17	0.13	0.10
w_{3E}^*	0.27	0.27	0.28	0.28	0.28	0.28	0.28	0.28	0.28	0.28
w_{4E}^*	0.34	0.33	0.33	0.33	0.36	0.39	0.44	0.49	0.54	0.59

Table 7.2: The efficiency of the E-optimal designs ξ_E^* in the linear regression model (7.48) on the interval $[0, \infty)$ with $b_1 = -1 - z$ and $b_2 = -1 + z$ and the efficiency of the E-optimal design $\xi_E^*(-1)$ given in (7.50) in the corresponding limiting model (7.49). The efficiencies $\mathrm{eff}_D(\xi)$, $d_i(\xi)$, and $C_E(\xi)$ are defined in (7.51), (7.52), and (7.53), respectively

z	0.1	0.3	0.4	0.5	0.6	0.7	0.8	0.9	0.95
$d_1(\xi_E^*)$	0.81	0.81	0.83	0.87	1.04	1.28	0.72	0.52	0.48
$d_2(\xi_E^*)$	0.80	0.79	0.78	0.76	0.74	0.71	0.68	0.63	0.59
$d_3(\xi_E^*)$	0.81	0.81	0.81	0.83	0.86	0.94	1.08	1.38	1.79
$d_4(\xi_E^*)$	0.82	0.84	0.85	0.89	0.97	1.12	1.36	1.89	2.53
$d_1(\xi_E^*(-1))$	0.81	0.82	0.83	0.87	0.93	0.95	0.92	1.14	1.37
$d_2(\xi_E^*(-1))$	0.80	0.80	0.82	0.86	0.94	1.09	1.38	2.04	2.81
$d_3(\xi_E^*(-1))$	0.81	0.81	0.83	0.86	0.93	1.09	1.51	3.42	10.00
$d_4(\xi_E^*(-1))$	0.82	0.84	0.85	0.88	0.94	1.08	1.49	3.48	10.59
$\mathrm{eff}_D(\xi_E^*)$	0.89	0.89	0.89	0.88	0.85	0.81	0.75	0.67	0.60
$\mathrm{eff}_D(\xi_E^*(-1))$	0.89	0.89	0.88	0.88	0.87	0.84	0.78	0.63	0.48
$C_E(\xi_E^*)$	1.23	1.23	1.25	1.27	1.32	1.39	1.47	1.61	1.75
$C_E(\xi_E^*(-1))$	1.23	1.23	1.22	1.16	1.08	0.92	0.72	0.50	0.38

D-optimal design for various values of the nonlinear parameters b_1 and b_2 and are defined by

$$\mathrm{eff}_D(\xi) = \left(\frac{\det M(\xi, b)}{\det M(\xi_D^*, b)} \right)^{1/m} \tag{7.51}$$

$$d_i(\xi) = \frac{e_i^T M^{-1}(\xi, b) e_i}{e_i^T M^{-1}(\xi_D^*, b) e_i} \tag{7.52}$$

(in other words, we compare the performance of the design ξ for estimating individual coefficients with respect to the D-optimal design) and

$$C_E(\xi) = \frac{\lambda_{\min}(M(\xi, b))}{\lambda_{\min}(M(\xi_D^*, b))}. \tag{7.53}$$

Again, we observe a very good performance of the E-optimal designs. These designs produce a reasonable D-efficiency for a moderate size of the difference $|b_1 - b_2|$, but are in many cases substantially more efficient than the D-optimal designs for estimating the individual coefficients. The behavior of the design $\bar{\xi}_E$ in the limiting regression model (7.48) is interesting from a practical point of view because it is very similar to the performance of the E-optimal design for a broad range of b_1 and b_2 values. Consequently, this design might be appropriate if rather unprecise prior information for the nonlinear parameters is available. For example, if it is known (by scientific background) that $b_1 \in [\underline{b}_1, \bar{b}_1]$ and $b_2 \in [\underline{b}_2, \bar{b}_2]$, the design

$$\bar{\xi}_E\left(\frac{\underline{b}_1 + \bar{b}_2}{2}\right)$$

might be a robust choice for practical experiments.

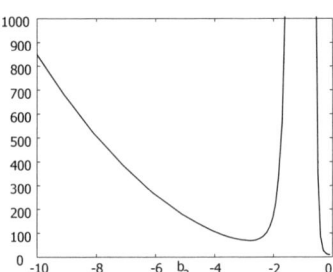

Figure 7.3: The ratio $\lambda_{(2)}/\lambda_{c^*}$ for the design $\xi_E^*(b)$, where $b = (-1, b_2)$. The designs are E-optimal if this ratio is larger or equal than 1.

Example 7.5.4 Our final example discusses the rational model (7.47) with an additional term for the intercept:

$$Y = a_1 + \frac{a_2}{t - b_1} + \frac{a_3}{t - b_2} + \varepsilon, \quad t \in [-1, 1], \tag{7.54}$$

where $|b_i| > 1$ $(i = 1, 2)$ and $|b_2 - b_2| > 0$ (this corresponds to the case $k = 2, s = 1$ in the general model (7.4)). The limiting model is given by

$$Y = \beta_1 + \frac{\beta_2}{t - x} + \frac{\beta_3}{(t - x)^2} + \frac{\beta_4}{(t - x)^3} + \frac{\beta_5}{(t - x)^4} + \varepsilon. \tag{7.55}$$

The notation is essentially the same as in the previous example. Our numerical study showed that the locally E-optimal design for model (7.54) is supported at the Chebyshev points for all choices of the parameters (b_1, b_2) ($|b_i| > 1, b_1 \neq b_2$). In Tables 7.3 and 7.4 we display the main features of the locally E- and D-optimal designs ξ_E^* and ξ_D^* and the E-optimal design $\bar{\xi}_E\left(\frac{b_1+b_2}{2}\right)$ in the limiting regression model (7.55), which is given by

$$\bar{\xi}_E(-3) = \begin{pmatrix} -1 & -0.84 & -0.33 & 0.49 & 1 \\ \frac{1}{8} & \frac{1}{4} & \frac{1}{4} & \frac{1}{4} & \frac{1}{8} \end{pmatrix}. \tag{7.56}$$

The conclusions are very similar to those in Example 7.5.3. This indicates that the observations from this example are, in some sense, representative.

7.6 Exponential Models

In the present section we will consider the models of the form

$$Y = \sum_{i=1}^{k} a_i e^{-b_i t} + \varepsilon, \; 0 < b_1 < \cdots < b_k, \tag{7.57}$$

where $a_1, \ldots, a_k$ and $b_1, \ldots, b_k$ are parameters to be estimated. These models can be obtained from (7.1) by substituting $s = 0$ and $\varphi(t, b) = \exp(-bt)$.

The corresponding linearized model assumes the form

$$Y = \beta^T f(t, b) + \varepsilon,$$

where $\beta = (\beta_1, \ldots, \beta_{2k})^T$ is the parameter vector to be estimated,

$$f(t, b) = (f_1(t, b), \ldots, f_{2k}(t, b))^T, \; b = (b_1, \ldots, b_k)^T,$$

$$f_{2i-1}(t, b) = \exp(-b_i t), \; i = 1, \ldots, k, \tag{7.58}$$

$$f_{2i}(t, b) = t \exp(-b_i, t), \; i = 1, \ldots, k.$$

The limiting model assumes the form

$$Y = \beta^T \bar{f}(t, \gamma) + \varepsilon,$$

where

$$\bar{f}(t, \gamma) = (\bar{f}_1(t, \gamma), \ldots, \bar{f}_{2k}(t, \gamma))^T,$$

$$\bar{f}_i(t, \gamma) = e^{-\gamma t} t^{i-1}, \; i = 1, \ldots, 2k. \tag{7.59}$$

As was pointed out in Section 1.8, the system (7.58) with $b_1 < \cdots < b_k$ and the system (7.59) with arbitrary γ are Chebyshev systems on an arbitrary finite interval, and for the the Chebyshev points are uniquely

Table 7.3: Locally D- and E-optimal designs for the rational regression model (7.54) on the interval $[-1, 1]$, where $b_1 = -3 - z$, $b_2 = -3 + z$, and $a_3 = a_2 = 1$. Note that the largest and smallest support points, and of the locally E- and D-optimal design satisfy $t_{5E}^* = t_{5D}^* = 1$ and $t_{1E}^* = t_{1D}^* = -1$, respectively, and the masses of the locally D-optimal design are all equal.

z	0.1	0.2	0.3	0.5	1	1.5	1.9
t_{2D}^*	-0.81	-0.81	-0.81	-0.82	-0.83	-0.87	-0.95
t_{3D}^*	-0.32	-0.34	-0.34	-0.34	-0.38	-0.47	-0.70
t_{4D}^*	0.41	0.41	0.41	0.40	0.37	0.29	0.08
t_{2E}^*	-0.84	-0.84	-0.84	-0.85	-0.86	-0.89	-0.96
t_{3E}^*	-0.33	-0.33	-0.34	-0.34	-0.38	-0.47	-0.70
t_{4E}^*	0.49	0.49	0.49	0.48	0.45	0.38	0.17
w_{1E}^*	0.13	0.13	0.13	0.12	0.11	0.09	0.05
w_{2E}^*	0.25	0.25	0.25	0.25	0.22	0.20	0.14
w_{3E}^*	0.25	0.25	0.25	0.25	0.25	0.25	0.25
w_{4E}^*	0.25	0.25	0.25	0.25	0.28	0.30	0.36
w_{5E}^*	0.12	0.12	0.12	0.13	0.14	0.16	0.20

Table 7.4: The efficiency of the E-optimal designs ξ_E^* in the rational regression model (7.54) on the interval $[-1, 1]$ with $b_1 = -3 - z$, $b_2 = -3 + z$, and $a_3 = a_2 = 1$ and the efficiency of the E-optimal design $\bar{\xi}_E(-1)$ given in (7.56) in the corresponding limiting model (7.55). The efficiencies $\mathrm{eff}_D(\xi)$, $d_i(\xi)$, and $C_E(\xi)$ are defined in (7.51), (7.52), and (7.53), respectively.

z	0.1	0.2	0.3	0.5	1	1.5	1.9
$d_1(\xi_E^*)$	0.86	0.87	0.87	0.87	0.84	0.82	0.75
$d_2(\xi_E^*)$	0.83	0.84	0.84	0.84	0.85	0.90	1.21
$d_3(\xi_E^*)$	0.83	0.84	0.84	0.84	0.87	0.97	1.53
$d_4(\xi_E^*)$	0.83	0.84	0.84	0.83	0.88	0.81	0.74
$d_5(\xi_E^*)$	0.83	0.84	0.84	0.84	0.83	0.82	0.76
$d_1(\bar{\xi}_E^*(-3))$	0.86	0.88	0.88	0.89	0.96	1.31	3.62
$d_2(\bar{\xi}_E^*(-3))$	0.83	0.84	0.84	0.84	0.85	1.05	5.74
$d_3(\bar{\xi}_E^*(-3))$	0.83	0.84	0.84	0.84	0.84	1.01	5.72
$d_4(\bar{\xi}_E^*(-3))$	0.83	0.84	0.84	0.83	1.08	1.28	3.74
$d_5(\bar{\xi}_E^*(-3))$	0.83	0.84	0.84	0.84	0.88	1.21	3.94
$\mathrm{eff}_D(\xi_E^*)$	0.93	0.93	0.93	0.93	0.93	0.91	0.83
$\mathrm{eff}_D(\bar{\xi}_E^*(-3))$	0.93	0.93	0.93	0.93	0.93	0.91	0.66
$C_E(\xi_E^*)$	1.20	1.19	1.19	1.19	1.20	1.22	1.33
$C_E(\bar{\xi}_E^*(-3))$	1.20	1.19	1.19	1.19	1.14	0.82	0.26

determined. Thus, Theorem 7.3.2 holds for the regression models (7.57) on an arbitrary finite interval.

It is also not difficult to check (and it is left to the reader) that condition (d) of Lemma 7.4.1 is satisfied for the function system (7.58) and, thus, this lemma can be applied.

Let Ω be the set of values b for which the minimal eigenvalue of the information matrix $M(\xi_{c^*})$ is equal to $b/c^{*T}c^*$. Due to Lemma 7.2.2, in this case the Chebyshev design ξ_{c^*} is the unique locally E-optimal design. Thus, in the case $b \in \Omega$ for constructing locally E-optimal designs, it will do to find Chebyshev points for the system (7.58). Also, if all coefficients determined by (7.12) are non-negative, then for constructing c-optimal designs, it also will do to find the Chebyshev points. Other cases appears to be more complex and are not covered by the theory developed above. However, the examples given below show that the theory for many models cover either the whole set of possible values of the parameters or at least its significant part.

The following remark is useful for understanding the behavior of optimal designs.

Remark 7.6.1 Let Ω denote the set of all vectors $b = (b_1, \ldots, b_k)^T \in \mathbb{R}^k$ with $b_i \neq b_j$, $i \neq j$, $b_i > 0$, $i = 1, \ldots, k$, such that the minimum eigenvalue of the information matrix of the local E-optimal design (with respect to the vector b) is simple. The following properties of local E-optimal designs follow by standard arguments from general results on E-optimal designs (see Dette and Studden (1993), Pukelsheim (1993), as well as Chapters 3 and 4 of the present book) and simplify the construction of local E-optimal designs substantially.

1. For any $b \in \Omega$, the local E-optimal design for the exponential regression model (7.57) (with respect to the parameter b) is unique.

2. For any $b \in \Omega$, the support points of the local E-optimal design for the exponential regression model (7.57) (with respect to the parameter b) do not depend on the parameters $a_1, \ldots, a_k$.

3. For any $b \in \Omega$, the local E-optimal design for the exponential regression model (7.57) (with respect to the parameter b) has $2k$ support points; moreover, the point d is always a support point of the local E-optimal design. The support points of the E-optimal design are the extremal points of the Chebyshev function $p^T f(x)$, where p is an eigenvector corresponding to the minimal eigenvalue of the information matrix $M(\xi_E^*(b))$.

4. For any $b \in \Omega$, the weights of the local E-optimal design for the exponential regression model (7.57) (with respect to the parameter b) are given by

$$w^* = \frac{JF^{-1}p}{p^T p}, \tag{7.60}$$

where $p^T = \mathbf{1}_{jk}^T JF^{-1}$, $J = \mathrm{diag}(1, -1, 1, \ldots, 1, -1)$,

$$F = (f(x_1^*), \cdots, f(x_m^*)) \in \mathbf{R}^{2k \times 2k}$$

and $x_1^* < \ldots < x_{2k}^*$ denote the support points of the local E-optimal design.

5. If $b \in \Omega$, let $x_{1;d}^*(b), \ldots, x_{2k;d}^*(b)$ denote the support points of the local E-optimal design for the exponential regression model (7.57) with design space $\mathfrak{X} = [d, +\infty)$. Then $x_{1;0}^*(b) \equiv 0$,

$$x_{i;d}^*(b) = x_{i;0}^*(b) + d, \ i = 2, \ldots, 2k,$$

$$x_{i;0}^*(\nu b) = x_{i;0}^*(b/\nu), \ i = 2, \ldots, 2k$$

for any $\nu > 0$.

Consider now a few examples.

Note that the methods for constructing locally E-optimal designs applied below can be used only in the case when the support points of the designs are the Chebyshev points for the system (7.58).

For this reason, all designs obtained by the Taylor expansion were checked for optimality by means of the equivalence theorem for E-criterion (Theorem 3.3.1). In all cases considered in our numerical study, the Equivalence Theorem confirmed our designs to be locally E-optimal and we did not find cases where the multiplicity of the minimum eigenvalue of the information matrix in the exponential regression model (7.57) was larger than 1.

Example 7.6.1 Consider the exponential model $\mathbf{E}(Y(x)) = a_1 e^{-\lambda_1 x}$ corresponding to the case $k = 1$. In this case, the Chebyshev function $\phi(x) = (1 + q_2^* x) e^{-\lambda_1 x}$ minimizing

$$\sup_{x \in [0, \infty)} |(1 + ax) e^{-\lambda_1 x}|$$

with respect to the parameter $a \in \mathbb{R}$ and the corresponding extremal point x_2^* are determined by the equations $\phi(x_2^*) = -\phi(0)$ and $\phi'(x_2^*) = 0$, which are equivalent to

$$e^{-\lambda_1 x_2} - \lambda_1 x_2 + 1 = 0, \quad p_1 e^{-\lambda_1 x_2} + \lambda_1 = 0.$$

Therefore, the second point of the locally E-optimal design is given by $x_2^* = t^*/\lambda_1$, where t^* is the unique solution of the equation $e^{-t} = t - 1$ (the other support point is 0) and the locally E-optimal design is given by $\{0, x_2^*; w_1^*, w_2^*\}$, where the weights are calculated by the formula given in Remark 7.6.1; that is,

$$w_1^* = \frac{x_2^* e^{-\lambda_1 x_2^*} + \lambda_1}{x_2^* e^{-\lambda_1 x_2^*} + \lambda_1 + \lambda_1 e^{\lambda_1 x_2^*}}, \quad w_2^* = \frac{\lambda_1 e^{\lambda_1 x_2^*}}{x_2^* e^{-\lambda_1 x_2^*} + \lambda_1 + \lambda_1 e^{\lambda_1 x_2^*}}. \quad (7.61)$$

Example 7.6.2 For the exponential model

$$\mathbf{E}(Y(x)) = a_1 e^{-\lambda_1 x} + a_2 e^{-\lambda_2 x} \qquad (7.62)$$

corresponding to the case $k = 2$, the situation is more complicated and the solution of the locally E-optimal design problem cannot be determined directly. In this case, we used the Taylor expansion introduced in Section 2.4 for the construction of the locally E-optimal design, where the point $\lambda_{(0)}$ in this expansion was given by the vector $\lambda_{(0)} = (1.5, 0.5)^T$. By Remark 7.6.1(5) we can restrict ourselves to the case $\lambda_1 + \lambda_2 = 2$. Locally E-optimal designs for arbitrary values of $\lambda_1 + \lambda_2$ can be easily obtained by rescaling the support points of the locally E-optimal design found under the restriction $\lambda_1 + \lambda_2 = 2$, whereas the weights have to be recalculated using Remark 7.6.1(4). We consider the parametrization $\lambda_1 = 1 + z, \lambda_2 = 1 - z$ and study the dependence of the optimal design on the parameter z. Because $\lambda_1 > \lambda_2 > 0$, an admissible set of values z is the interval $(0, 1)$. We choose the center of this interval as the origin for the Taylor expansion. Table 7.5 contains the coefficients in the Taylor expansion for the points and weights of the locally E-optimal design; that is,

$$x_i^* = x_i(z) = \sum_{j=0}^{\infty} x_{i(j)}(z - 0.5)^j, \; w_i^* = w_i(z) = \sum_{j=0}^{\infty} w_{i(j)}(z - 0.5)^j$$

(note that $x_1^* = 0$ and $w_1^* = 1 - w_2^* - w_3^* - w_4^*$).

Table 7.5: The coefficients of the Taylor expansion for the support points and weights of the locally E-optimal designs

j	0	1	2	3	4	5	6
$x_{2(j)}$	0.4151	0.0409	0.0689	0.0810	0.1258	0.1865	0.2769
$x_{3(j)}$	1.8605	0.5172	0.9338	1.2577	2.1534	3.6369	6.3069
$x_{4(j)}$	5.6560	4.4313	10.505	20.854	44.306	90.604	181.67
$w_{2(j)}$	0.1875	0.2050	0.6893	0.3742	-1.7292	-1.2719	7.0452
$w_{3(j)}$	0.2882	0.2243	-0.0827	-0.8709	-0.1155	2.7750	1.8101
$w_{4(j)}$	0.4501	-0.4871	-0.9587	0.2323	2.9239	-0.2510	-12.503

The support points are depicted as a function of the parameter z in Figure 7.4. We observe for a broad range of the interval $(0, 1)$ only a weak dependence of the locally E-optimal design on the parameter z. Consequently, it is of some interest to investigate the robustness of the locally E-optimal design for the parameter $z = 0$, which corresponds to the vector $\lambda = (1, 1)$.

This vector yields to the limiting model (7.18), and by Theorem 7.7.1 given in the appendix, the locally E-optimal designs converge weakly to the design $\bar{\xi}_E^* := \bar{\xi}_{e_{2k}}^*$. The support points of this design can obtained from

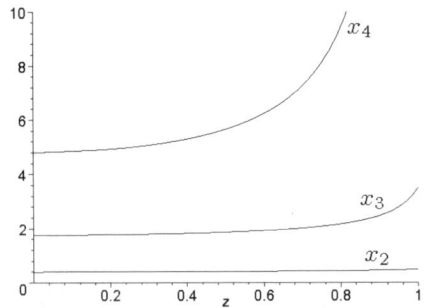

Figure 7.4: Support points of the locally E-optimal design $\xi_E^*(\lambda)$ in the exponential regression model (7.5), where $k = 2$ and $b \in (1 + z, 1 - z)^T$.

the corresponding Chebyshev problem:

$$\inf_{a_1, a_2, a_3} \sup_{x \in [0, \infty)} |(1 + a_1 x + a_2 x^2 + a_3 x^3) e^{-x}|$$

The solution of this problem can be found numerically using the Remez algorithm (see Studden and Tsay (1976)); that is,

$$P_3(x) = (x^3 - 3.9855 x^2 + 3.15955 x - 0.27701) e^{-x}.$$

The extremal points of this polynomial are given by

$$x_1^* = 0, \ x_2^* = 0.40635, \ x_3^* = 1.75198, \ x_4^* = 4.82719;$$

the weights of design $\bar{\xi}_E^*$ defined in Theorem 7.7.1 are calculated using formula (7.12); that is

$$w_1^* = 0.0767, \ w_2^* = 0.1650, \ w_3^* = 0.2164, \ w_4^* = 0.5419.$$

Table 7.6: Efficiencies of locally D- and E-optimal designs in the exponential regression model (7.62) ($\lambda_1 = 1 + z, \lambda_2 = 1 - z$). The locally D- and E-optimal design are denoted by $\xi_D^*(\lambda)$ and $\xi_E^*(\lambda)$, respectively, and $\bar{\xi}_D^*$ and $\bar{\xi}_E^*$ denote the weak limit of the locally D- and E-optimal design as $\lambda \to (1, 1)$, respectively

z	0.1	0.2	0.3	0.4	0.5	0.6	0.7	0.8	0.9
$I_D(\xi_D^*)$	1.00	1.00	1.00	0.99	0.98	0.95	0.90	0.80	0.61
$I_D(\xi_E^*(\lambda))$	0.75	0.74	0.75	0.75	0.78	0.82	0.87	0.90	0.89
$I_D(\bar{\xi}_E^*)$	0.74	0.74	0.76	0.77	0.78	0.79	0.78	0.72	0.58
$I_E(\xi_E^*)$	1.00	1.00	0.98	0.94	0.85	0.72	0.58	0.45	0.33
$I_E(\xi_D^*(\lambda))$	0.66	0.66	0.66	0.67	0.70	0.74	0.79	0.82	0.80
$I_E(\bar{\xi}_D^*)$	0.65	0.64	0.62	0.59	0.56	0.52	0.47	0.41	0.33

Table 7.7: Efficiencies (7.65) of the designs $\bar{\xi}_D^*$ and $\bar{\xi}_E^*$ [obtained as the weak limit of the corresponding locally optimal designs as $\lambda \to (1,1)$] for estimating the individual coefficients in the exponential regression model (7.62) ($\lambda_1 = 1 + z, \lambda_2 = 1 - z$)

z	0.1	0.2	0.3	0.4	0.5	0.6	0.7	0.8	0.9
$I_1(\bar{\xi}_E^*, \lambda)$	1.00	1.00	0.98	0.93	0.84	0.69	0.53	0.40	0.27
$I_1(\bar{\xi}_D^*, \lambda)$	0.65	0.64	0.61	0.57	0.50	0.41	0.32	0.26	0.19
$I_2(\bar{\xi}_E^*, \lambda)$	0.99	0.97	0.92	0.85	0.76	0.65	0.55	0.44	0.34
$I_2(\bar{\xi}_D^*, \lambda)$	0.68	0.70	0.70	0.68	0.65	0.60	0.54	0.46	0.37
$I_3(\bar{\xi}_E^*, \lambda)$	1.00	1.00	0.98	0.93	0.85	0.73	0.56	0.38	0.20
$I_3(\bar{\xi}_D^*, \lambda)$	0.65	0.64	0.62	0.58	0.52	0.45	0.35	0.24	0.13
$I_4(\bar{\xi}_E^*, \lambda)$	1.00	0.99	0.97	0.94	0.88	0.76	0.57	0.33	0.10
$I_4(\bar{\xi}_D^*, \lambda)$	0.63	0.59	0.54	0.49	0.42	0.34	0.24	0.13	0.04

Some E-efficiencies

$$I_E(\xi, \lambda) = \frac{\lambda_{\min}(M(\xi))}{\lambda_{\min}(M(\xi_E^*(\lambda)))} \tag{7.63}$$

of the limiting design $\bar{\xi}_E^*$ are given in Table 7.6 and we observe that this design yields rather high efficiencies whenever $z \in (0, 0.6)$. In this table, we also display the E-efficiencies and the locally D-optimal design $\xi_D^*(\lambda)$, the D-efficiencies

$$I_D(\xi, \lambda) = \left(\frac{\det M(\xi)}{\sup_\eta \det M(\eta)}\right)^{1/2k} \tag{7.64}$$

of the locally E-optimal design $\xi_E^*(\lambda)$, and the corresponding efficiencies of the the the weak limit of the locally D-optimal designs $\bar{\xi}_D^*$. We observe that the design $\bar{\xi}_D^*$ is very robust with respect to the D-optimality criterion. On the other hand, the D-efficiencies of the E-optimal designs $\xi_E^*(\lambda)$ and its corresponding limit $\bar{\xi}_E^*$ are substantially higher than the E-efficiencies of the $\xi_D^*(\lambda)$ and $\bar{\xi}_D^*$.

Finally, investigate the efficiencies

$$I_i(\xi, \lambda) = \frac{\inf_\eta e_i^T M^{-1}(\eta) e_i}{e_i^T M^{-1}(\xi) e_i}, \quad i = 1, \ldots, 2k, \tag{7.65}$$

of the optimal designs $\bar{\xi}_D^*$ and $\bar{\xi}_E^*$ for the estimation of the individual parameters. These efficiencies are shown in Table 7.7. Note that in most cases the design $\bar{\xi}_E^*$ is substantially more efficient for estimating the individual parameters than the design $\bar{\xi}_D^*$. The design $\bar{\xi}_E^*$ can be recommended for a large range of possible values of z.

Example 7.6.3 For the exponential model

$$\mathbf{E}(Y(x)) = a_1 e^{-\lambda_1 x} + a_2 e^{-\lambda_2 x} + a_3 e^{-\lambda_3 x} \tag{7.66}$$

corresponding to the case $k = 3$, the locally E-optimal designs can be calculated by similar methods. For the sake of brevity, we present only the limiting design [obtained form the locally D- and E-optimal designs if $\lambda \to (1,1,1)$] and investigate the robustness with respect to the D- and E-optimality critera. The support points of the e_6-optimal designs in the heteroscedastic polynomial regression model (7.18) (with $\gamma = 1$) can be found as the extremal points of the Chebyshev function

$$P_5(x) = (x^5 - 11.7538x^4 + 42.8513x^3 - 55.6461x^2 + 21.6271x - 1.1184)e^{-x},$$

which are given by

$$x_1^* = 0, \ x_2^* = 0.2446, \ x_3^* = 1.0031,$$
$$x_4^* = 2.3663, \ x_5^* = 4.5744, \ x_6^* = 8.5654.$$

For the weights of the limiting design $\bar{\xi}_E^* := \bar{\xi}_{e_6}^*$, we obtain from the results of Section 7.3

$$w_1^* = 0.0492, \ w_2^* = 0.1007, \ w_3^* = 0.1089,$$
$$w_4^* = 0.1272, \ w_5^* = 0.1740, \ w_6^* = 0.4401.$$

For the investigation of the robustness properties of this design, we note by Remark 7.6.1(5) that we can restrict ourselves to the case $\lambda_1 + \lambda_2 + \lambda_3 = 3$. The support points in the general case are obtained by a rescaling, and the weights have to be recalculated using Remark 7.6.1(4). For the sake of brevity, we do not present the locally E-optimal designs, but restrict ourselves to some efficiency considerations. For this, we introduce the parametrization $\lambda_1 = 1 + u + v$, $\lambda_2 = 1 - u$, and $\lambda_3 = 1 - v$, where the restriction $\lambda_1 > \lambda_2 > \lambda_3 > 0$ yields

$$u < v, v < 1 \ , \ \ u > -v/2.$$

In Table 7.8 we show the E-efficiency defined in (7.63) of the design $\bar{\xi}_E^*$, which is the weak limit of the locally E-optimal design $\xi_E^*(\lambda)$ as $\lambda \to (1,1,1)$ (see Theorem 7.7.1). Two conclusions can be drawn from our numerical results. On the one hand, we observe that the optimal design $\bar{\xi}_E^*$ is robust in a neighborhood of the point $(1,1,1)$. On the other hand, we see that the locally E-optimal design $\xi_E^*(\lambda)$ is also robust if the nonlinear parameters λ_1, λ_2, and λ_3 differ not too substantially (i.e., the "true" parameter is contained in a moderate neighborhood of the point $(1,1,1)$). Table 7.8 also contains the D-efficiencies of the E-optimal designs defined in (8.46) and the E-efficiencies of the locally D-optimal design $\xi_D^*(\lambda)$ and its corresponding weak limit as $\lambda \to (1,1,1)$. Again, the D-efficiencies of the E-optimal designs are higher than the E-efficiencies of the D-optimal designs.

Finally, compare briefly the limits of the locally E- and D-optimal designs if $\lambda \to (1,1,1)$ with respect to the criterion of estimating the individual coefficients in the exponential regression model (7.66). In Table 7.9, we

Table 7.8: Efficiencies of locally and limiting D- and E-optimal design, with $k = 3$ ($\lambda_1 = 1 + u + v, \lambda_2 = 1 - u, \lambda = 1 - v$)

u	0	0	0	-0.2	-0.2	0.2	0.2	0.4	0.4
v	0.2	0.5	0.8	0.6	0.8	0.3	0.8	0.5	0.8
$I_D(\xi_D^*)$	1.00	0.98	0.83	0.97	0.86	0.99	0.79	0.92	0.70
$I_D(\xi_E^*(\lambda))$	0.78	0.85	0.90	0.86	0.90	0.66	0.90	0.61	0.86
$I_D(\bar{\xi}_E^*)$	0.75	0.78	0.74	0.78	0.75	0.77	0.71	0.78	0.65
$I_E(\bar{\xi}_E^*)$	0.98	0.76	0.43	0.71	0.48	0.93	0.36	0.53	0.19
$I_E(\xi_D^*(\lambda))$	0.65	0.73	0.79	0.74	0.79	0.55	0.79	0.52	0.74
$I_E(\xi_D^*)$	0.63	0.57	0.37	0.53	0.40	0.46	0.31	0.23	0.09

show the efficiencies of these designs for estimating the parameters in a_1, b_1, a_2, b_2, a_3, and b_3 in the model (7.66). We observe that, in most cases, the limit of the locally E-optimal designs $\bar{\xi}_E^*$ yields substantially larger efficiencies than the corresponding limit of the locally D-optimal design $\bar{\xi}_D^*$. Moreover, this design is robust for many values of the parameter (u, v).

Table 7.9: Efficiencies (7.65) of the designs $\bar{\xi}_D^*$ and $\bar{\xi}_E^*$ (obtained as the weak limit of the corresponding locally optimal designs as $\lambda \to (1, 1, 1)$ for estimating the individual coefficients in the exponential regression model (7.66) ($\lambda_1 = 1 + u + v, \lambda_2 = 1 - u, \lambda = 1 - v$)

u	0	0	0	-0.2	-0.2	0.2	0.2	0.4	0.4	0.7
v	0.2	0.5	0.8	0.6	0.8	0.3	0.8	0.5	0.8	0.8
$I_1(\bar{\xi}_E^*)$	0.98	0.77	0.43	0.71	0.48	0.86	0.35	0.52	0.26	0.11
$I_1(\bar{\xi}_D^*)$	0.63	0.56	0.36	0.53	0.40	0.59	0.30	0.41	0.22	0.10
$I_2(\bar{\xi}_E^*)$	0.97	0.74	0.43	0.70	0.48	0.80	0.37	0.49	0.29	0.19
$I_2(\bar{\xi}_D^*)$	0.65	0.59	0.42	0.55	0.43	0.63	0.38	0.48	0.33	0.23
$I_3(\bar{\xi}_E^*)$	0.90	0.73	0.43	0.71	0.48	0.93	0.38	0.53	0.16	0.02
$I_3(\bar{\xi}_D^*)$	0.71	0.59	0.38	0.53	0.40	0.46	0.47	0.23	0.04	0.01
$I_4(\bar{\xi}_E^*)$	0.99	0.82	0.41	0.73	0.47	0.93	0.31	0.53	0.17	0.02
$I_4(\bar{\xi}_D^*)$	0.60	0.50	0.29	0.51	0.36	0.48	0.20	0.25	0.10	0.01
$I_5(\bar{\xi}_E^*)$	0.99	0.85	0.30	0.76	0.35	0.93	0.21	0.53	0.11	0.02
$I_5(\bar{\xi}_D^*)$	0.55	0.39	0.12	0.33	0.14	0.46	0.09	0.23	0.05	0.01
$I_6(\bar{\xi}_E^*)$	0.99	0.84	0.26	0.75	0.31	0.93	0.18	0.53	0.09	0.02
$I_6(\bar{\xi}_D^*)$	0.53	0.34	0.08	0.27	0.10	0.45	0.06	0.22	0.03	0.01

7.7   Appendix: Some Auxiliary Results

Recall the notation in Sections 7.2 and 7.3

$$f_i(t) = h_i(t), \quad t = 1, \ldots, s,$$

$$f_{s+2i-1}(t) = f_{s+2i-1}(t, b) = \varphi(t, b_i,) \quad i = 1, \ldots, k, \tag{7.67}$$

$$f_{s+2i}(t) = f_{s+2i}(t, b) = \varphi'(t, b_i), \quad i = 1, \ldots, k$$

$$\bar{f}_i(t) = h_i(t), \quad i = 1, \ldots, s,$$

$$\bar{f}_{s+i}(t) = \bar{f}_{s+i}(t, x) = \varphi^{(i)}(t, x), \quad i = 1, \ldots, 2k. \tag{7.68}$$

Let $f(t, b) = (f_1(t), \ldots, f_m(t))^T$ and $\bar{f}(t, x) = (\bar{f}_1(t), \ldots, \bar{f}_m(t))^T$ denote the corresponding vectors of regression functions ($m = s + 2k$) and consider a design ξ on the interval I with at least m support points. In this appendix, we investigate the relation between the information matrices

$$M(\xi, b) = \int_I f(t, b) f^T(t, b) \, d\xi(t)$$

and

$$\bar{M}(\xi, b) = \int_I \bar{f}(t, x) \bar{f}^T(t, x) \, d\xi(t)$$

defined by (7.7) and (7.21), respectively, if

$$\delta_i = r_i \delta = b_i - x \to 0, \quad i = 1, \ldots, k, \tag{7.69}$$

where the components of the vector $r = (r_1, \ldots, r_k)$ are different and ordered.

Theorem 7.7.1 *Assume that $\varphi \in C^{0,2k-1}$ and ξ is an arbitrary design, such that the matrix $\bar{M}(\xi, b)$ is nonsingular. If assumption (7.69) is satisfied, it follows that for sufficiently small δ, the matrix $M(\xi, b)$ is invertible, and if $\delta \to 0$,*

$$M^{-1}(\xi, b) = \delta^{-4k+4} T(\delta) \begin{pmatrix} \bar{M}^{(1)}(\xi) & \bar{M}^{(2)}(\xi) F \\ F^T \bar{M}^{(2)^T}(\xi) & \gamma \gamma^T h + o(1) \end{pmatrix} T(\delta) + o(1),$$

$$\tag{7.70}$$

where the matrices $T(\delta) \in \mathbb{R}^{m \times m}$, $\bar{M}^{(1)}(\xi) \in \mathbb{R}^{s \times s}$, $\bar{M}^{(2)}(\xi) \in \mathbb{R}^{s \times 2k}$, and $\bar{M}^{(3)}(\xi) \in \mathbb{R}^{2k \times 2k}$ are defined by

$$T(\delta) = \mathrm{diag}\Big(\underbrace{\delta^{2k-2}, \ldots, \delta^{2k-2}}_{s}, \underbrace{\frac{1}{\delta}, 1, \frac{1}{\delta}, 1, \ldots, \frac{1}{\delta}, 1}_{2k}\Big),$$

$$\begin{pmatrix} \bar{M}^{(1)} & \bar{M}^{(2)}(\xi) \\ \bar{M}^{(2)^T}(\xi) & \bar{M}^{(3)}(\xi) \end{pmatrix} = \bar{M}^{-1}(\xi, x),$$

the vector $\gamma = (\gamma_1, \ldots, \gamma_{2k})^T$ and $h \in \mathbb{R}$ are given by $h = [(2k - 1)!]^2 e_m^T \bar{M}^{-1}(\xi, x) e_m$,

$$\gamma_{2i} = \prod_{j \neq i} (r_i - r_j)^{-2}, \quad i = 1, \ldots, k,$$

$$\gamma_{2i-1} = -\gamma_{2i} \sum_{j \neq i} \frac{2}{r_i - r_j}, \quad i = 1, \ldots, k, \tag{7.71}$$

and the matrix $F \in \mathbb{R}^{2k \times 2k}$ is defined by

$$F = \begin{pmatrix} 0 & \cdots & 0 & \frac{\gamma_1}{0!} \\ \vdots & & & \\ 0 & \cdots & 0 & \frac{\gamma_{2k}}{(2k-1)!} \end{pmatrix}.$$

Proof. Define $\psi(\delta) = (1, \delta, \ldots, \delta^{2k-1})^T$ and introduce the matrices

$$L = (\ell_1, \ldots, \ell_{2k})^T \in \mathbb{R}^{2k \times 2k}, \tag{7.72}$$

$$U = \operatorname{diag}\left(1, \frac{1}{1!}, \frac{1}{2!}, \ldots, \frac{1}{(2k-1)!}\right) \in \mathbb{R}^{2k \times 2k}, \tag{7.73}$$

where $\ell_{2i-1} = \psi(\delta_i)$ and $\ell_{2i} = \psi'(\delta_i)$, $i = 1, \ldots, k$). For fixed $t \in I$, we use the Taylor expansions

$$\varphi(t, x + \delta) = \sum_{j=0}^{2k-1} \frac{\varphi^{(i)}(t, x)}{j!} \delta^j + o(\delta^{2k-1})$$

$$\varphi'(t, x + \delta) = \sum_{j=1}^{2k-1} \frac{\varphi^{(i)}(t, x)}{(j-1)!} \delta^{j-1} + o(\delta^{2k-2})$$

to obtain the representation

$$f(t, b + \delta r) = \begin{pmatrix} I_s & 0 \\ 0 & LU \end{pmatrix} \bar{f}(t, x) + \begin{pmatrix} 0 \\ \tilde{f}(t) \end{pmatrix}, \tag{7.74}$$

where $I_s \in \mathbb{R}^{s \times s}$ denotes the identity matrix and the vector $\tilde{f}$ is of order

$$\tilde{f}(t) = (o(\delta^{2k-1}), o(\delta^{2k-2}), o(\delta^{2k-1}), \ldots, o(\delta^{2k-2}))^T. \tag{7.75}$$

It follows from in Karlin and Studden (1966, pp. 127–129) that

$$\det L = \prod_{1 \leq i < j \leq k} (\delta_i - \delta_j)^4,$$

and, consequently, $V = (v_1, \ldots, v_{2k}) := L^{-1}$ exists. The equality $LV = I_m$ implies the equations

$$v_{2i}^T \psi(\delta_j) = 0, \qquad v_{2i}^T \psi'(\delta_j) = 0, \quad j \neq i,$$
$$v_{2i}^T \psi(\delta_i) = 0, \qquad v_{2i}^T \psi'(\delta_i) = 1,$$

which shows that $\delta_1, \ldots, \delta_{i-1}, \delta_{i+1}, \ldots, \delta_k$ are roots of multiplicity 2 of the polynomial $v_{2i}^T \psi(\delta)$ and δ_i is a root of multiplicity 1. Because this polynomial has degree $2k - 1$, it follows that

$$v_{2i}^T \psi(\delta) = (\delta - \delta_i) \prod_{j \neq i} \left(\frac{\delta - \delta_j}{\delta_j - \delta_i} \right)^2, \qquad (7.76)$$

and a similar argument shows

$$v_{2i-1}^T \psi(\delta) = \frac{\delta - \alpha_i}{\delta_i - \alpha_i} \prod_{j \neq i} \left(\frac{\delta - \delta_j}{\delta_i - \delta_j} \right)^2, \qquad (7.77)$$

where the constants $\alpha_1, \ldots, \alpha_k$ are given by

$$\alpha_i = \delta_i + \left(\sum_{j \neq i} \frac{2}{\delta_i - \delta_j} \right)^{-1}, \quad i = 1, \ldots, k. \qquad (7.78)$$

From (7.74) and (7.75) we obtain

$$f(t, b + \delta r) f^T(t, b + \delta r)$$

$$= \left(\begin{array}{cc} I_s & 0 \\ 0 & LU \end{array} \right) \bar{f}(t, x) \bar{f}^T(t, x) \left(\begin{array}{cc} I_s & 0 \\ 0 & LU \end{array} \right)^T + o(\delta^{2k-2}),$$

and integrating the right-hand side with respect to the design ξ shows that

$$M(\xi, b + \delta r) = \left(\begin{array}{cc} I_s & 0 \\ 0 & LU \end{array} \right) \bar{M}(\xi, x) \left(\begin{array}{cc} I_s & 0 \\ 0 & LU \end{array} \right)^T + o(\delta^{2k-2}). \qquad (7.79)$$

Now, define $H_1(\delta) = \mathrm{diag}(\delta^{2k-1}, \delta^{2k-2}, \delta^{2k-1}, \ldots, \delta^{2k-1}, \delta^{2k-2}) \in \mathbb{R}^{2k \times 2k}$ and

$$H(\delta) = \left(\begin{array}{cc} I_s & 0 \\ 0 & H_1(\delta) \end{array} \right) \in \mathbb{R}^{m \times m};$$

we obtain from (7.76) and (7.77) that

$$H_1(\delta)(L^{-1})^T = (0 \mid \gamma) + o(1) ,$$

where $\gamma = (\gamma_1, \ldots, \gamma_{2k})^T$ is defined by (7.71) and $0 \in \mathbb{R}^{2k \times 2k-1}$ denotes the matrix with all entries equal to zero. By (7.67), this implies for the inverse of the matrix $M(\xi, b + \delta r)$,

$$M^{-1}(\xi, b + \delta r)$$

$$= H^{-1}(\delta) \left\{ \left(\begin{array}{cc} I & 0 \\ 0 & F \end{array} \right) \bar{M}^{-1}(\xi, x) \left(\begin{array}{cc} I & 0 \\ 0 & F^T \end{array} \right) + o(1) \right\} H^{-1}(\delta)$$

$$= \delta^{-4k+4} T(\delta) \left\{ \left(\begin{array}{cc} \bar{M}^{(1)}(\xi) & \bar{M}^{(2)}(\xi) F^T \\ F \bar{M}^{(2)^T}(\xi) & F \bar{M}^{(3)}(\xi) F^T \end{array} \right) + +o(1) \right\} T(\delta),$$

where the matrix F is given by $F = (0|\gamma)U^{-1} \in \mathbb{R}^{2k \times 2k}$. The assertion now follows by a straightforward calculation that shows that

$$F\bar{M}^{(3)}(\xi)F^T = h\gamma\gamma^T.$$

■

Chapter 8

The Monod Model

In this chapter, the estimation problem and the problem of designing experiments in a nonlinear regression model, used in microbiology, are studied. The model is called Monod model, defined implicitly by a differential equation for the regression function and it has numerous applications in microbial growth kinetics, water research, pharmacokinetics, and plant physiology. It is proved that least squares parameter estimates are asymptotically unbiased and normally distributed. The asymptotic covariance matrix of the least squares estimator is the basis for construction of efficient designs of experiments.

In spite of the regression function determined only implicitly, it is established that the matrix corresponds to the covariance matrix of a certain explicitly given linear model. Also, the basis functions of the last generates a Chebyshev system. This allows one to apply the theory developed in the previous chapters for studying locally D-, E-, and c-optimal designs for the Monod model. Sensitivity of these designs to the choice of initial values for parameters is investigated numerically and on the basis of Taylor expansions for support points. For rather small deviations of the initial values from the proper ones, the designs remains very efficient and proves to be considerably better equidistant designs that usually used in practice. If the deviations are more serious, the maximin efficient designs, defined in Section 1.7, are recommended. Such designs are constructed numerically for a variety of intervals for individual parameters of the model.

The chapter is based on Dette, Melas, Pepelyshev, and Strigul (2003 and 2005). Note that Section 8.5 is completely new.

8.1   Introduction

The Monod model is widely applied for modeling biodegradation rates. It is used for describing microbial growth and substrate degradations in all kinds of applications (e.g., batch and continuous fermentation, activated sludge wastewater treatment, pharmacokinetics, plant physiology etc. (see, e.g.. Pirt (1975) and Holmberg (1982))). Much of the versatility of the Monod model is due to the fact that it can describe biodegradation rates following zero-one first-order kinetics with respect to the target substrate concentration (see Holmberg (1982)). Roughly speaking, the model consists of a first-order differential equation; that is,

$$\eta'(t) = \mu(t)\eta(t), \tag{8.1}$$

where the function μ is defined by

$$\mu(t) = \vartheta_1 \frac{s(t)}{s(t) + \vartheta_2} \tag{8.2}$$

and the function s is given by the expression

$$s(t) - s_0 = (\eta_0 - \eta(t))/\vartheta_3 \tag{8.3}$$

(here, $s_0 = s(0)$ and $\eta_0 = \eta(0)$ are given initial conditions). In microbiology, a traditional notation is used for the unknown parameters of the Monod model (see, e.g., Pirt (1975)). The parameter ϑ_1 denotes the maximum growth rate and is usually denoted by $\mu_{\max}$ or $V_{\max}$; ϑ_2 is the saturation of affinity constant and is often denoted by K_s. The parameter ϑ_3 is the yield coefficient (often denoted by Y) and $\eta(t)$ and $s(t)$ denote the concentration of microorganisms and the concentration of the substrate, respectively. The explanatory variable t usually denotes the time, which varies in a compact interval $[0, T]$, where the maximal time T can be of quite different size. The minimum is several hours for optimal microbiological media, and the maximum is 1 year or more for specialized groups of microorganisms. Due to natural biological conditions we can assume $\vartheta_i > 0$, $i = 1, 2, 3$, the initial conditions s_0 and η_0 are usually known and positive and the explanatory variable t varies in an interval, say $[0, T]$.

In a number of recent papers, the problem of parameter estimation and the problem of designing experiments for this model was discussed in an extensive empirical study (see Vanrolleghem, Van Daele, and Dochaine (1995), Merkel, Schwarz, Fritz, Reuss, and Krauth (1996), Ossenbruggen, Spanjers, and Klapwik (1996)) . The results of these authors indicate that the information quality of the experiments is highly dependent on the design, and major improvements can be obtained by choosing the observations at appropriate allocations (see Vanrolleghem, Van Daele, and Dochaine (1995)).

It it the purpose of the present chapter to provide some more theoretical background for statistical inference in the Monod model. To be precise, we assume that at experimental conditions $t_1, \ldots, t_n$, independent observations $y_{1(1)}, \ldots, y_{1(r_1)}, \ldots, y_{n(1)}, \ldots, y_{n(r_n)}$ are available, which are given by

$$y_{j(i)} = \eta(t_j, \theta) + \varepsilon_{j(i)}, \ i = 1, \ldots, r_j, \ j = 1, \ldots, n, \tag{8.4}$$

where $\eta = \eta(\cdot, \theta)$ is a solution of the Monod equation defined by (8.1). In other words, r_j denotes the number of observations at time point t_j ($j = 1, \ldots, n$), $\varepsilon_{j(i)}$ are independent and identically distributed random values with zero mean and constant variance $\sigma^2 > 0$, and $\theta = (\vartheta_1, \vartheta_2, \vartheta_3)^T$ denotes the vector of parameters. Throughout this section the "true" value for θ in model (8.1) will be denoted by $\theta^* = (\vartheta_1^*, \vartheta_2^*, \vartheta_3^*)^T$.

In Section 8.2*, we demonstrate that statistical inference in this model is closely related to analysis in an equivalent linear regression model and we establish consistency and asymptotic normality of the least squares estimator (LSE) $\hat{\theta} = (\hat{\vartheta}_1, \hat{\vartheta}_2, \hat{\vartheta}_3)^T$, which minimizes the expression

$$\sum_{j=1}^{n} \sum_{i=1}^{r_j} (y_{j(i)} - \eta(t_j, \theta))^2 \; . \tag{8.5}$$

For a sufficiently large sample size, we show that the covariance matrix of the LSE can be approximated by the inverse of the Fisher information matrix

$$\frac{N}{\sigma^2} \left(\sum_{j=1}^{n} \frac{r_j}{N} \frac{\partial}{\partial \vartheta_i} \eta(t_k, \theta) \Big|_{\theta=\theta^*} \frac{\partial}{\partial \vartheta_j} \eta(t_k, \theta) \Big|_{\theta=\theta^*} \right)^3_{i,j=1} , \tag{8.6}$$

where $N = \sum_{j=1}^{n} r_j$ denotes the total number of observations. In the second part of Section 8.2, we will study locally optimal designs (see Section 1.7) for estimating the parameters in the nonlinear regression function obtained as a solution of the Monod equation. These designs minimize an appropriate functional of the Fisher information matrix defined in (8.6). Although the regression function in the Monod model is only given implicitly, we are able to obtain an explicit representation of the information matrix (8.6), which can be used for the construction of locally optimal designs. Exemplarily we determine optimal designs with respect to the D-, E-, and c-optimality critera. Section 8.3 deals with the D-optimality criterion. We find the best three-point designs and show that these designs are D-optimal within the class of all designs if the design region $[0, T]$ is sufficiently large and the initial value η_0 is sufficiently close to 0. These results are used for the construction of efficient designs (with respect to the D-criterion) on arbitrary design spaces. In Section 8.4, we present some numerical results, compare the D-optimal designs with the uniform design (which is commonly used for this type of problem), and study the sensitivity of the locally D-optimal designs with respect to changes of the initial values for the parameters. Locally E-optimal designs and optimal designs for estimating the individual parameters are investigated in Section 8.5,

---

*Note that in Sections 8.2–8.4 and 8.6 some materials (theorems, tables, and figures) are taken from Dette, H., Melas, V. B., Pepelyshev, A., Strigul, N. (2003). Efficient design of experiments in the Monod model. *J. Roy. Statist. Soc.*, Series B, **65**, 725-742. ©2003 Royal Statistical Society.

which also contains some numerical results. Some conclusions and recommendations are given in Section 8.6, and all technical details were deferred to an appendix in Section 8.8. Finally, it must be stressed at this point that locally optimal designs are influenced by a preliminary "guess" for the parameter values. Our results demonstrate that without any prior information, locally optimal designs for the Monod model are not robust with respect to misspecification of the initial parameters and therefore give some arguments in favor of more robust optimality criteria (see Section 1.7). The results of Sections 8.2–8.6 provide a first step in the construction of optimal designs for the Monod model with respect to these more sophisticated criteria. However, if certain intervals for the nonlinear parameters can be specified based on a microbiological background, we are able to construct locally optimal designs that are robust with respect to misspecification of the initial parameters and allow efficient estimation of the parameters in the Monod model. In many applications of the Monod model, the statistical inference is made in two steps and some information regarding the parameters in the model is available from the first step, which can be used for the construction of efficient designs in the second step (see, e.g., Merkel, Schwarz, Fritz, Reuss, and Krauth (1996)). In such cases the application of the locally optimal designs determined in Sections 8.2–8.6 is well justified and yields a substantial improvement with respect to the accuracy of the parameter estimates. Moreover, such information can be used also for constructing maximin efficient designs defined in Section 1.7. In Section 8.7[†], we construct numerically such designs for the Monod model. These designs are based on given intervals for values of nonlinear parameters and have some advantages to be discussed in this section. The last section contains the proofs of the results.

8.2   Equivalent Regression Models

The analogue (up to the constant N/σ^2) of the Fisher information matrix (8.6) is the matrix

$$
M(\xi, \theta^*) = \left(\sum_{k=1}^{n} w_k \frac{\partial}{\partial \vartheta_i} \eta(t_k, \theta) \Big|_{\theta=\theta^*} \frac{\partial}{\partial \vartheta_j} \eta(t_k, \theta) \Big|_{\theta=\theta^*} \right)_{i,j=1}^{3}, \qquad (8.7)
$$

where

$$
\xi = \left(\begin{array}{ccc} x_1 & \cdots & x_n \\ \omega_1 & \cdots & \omega_n \end{array}, \right) \qquad (8.8)
$$

which is called *the information matrix* of the design ξ. Our first result shows that least squares estimates based on an approximate design and

[†]In this section some materials are taken from Dette, H., Melas, V.B., Pepelyshev, A., Strigul, N. (2005). Design of experiments for the Monod model – robust and efficient designs. *J. Theor. Biol.* **234**, 537–550. ©2005 Elsevier Ltd. with permission of Elsevier Publisher.

a simple rounding procedure are consistent and asymptotically normally distributed. The proof can be found in the appendix.

Theorem 8.2.1 *Let ξ denote an arbitrary design of the form (8.8) on the interval $[0, T]$, $n \geq 3$, and assume that r_j observations are taken at the points t_j, where the values r_j are obtained by rounding the values $w_j N$ to integers such that $\sum_{j=1}^{n} r_j = N$. If $\eta_0, s_0 > 0$ and*

$$\theta^* \in \Omega = \{\theta = (\vartheta_1, \vartheta_2, \vartheta_3)^T : \vartheta_i > 0, \ i = 1, 2, 3\}, \tag{8.9}$$

then the nonlinear LSE $\hat{\theta}$ of the parameter θ^* minimizing (8.5) satisfies

$$\sqrt{N}(\hat{\theta} - \theta^*) \overset{\mathcal{D}}{\Longrightarrow} \mathcal{N}(0, \sigma^2 M^{-1}(\xi, \theta^*)), \quad N \to \infty .$$

In other words, the vector $\sqrt{N}(\hat{\theta} - \theta^*)$ has asymptotically a normal distribution with mean zero and covariance matrix $\sigma^2 M^{-1}(\xi, \theta^*)$.

Note that Theorem 8.2.1 provides asymptotic unbiasedness and normality of the LSE $\hat{\theta}$ in the Monod model (8.1). Moreover, asymptotically the covariance matrix of the vector $\sqrt{N}\hat{\theta}_N$ is given by

$$\sigma^2 M^{-1}(\xi, \theta^*),$$

where the information matrix is defined in (8.7). Following Chernoff (1953), we assume that θ^0 is a prior guess of the "true" parameter θ^* and call a design $\xi_{\theta^0}^*$ locally D-optimal design if

$$\det M(\xi_{\theta^0}^*, \theta^0) = \max_{\xi} \det M(\xi, \theta^0),$$

where the maximum is taken over all designs on the interval $[0, T]$. The concept of local optimality for nonlinear regression models requires two assumptions:

- The number of observations is sufficiently large such that the asymptotic theory is applicable.

- If the design is optimal for prior guess θ^0, then it is also efficient for the "true" (but unknown) parameter θ^*; that is,

$$I_{\theta^*}(\xi_{\theta^0}^*) = \left(\frac{\det M(\xi_{\theta^0}^*, \theta^*)}{\det M(\xi_{\theta^*}^*, \theta^*)} \right)^{1/m} > 1 - \delta,$$

  where m is the number of parameters ($m = 3$ in this case) and δ is a small positive constant.

The quantity $I_{\theta^*}(\xi_{\theta^0}^*)$ is called "relative efficiency" of the design $\xi_{\theta^0}^*$ with respect to the (unknown) design $\xi_{\theta^*}^*$, and $I_{\theta^*}^{-1}(\xi_{\theta^0}^*)$ represents the (relative) additional amount of observations that is necessary to obtain the same accuracy with the design $\xi_{\theta^0}^*$ compared to the "ideal" (but unknown) design $\xi_{\theta^*}^*$. The first assumption can be verified by simulations, and the second assumption can be verified by a robustness study if the locally D-optimal designs are known (see Section 8.4 for an example). If these assumptions can be justified, the problem of searching a locally D-optimal design for the Monod model (8.1) is equivalent to the problem of searching a D-optimal design for the linear regression model

$$\theta^T f(t) \;=\; \theta^T \frac{\partial}{\partial \theta} \eta(t, \theta)\Big|_{\theta=\theta^0}\,,$$

and all results from the classical linear theory can be transferred to local optimality criteria. For example, the local D-optimality criterion guarantees (asymptotically) a minimum volume of the confidence ellipsoid if the random errors in the model (8.4) follow a normal distribution (see Karlin and Studden (1966, Chap. X)) and the local D-optimality can be characterized by the celebrated equivalence theorem of Kiefer and Wolfowitz (1960), which shows that a design $\xi_{\theta^0}^*$ is locally D-optimal if and only if

$$f^T(t) M^{-1}(\xi_{\theta^0}^*, \theta^0) f(t) \leq m, \quad \forall t \in [0, T], \qquad (8.10)$$

where m is the number of parameters in the model. Moreover, the D-efficiency of a given design with respect to a locally D-optimal design can be evaluated by Kiefer's inequality without an explicit construction of a locally D-optimal design. This inequality yields for the D-efficiency,

$$\left(\frac{\det M(\xi, \theta^0)}{\max_\xi \det M(\xi, \theta^0)} \right)^{1/m} \geq e^{1 - v/m},$$

where the constant v is defined by

$$v = \max_{t \in [0,T]} f^T(t) M^{-1}(\xi, \theta^0) f(t)$$

(see Pukelsheim (1993)). Two further optimality criteria will be discussed in Section 8,5, which have been proposed as alternative for the Monod model (see, e.g., Vanrolleghem, Van Daete, and Dochain (1995) or Versyck, Bernaerts, Geeraerd, and Van Impe (1999)). A design is called locally E-optimal if it maximizes the minimum eigenvalue

$$\lambda_{\min}\left(M(\xi, \theta^0) \right) \qquad (8.11)$$

of the information matrix in the class of all (approximate) designs. An E-optimal design minimizes the worst variance

$$\max_{p^T p = 1} \mathrm{Var}(p^T \hat\theta),$$

taken over the variances of all (normalized) linear combinations of the parameter estimates $p^T\hat{\theta} = \sum_{i=1}^m p_i \hat{\vartheta}_i$ for the specific value θ^0. If only one linear combination is of interest, say $c^T\theta$, the locally c-optimality criterion might be useful, which determines a design minimizing the quantity

$$c^T M^-(\xi, \theta^0)c, \tag{8.12}$$

where A^- denotes the generalized inverse of the matrix A and the minimum is taken over the class of all designs ξ for which the linear combination $c^T\theta$ is estimable (i.e., $c \in \text{range}(M(\xi, \theta^0)))$ (see Pukelsheim (1993)). Note that for the special choice of a unit vector $c = e_k = (0, \ldots, 0, 1, 0, \ldots, 0)^T$, the c-optimal design minimizes the variance of the LSE for the parameter ϑ_k, $k = 1, \ldots, m$, for the specific value θ^0.

Locally optimal designs have been constructed for several nonlinear regression models (see Box and Lucas (1959), Melas (1978), Rasch (1990), Ford, Torsney, and Wu (1992), Haines (1992, 1993), Sitter and Torsney (1995), He, Studden and Sun (1996), Dette and Wong (1999) among many others). The problem of determining optimal designs for the Monod model under consideration is substantially more complex because the regression function in model (8.1) is only defined implicitly and the model has two nonlinear parameters. The main step in the solution of this design problem consists in a derivation of an alternative representation of the information matrix defined in (8.7). For this purpose, we introduce the notation

$$\begin{aligned} c &= c(\theta) = s_0\vartheta_3 + \eta_0, \\ b &= b(\theta) = \vartheta_2\vartheta_3/c, \end{aligned} \tag{8.13}$$

Combining (8.2) and (8.3), we obtain from (8.1) the differential equation

$$\eta'(t) = \vartheta_1 \frac{s_0\vartheta_3 + \eta_0 - \eta(t)}{s_0\vartheta_3 + \eta_0 - \eta(t) + \vartheta_2\vartheta_3}\eta(t) = \vartheta_1 \frac{c - \eta(t)}{c(1+b) - \eta(t)}\eta(t). \tag{8.14}$$

From the initial conditions $s_0 > 0$, $\eta_0 > 0$, and $\vartheta_i > 0$, $i = 1, 2, 3$, it follows that

$$\eta'(0) = \vartheta_1 \frac{s_0\vartheta_3}{s_0\vartheta_3 + \vartheta_2\vartheta_3} > 0,$$

and the following lemma describes some general properties of the regression function η.

Lemma 8.2.1 *Let η denote a nonconstant solution of the Monod equation (8.1) and $T > 0$; then*

$$\eta(t) > 0, \quad \eta'(t) > 0, \quad t \in [0, T],$$
$$\eta_0 \le \eta(t) < c, \quad t \in [0, T],$$
$$\lim_{t \to \infty} \eta(t) = c,$$

Note that the function η is strictly increasing on the interval $[0, T]$ and consequently, for a fixed vector θ, the inverse

$$t(x) = t(x, \theta) = \eta^{-1}(x, \theta) \tag{8.15}$$

exists on the interval $[\eta_0, \eta(T)]$ and satisfies $t(\eta_0) = 0$. From (8.14) it follows by a straightforward calculation

$$\frac{d\eta}{dt} = \frac{d\eta}{dt}\bigg|_{t=\eta^{-1}(u)} = \vartheta_1 \left(\frac{1+b}{u} + \frac{b}{c-u}\right)^{-1} ,$$

$$\frac{dt}{du} = \frac{d\eta}{du}^{-1} = \frac{1}{\vartheta_1}\left(\frac{1+b}{u} + \frac{b}{c-u}\right) ,$$

and integrating the last formula, we obtain

$$t(x) = \int_{\eta_0}^{x} \frac{1}{\vartheta_1}\left(\frac{1+b}{u} + \frac{b}{c-u}\right) du = \frac{1}{\vartheta_1}\left((1+b)\ln\frac{x}{\eta_0} + b\ln\frac{c-x}{c-\eta_0}\right). \tag{8.16}$$

Throughout this chapter let

$$\mathcal{X} = \{\eta(t, \theta^0) \mid t \in [0, T]\} \tag{8.17}$$

denote the induced design space; then any design ξ of the form (8.8) on the interval $[0, T]$ with $0 \le t_1 < t_2 < \cdots < t_n \le T$ induces a design ζ on $\mathcal{X}$ by the transformation

$$\eta_i = \eta(t_i, \theta^0), \quad i = 1, \ldots, n. \tag{8.18}$$

In the following, we will prove that the matrix $M(\xi, \theta^0)$ can be represented as a function of the points $\eta_1, \ldots, \eta_n$. Note that such a relation is obvious for linear models (see Pukelsheim (1993, p. 3)) but, in general not clear for the Monod model under consideration, because the information matrix for this model contains the partial derivatives of the regression function η evaluated at the points t_i. In order to prove this dependency, we differentiate the identity

$$\eta(t(x, \theta), \theta) \equiv x , \quad \theta \in \Omega , \ x \in [\eta_0, c], \tag{8.19}$$

with respect to the parameters ϑ_i $(i = 1, 2, 3)$ and obtain

$$\frac{\partial \eta(t, \theta)}{\partial t}\frac{\partial t(x, \theta)}{\partial \vartheta_i} + \frac{\partial \eta(t, \theta)}{\partial \vartheta_i} = 0 , \quad i = 1, 2, 3, \tag{8.20}$$

with $t = t(x, \theta)$ and $x = \eta(t, \theta)$. Now, observing (8.16) and (8.20), it follows by a direct computation that for any $t \ge 0$,

$$\frac{\partial \eta(t, \theta)}{\partial \theta} = K\varphi(x), \tag{8.21}$$

where $x = \eta(t, \theta)$ and the matrix $K \in \mathbb{R}^{3 \times 3}$ is defined by

$$
K = \begin{pmatrix}
\dfrac{1+b}{\vartheta_1} & \dfrac{b}{\vartheta_1} & 0 \\[2mm]
-\dfrac{b}{\vartheta_2} & -\dfrac{b}{\vartheta_2} & 0 \\[2mm]
-\dfrac{b\eta_0}{c\vartheta_3} & -\dfrac{b\eta_0}{c\vartheta_3} & -\dfrac{b}{\vartheta_3}
\end{pmatrix} .
\tag{8.22}
$$

Here,

$$
\varphi(x) = (\varphi_1(x), \varphi_2(x), \varphi_3(x))^T
$$

denotes a vector of regression functions with components

$$
\begin{aligned}
\varphi_1(x) &= \varphi_1(x, \theta) = v(x) \ln \frac{x}{\eta_0}, \\
\varphi_2(x) &= \varphi_2(x, \theta) = v(x) \ln \frac{c - x}{c - \eta_0}, \\
\varphi_3(x) &= \varphi_3(x, \theta) = v(x) \frac{x - \eta_0}{c - x},
\end{aligned}
\tag{8.23}
$$

and

$$
v(x) = v(x, \theta) = \frac{x(c - x)}{(1 + b)c - x} .
\tag{8.24}
$$

Note that for each θ, the function $\eta(t) = \eta(t, \theta)$ is strictly increasing with limit $\lim_{t \to \infty} \eta(t) = c$ (see Lemma 8.2.1). Thus, it is possible to extend this function by the definition $\eta(\infty) = c$. By the above discussion, we can now transfer the original (locally) optimal design problem for the Monod model to a design problem for a linear model on the induced design space $\mathcal{X}$. To this end, let

$$
\zeta = \begin{pmatrix} x_1 & \cdots & x_n \\ w_1 & \cdots & w_n \end{pmatrix}, \quad \eta_0 \le x_1 < x_2 < \cdots < x_n \le \bar{c}
\tag{8.25}
$$

denote an arbitrary design on the interval $[\eta_0, \bar{c}]$, where $\bar{c} \le c$, and define

$$
\xi_\zeta = \begin{pmatrix} t_1 & \cdots & t_n \\ w_1 & \cdots & w_n \end{pmatrix},
\tag{8.26}
$$

with $t_i = t(x_i, \theta^0)$, $i = 1, \ldots, n$, as the corresponding design on the original design space $[0, T]$ with $T = t(\bar{c}, \theta^0)$. We define

$$
\bar{M}(\zeta) = \bar{M}(\zeta, \theta^0) = \sum_{j=1}^{n} w_j \varphi(x_j) \varphi(x_j)^T
$$

as the information matrix of the design ζ in the (homoscedastic) linear regression model

$$
\beta^T \varphi(x) ,
\tag{8.27}
$$

where $\beta = (\beta_1, \beta_2, \beta_3)$ denotes the vector of parameters. It follows from (8.7) and (8.21) that

$$M(\xi_\zeta, \theta^0) = K\bar{M}(\zeta, \theta^0)K^T,$$

$$\det M(\xi_\zeta, \theta^0) = \frac{b^4}{(\vartheta_1^0 \vartheta_2^0 \vartheta_3^0)^2} \det\bar{M}(\zeta, \theta^0). \tag{8.28}$$

The following results are now obvious from these considerations.

Theorem 8.2.2 *A design ξ_ζ is a locally D-optimal design for the Monod model (8.1) on the interval $[0, T]$ if and only if the induced design ζ is D-optimal for the regression model (8.27) on the interval $[\eta_0, \bar{c}]$, $\bar{c} = \eta(T)$, under the standard assumptions about the measurement errors.*

Theorem 8.2.3 *A design ξ_ζ is a locally E- (e_k)-optimal design for the Monod model (8.1) on the interval $[0, T]$ if and only if the design ζ is E- (e_k)-optimal for regression function $\beta^T K\varphi(x)$ on the interval $[\eta_0, \bar{c}]$, $\bar{c} = \eta(T)$, under the standard assumptions on the measurement errors.*

Consequently, it is sufficient to construct locally D-optimal designs for the regression model (8.27) and locally E- and e_k-optimal designs for the regression model $\beta^T K\varphi(x)$. The locally optimal designs for the Monod model (8.1) are simply obtained by transforming the design η in (8.25) to the design ξ_ζ in (8.26). We will illustrate this method in the following sections, discussing the different optimality criteria separately.

8.3   Locally D-Optimal Designs

Due to Theorem 8.2.2, it will be sufficient to study D-optimal designs for the linear regression model (8.27) under the standard assumptions about measurements errors. Recall the notation $c = \eta(\infty, \theta^0)$ and $\bar{c} = \eta(T, \theta^0) \leq c$; then it is easy to see that the definition of the vector of regression functions φ in model (8.27) can be continuously extended by putting

$$\varphi_i(c) = 0, \; i = 1, 2$$
$$\varphi_3(c) = \frac{c - \eta_0}{b} = \frac{s_0 c}{\vartheta_2}.$$

In other words, the vector of regression functions in model (8.27) is well defined on intervals $[\eta_0, \bar{c}]$, where $\bar{c} = \eta(T, \theta^0)$, $0 < T \leq \infty$, $\bar{c} \leq c$. Let us denote designs that are D-optimal in the class of all designs supported at k support points as D-optimal k-point designs (E- and e_k-optimal k-point designs are defined similarly). It is well known (see Karlin and Studden (1966, Chap. X)) that if the number of design points is less than the number of estimated parameters in the regression model, then the information matrix is singular. Thus, the D-optimal design has at least three support points. The following result determines the best three-point design for the regression model (8.27). The proof is deferred to the appendix.

Lemma 8.3.1 *Assume that $\theta^0 \in \Omega$ and that x_1^*, x_2^* are determined by the relation*

$$\Phi(x_1^*, x_2^*, \theta^0) = \max\left\{\Phi(x_1, x_2, \theta^0) \;\middle|\; \eta_0 \le x_1 < x_2 \le \bar{c}\right\}, \qquad (8.29)$$

where the function Φ is defined by

$$\Phi(x_1, x_2, \theta^0) = \det(\varphi_i(x_j))_{i,j=1}^3, \qquad (8.30)$$

with $x_3 = \bar{c}$. The design

$$\zeta_{\bar{c}}^* = \begin{pmatrix} & x_1^* & x_2^* & \bar{c} \\ p & 1/3 & 1/3 & 1/3 \end{pmatrix} \qquad (8.31)$$

is a D-optimal three-point design for the regression model (8.27) on the interval $[\eta_0, \bar{c}]$ for any $\bar{c}$ with $\eta_0 < \bar{c} \le c$.

In general, it is not clear if there exist better designs (with respect to the D-criterion) with more than three support points. In our numerical study, we did not find D-optimal designs for the regression model (8.27) with more than three support points, but a general proof of this property for arbitrary design regions seems to be difficult. However, it is possible to obtain theoretical results in this direction if the right endpoint $\bar{c}$ of the design space is sufficiently large and the initial condition η_0 is sufficiently small. For this purpose, we consider at first the design problem for the regression model (8.27) on the interval $[0, c]$. In this case, the vector of the regression functions can be rewritten in a more convenient form by the substitution $x \to cx$:

$$\hat{\varphi}(x) = cv(x) \left(\ln \frac{x}{\tilde{\eta}_0}, \ln \frac{1-x}{1-\tilde{\eta}_0}, \frac{x-\tilde{\eta}_0}{1-x}\right)^T,$$

where $\tilde{\eta}_0 = \eta_0/c$. Thus, we can assume without loss of generality that $c = 1$ and the D-optimal designs on the general interval $[0, c]$ are obtained by a rescaling from the D-optimal designs on the interval $[0, 1]$ (calculated for $c = 1$ and initial condition $\tilde{\eta}_0 = \eta_0/c$). Because D-optimal designs are not changed under nonsingular transformations of the regression functions, the problem is reduced to the investigation of D-optimal designs for the regression model $\beta^T \tilde{\varphi}(x, \tilde{\eta}_0)$, where the vector $\tilde{\varphi}$ is defined by

$$\tilde{\varphi}(x, \tilde{\eta}_0) = (\tilde{\varphi}_1(x), \tilde{\varphi}_2(x), \tilde{\varphi}_3(x))^T$$

$$= v(x) \left(\frac{\ln(x/\tilde{\eta}_0)}{-\ln \tilde{\eta}_0}, \ln \frac{1-x}{1-\tilde{\eta}_0}, \frac{x-\tilde{\eta}_0}{1-x}\right)^T. \qquad (8.32)$$

A direct computation shows that for $\tilde{\eta}_0 \to 0$,

$$\tilde{\varphi}(x, \tilde{\eta}_0) \to \psi(x)$$

$$\tilde{\Phi}(x_1, x_2, \theta^0) = \det(\tilde{\varphi}_i(x_j))_{i,j=1}^3 \to \Psi(x_1, x_2)$$

(in the last equation, we put $x_3 = 1$), where the functions ψ and Ψ are defined by

$$\psi(x) = v(x)\left(1, \ln(1-x), \frac{x}{1-x}\right)^T,$$

$$\Psi(x_1, x_2) = \frac{1}{b}v(x_1)v(x_2)[\ln(1-x_1) - \ln(1-x_2)] \qquad (8.33)$$

$$= \frac{1}{b}\frac{\ln(1-x_1) - \ln(1-x_2)}{(1-x_1+b)(1-x_2+b)}x_1 x_2(1-x_1)(1-x_2).$$

Lemma 8.3.2

(1) For any $b = \vartheta_2\vartheta_3 > 0$, the unique D-optimal design for the regression model $\beta^T\psi(x)$ on the interval $[0, 1]$ is given by

$$\tilde{\zeta} = \left(\begin{array}{ccc} \tilde{x}_1 & \tilde{x}_2 & 1 \\ 1/3 & 1/3 & 1/3 \end{array}\right),$$

where the points $\tilde{x}_1 = \tilde{x}_1(b)$ and $\tilde{x}_2 = \tilde{x}_2(b)$ are determined by the relation

$$\Psi(\tilde{x}_1, \tilde{x}_2) = \max\left\{ \Psi(x_1, x_2) \mid 0 \le x_1 \le x_2 \le 1 \right\}$$

and the function Ψ is defined in (8.33).

(2) There exist positive numbers $\varepsilon > 0$ and $\delta > 0$ such that for any $\eta_0 < \varepsilon$ and any $\bar{c} > c - \delta$, the design given in Lemma 8.3.1 by formula (8.31) is the unique D-optimal design for the regression model (8.27) with design region $[\eta_0, \bar{c}]$. Moreover, if $\eta_0 \to 0$ and $\bar{c} \to 1$, we have

$$V^T \bar{M}^{-1}(\zeta_{\bar{c}}^*)V \to \tilde{M}^{-1}(\tilde{\zeta}),$$

where $\tilde{M}(\zeta)$ is the information matrix of design ζ in the regression model $\beta^T\psi(x)$, $\bar{M}(\zeta)$ is the information matrix of design ζ in the regression model $\beta^T\varphi(x)$, and $V = \mathrm{diag}(-1/\ln\eta_0, 1, 1)$. In particular, it follows that

$$e_3^T \bar{M}^{-1}(\zeta_{\bar{c}}^*)e_3 \to 3b^2.$$

The following theorem is now obtained by the combination of Lemmas 8.3.1 and 8.3.2 and Theorem 8.2.2.

Theorem 8.3.1 *Consider the Monod model defined by (8.1) - (8.3) on the interval $[0, T]$.*

(1) For any $\theta^0 \in \Omega$ and $0 < T \le \infty$, a locally D-optimal three-point design is given by

$$\xi_T^* = \left(\begin{array}{ccc} t_1^* & t_2^* & T \\ 1/3 & 1/3 & 1/3 \end{array}\right), \qquad (8.34)$$

where

$$t_i^* = \frac{1}{\vartheta_1} \left[(1+b) \ln \left(\frac{x_i^*}{\eta_0} \right) + b \ln \frac{c - \eta_0}{c - x_i^*} \right] \, , \quad i = 1, 2,$$

and the points x_i^* are determined by the relation (8.29).

(2) *The design ξ_T^* determined by (8.34) is the unique locally D-optimal design for sufficiently large T and sufficiently small η_0.*

(3) *For $N \to \infty$, $T \to \infty$, and $\eta_0 \to 0$ the following relation holds:*

$$\frac{N}{\sigma^2} D(\hat{\vartheta}_3) \to 3 \vartheta_1^{-2}.$$

8.4 A Numerical Study

We begin with a numerical construction of the design ζ_c^* defined by (8.31), which is D-optimal for the regression model (8.27) on the interval $[0, c]$ according to the second part of Lemma 8.3.2. The function $\Phi(x_1, x_2, \theta^0)$ defined in (8.30) for $x_3 = c$ is of the from

$$\Phi(x_1, x_2, \theta^0) = v(x_1)v(x_2) \left[\ln \frac{x_1}{\eta_0} \ln \frac{c - x_2}{c - \eta_0} - \ln \frac{x_2}{\eta_0} \ln \frac{c - x_1}{c - \eta_0} \right] \cdot \frac{c - \eta_0}{b} \, ,$$

where the function v is defined by (8.24). The maximum of Φ on the set $\{(x_1, x_2) \mid \eta_0 \le x_1 < x_2 \le c\}$ can be calculated by a standard gradient method. Define $\tilde{\eta}_0 = \eta_0/c$: then it is again sufficient to consider only the case $c = 1$ (see the discussion in Section 8.3). The designs ζ_c^* on the interval $[\eta_0, c]$ can be simply obtained from the designs on the interval $[\tilde{\eta}_0, 1]$ multiplying the support points with c. Table 8.1 shows locally D-optimal three-point designs on the interval $[\tilde{\eta}_0, 1]$ and the diagonal elements of the corresponding covariance matrix for various values of the parameters $b = \vartheta_2 \vartheta_3$ and $\tilde{\eta}_0$. According to Lemma 8.3.1, these designs are locally D-optimal among all designs with three support points. The optimality in the class of all designs was checked by an application of Kiefer's equivalence theorem (see Kiefer and Wolfowitz (1960) or equation (8.10)). We observed in all considered cases that the checking condition was satisfied and our numerical study shows that the D-optimal three-point designs are in fact D-optimal within the class of all approximate designs on interval $[\eta_0, c]$.

We will now discuss the corresponding designs for the Monod model defined by the differential equation (8.1), which are related to the D-optimal designs for the linear regression model (8.27) by the relations (8.25) and (8.26). Because $\eta(\infty) = c$, the corresponding designs for the Monod model have a support point at $T = \infty$ and cannot be realized in practice. This fact was also observed empirically by Vanrolleghem, Van Daele, and Dochain (1995), who showed that for the commonly used optimality criteria (including the D-, c-, and E-criterion), the optimal strategies yield to prohibitively

Table 8.1: Locally D-optimal three-point designs ζ_1^* for the equivalent regression model (8.27) on the interval $[\tilde{\eta}_0, 1]$ for various values of $\tilde{\eta}_0$ and $b = \vartheta_2 \vartheta_3$. These designs are of the form (8.31) and determined by Lemma 8.3.1. Also shown are the diagonal elements $\bar{m}^{ii}$ ($i = 1, 2$) of the matrix $\bar{M}^{-1}(\zeta_1^*)$ rounded to integers. Optimal designs on the interval $[\tilde{\eta}_0 c, c]$ can be obtained by rescaling the design ζ_1^* with the factor c. The value of $\bar{m}^{33} = (\bar{M}^{-1}(\zeta_1^*))_{33}$ is given by $3b^2/(1 - \tilde{\eta}_0)^2$.

b	x_1^*	x_2^*	$\bar{m}^{11}$	$\bar{m}^{22}$	$\bar{m}^{33}$	x_1^*	x_2^*	$\bar{m}^{11}$	$\bar{m}^{22}$	$\bar{m}^{33}$
			$\tilde{\eta}_0 = 0.2$					$\tilde{\eta}_0 = 0.1$		
0.1	0.70	0.95	31	23	0.05	0.67	0.95	13	19	0.04
0.25	0.65	0.93	71	70	0.29	0.61	0.92	28	57	0.23
0.75	0.59	0.91	279	367	2.64	0.55	0.89	101	284	2.08
1	0.58	0.90	428	593	0.54	4.69	0.89	153	456	3.70
2	0.56	0.89	1321	2013	18.7	0.51	0.88	458	1521	14.8
			$\tilde{\eta}_0 = 0.05$					$\tilde{\eta}_0 = 0.01$		
0.1	0.65	0.94	7	17	0.03	0.62	0.94	2	15	0.03
0.25	0.59	0.92	14	50	0.21	0.56	0.91	5	43	0.19
0.75	0.52	0.89	49	244	1.87	0.49	0.88	16	205	1.72
1	0.51	0.88	74	389	3.32	0.48	0.87	24	325	3.06
2	0.48	0.87	217	1286	13.3	0.45	0.86	69	1065	12.24

long experiments. However, even if the D-optimal designs are not directly implementable, they can be used as a basis for evaluating the efficiency of other designs, which are used in practice. For example, consider the design

$$\hat{\xi} = \begin{pmatrix} t_1^* & t_2^* & t_3 \\ 1/3 & 1/3 & 1/3 \end{pmatrix}, \qquad (8.35)$$

where $t_i^* = t_i^*(\theta^0)$, $i = 1, 2$, are points of the design

$$\xi_\infty^* = \begin{pmatrix} t_1^* & t_2^* & \infty \\ 1/3 & 1/3 & 1/3 \end{pmatrix}$$

on the infinite design space $[0, \infty]$, which was determined in Theorem 8.3.1. We will now choose the point t_3 such that the efficiency of the design $\hat{\xi}$ with respect to the design ξ_∞^*,

$$I = \left(\frac{\det M(\hat{\xi})}{\det M(\xi_\infty^*)} \right)^{1/3} = \left(\frac{\det \bar{M}(\zeta_{\hat{\xi}})}{\det \bar{M}(\zeta_{\xi_\infty^*})} \right)^{1/3} \qquad (8.36)$$

is equal to a given value $1 - \delta$ (note that $\zeta_{\xi_\infty^*} = \zeta_c^*$). The corresponding values of the point t_3 are represented in Table 8.2 for various values of δ, b, and η_0. From this table, we can conclude that for $t_3 = 2t_2^*$ the efficiency of the design $\hat{\xi}$ is at least 0.98 for all considered parameter combinations. Thus, if $t_2^* < T/2$, the designs

$$\xi_D^* = \left(\begin{array}{ccc} t_1^* & t_2^* & 2t_2^* \\ 1/3 & 1/3 & 1/3 \end{array} \right) \qquad (8.37)$$

are close to the locally D-optimal designs on the interval $[0, \infty]$ and they can be realized in practice. Note that the true efficiency of these designs on the interval $[0, T]$ is usually larger than $1 - \delta$, because our comparison is based on a design for an infinite design space, and the locally D-optimal design on the interval $[0, T]$ has always a smaller determinant than the locally D-optimal design ξ_∞^* on the infinite design space $[0, \infty]$. In other words, the designs given in (8.37) have at least D-efficiency $1 - \delta$ for the concrete interval $[0, T]$, whenever $t_2^* \leq T/2$.

Table 8.2: The third support point t_3 of the design $\hat{\xi}$ defined in (8.35), such that the relative D-efficiency (8.55) is at least $1 - \delta$. The last two columns show the ratio $k(1 - \delta) = t_3/t_2^*$.

b	η_0	t_1^*	t_2^*	$1 - \delta$				$k(0.95)$	$k(0.99)$
				0.9	0.95	0.98	0.99		
				t_3					
0.25	0.1	2.5	3.4	4.2	4.4	4.7	4.9	1.3	1.4
0.5	0.1	3.0	4.4	6.0	6.5	7.1	7.4	1.5	1.7
0.75	0.1	3.5	5.4	7.8	8.5	9.3	9.9	1.6	1.8
1	0.1	4.0	6.5	9.7	10.5	11.6	12.4	1.6	1.9
1.5	0.1	5.1	8.5	13.3	14.6	16.2	17.4	1.7	2.1
2	0.1	6.1	10.5	17.0	18.7	20.8	22.5	1.8	2.1
0.25	0.2	1.7	2.5	3.4	3.6	3.9	4.0	1.4	1.6
0.25	0.05	3.3	4.2	5.1	5.3	5.6	5.8	1.2	1.4
0.25	0.01	5.2	6.2	7.1	7.3	7.6	7.8	1.2	1.2

We now compare the design ξ_D^* defined in (8.37) with the uniform design

$$\xi_u = \left(\begin{array}{cccc} T/n & 2T/n & \dots & T \\ 1/n & 1/n & \dots & 1/n \end{array} \right). \qquad (8.38)$$

A numerical study indicates that the information matrix of the uniform design for $n \geq 10$, $T > 1.5t_2^*$ does not depend sensitively on the parameters n and T (these results are not displayed for the sake of brevity). Exemplarily, we consider the case $n = 10$, $T = 2t_2^*$; other situations give similar results. A comparison of the uniform design ξ_u and the design ξ_D^* given in (8.37) with respect to the D-criterion shows that the D-efficiencies,

$$I(\xi_u) = \left(\frac{\det M(\xi_u)}{\det M(\xi_D^*)} \right)^{\frac{1}{3}}, \qquad (8.39)$$

of the uniform design vary between 150% and 200% (see Table 8.3). This
indicates a rather poor performance of the uniform design ξ_u. A more
refined comparison is obtained by looking at the the asymptotic variances
of the estimators for the parameters ϑ_1, ϑ_2, and ϑ_3. Note that the ratio of
these variances is given by

$$d_i = \frac{(M^{-1}(\xi_u))_{ii}}{(M^{-1}(\xi_D^*))_{ii}}, \quad i = 1, 2, 3. \tag{8.40}$$

We observe from Table 8.3 that the uniform design produces a smaller
(asymptotic) variance of the estimator for the parameter ϑ_3 compared to the
design (8.37). On the other hand, the asymptotic variances obtained for the
estimators for the parameters ϑ_1 and ϑ_2 are substantially smaller than the
corresponding variances obtained from the uniform design (8.50). However,
in realistic situations (see Pirt (1975) and Blok (1994) and the simulations
of the following paragraph), an efficient estimation of ϑ_1 and ϑ_2 is more
important, because the parameter ϑ_3 is usually estimated with much higher
precision than the parameters ϑ_1 and ϑ_2. Moreover, the loss of efficiency
using the uniform design for estimating ϑ_1 and ϑ_2 is substantially larger
than the loss of efficiency using the design ξ_D^* for estimating the parameter
ϑ_3. Thus, if we consider estimation of the parameters ϑ_1 and ϑ_2 as more
important, the design ξ_D^* yields a reduction of approximately 50% of the
variance compared to the uniform design ξ_u (see the results in Table 8.3),
provided that the sample size N is sufficiently large. In order to investigate

Table 8.3: The efficiency of the design ξ_D^* defined in (8.37) in relation to
the uniform design ξ_u defined in (8.50) in the Monod model (8.1). Here,
the efficiency d_i defined in (8.40) ($i = 1, 2, 3$) corresponds to the estimation of the individual parameters ϑ_1, ϑ_2, and ϑ_3 and depends only on the
parameters $\tilde{\eta}_0 = \eta_0/c$ and b. $I(\xi_u)$ corresponds to the D-criterion and is
defined in (8.39).

$\tilde{\eta}_0$	0.2					0.1				
b	0.1	0.25	0.75	1	2	0.1	0.25	0.75	1	2
d_1	2.0	2.0	2.2	2.3	2.5	1.9	2.1	2.2	2.3	2.3
d_2	2.2	2.0	2.2	2.3	2.4	2.1	2.2	2.2	2.3	2.3
d_3	1.0	0.7	0.6	0.6	0.6	1.2	0.8	0.6	0.6	0.6
$I(\xi_u)$	1.6	1.4	1.4	1.5	1.5	1.7	1.6	1.5	1.5	1.5
$\tilde{\eta}_0$	0.05					0.01				
b	0.1	0.25	0.75	1	2	0.1	0.25	0.75	1	2
d_1	3.0	2.1	2.3	2.3	2.4	2.8	2.5	2.4	2.5	2.4
d_2	3.4	2.2	2.2	2.3	2.4	3.0	2.6	2.4	2.5	2.4
d_3	1.4	0.9	0.7	0.7	0.6	1.7	1.2	0.8	0.7	0.7
$I(\xi_u)$	2.1	1.6	1.6	1.5	1.6	2.4	2.0	1.7	1.7	1.7

how these asymptotic observations can be transferred to realistic sample sizes, a small simulation study was conducted. We simulated observations according to the model (8.4) where the errors are normally distributed with variance $\sigma^2 = 0.01^2$ and $\sigma^2 = 0.02^2$. The parameter $\theta = (\vartheta_1, \vartheta_2, \vartheta_3)^T$ was estimated by the least squares technique using the Nelder–Mead simplex method. For the parameter θ^* we fixed the value $(0.25, 0.5, 0.25)^T$, which corresponds to parameters observed in studies of microbial growth (see Pirt (1975) or Blok (1994)). The simulation was repeated 400 times for the sample sizes $N = 20, 30, 40, 60$, and the designs ξ_D^* and ξ_u defined in (8.37) and (8.50), respectively. The simulated variances of the estimators for the parameters ϑ_1, ϑ_2, and ϑ_3 are represented in Tables 8.4 and 8.5 corresponding to the choices $\sigma = 0.01$ and $\sigma = 0.02$, respectively.

Table 8.4: Simulated and asymptotic variances of the estimates for the parameters in the Monod model (8.1) for different sample sizes (N). The variances are multiplied with N/σ^2 ($\sigma = 0.01$) and are presented for the uniform design ξ_u and the design ξ_D^* in (8.37) obtained from the D-optimal design on an infinite design space.

N	20	30	40	60	∞
	uniform design				
θ_1	771	606	575	529	551
θ_2	17693	13637	12336	11944	12500
θ_3	2.1	2.1	2.3	2.1	2.3
	optimal design				
θ_1	432	295	195	180	269
θ_2	10101	6789	4357	4070	6055
θ_3	3.3	2.9	2.9	2.6	3.0

For the sake of transparency these values are multiplied by N/σ^2 (and rounded to integers). The results confirm our asymptotic findings for realistic sample sizes. The design ξ_D^* defined in (8.37) allows a substantially more precise estimation of the parameters ϑ_1 and ϑ_2 compared to the design ξ_u. Note that the variances for the estimation of the parameter ϑ_3 are substantially smaller (independent of the design) than the corresponding variances of the estimators for ϑ_1 and ϑ_2. Finally, it is worthwhile to mention that the asymptotic considerations of the previous paragraph are applicable if $N \geq 30$. Thus, our simulation study confirmed that the design ξ_D^* performs much better than the uniform design ξ_u.

8.5   Taylor Expansions

As was shown in the previous section, designs very close to locally D-optimal designs for model (8.1) on the interval $\mathfrak{X} \in [0, T]$ can be easily constructed on the basis of saturated locally D-optimal designs on $[0, \infty]$.

Table 8.5: Simulated and asymptotic variances of the estimates for the parameters in the Monod model (8.1) for different sample sizes (N). The variances are multiplied with N/σ^2 ($\sigma = 0.02$) and are presented for the uniform design ξ_u and the design ξ_D^* in (8.37) obtained from the D-optimal design on an infinite design space.

N	20	30	40	100	∞
		uniform design			
θ_1	2936	2080	1791	564	551
θ_2	72541	49874	43325	13494	12500
θ_3	4.0	3.8	3.7	3.6	2.3
		optimal design			
θ_1	1056	867	798	278	269
θ_2	23024	20390	14880	6359	6055
θ_3	2.6	2.9	2.9	3.0	3.0

The dependence of the last designs on θ_1^0 was found in the explicit form. Let us now study the dependence of the inner points of these designs on initial values for θ_2 and θ_3 on the basis of the functional approach.

In the appendix, it will be shown that the basis functions of model (8.38) generate a Chebyshev system on $[0, c]$ and the saturated locally D-optimal designs for this model have one and the same type $(0, 2, 1)$ for any parameter values. The support points of the designs for model (8.1) are obtained from that for model (8.38) by explicit formulas. Thus, using the theory of Chapter 2, it is not difficult to establish the following results.

Theorem 8.5.1 *The inner points of the saturated locally D-optimal design ξ_∞^*, t_1^*, and t_2^*, are real analytic functions of θ_2^0 and θ_3^0 under the condition $\theta_1^0, \theta_3^0 > 0$. These functions are increasing for any of the parameters.*

Due to Theorem 8.5.1, the functions $t_i^*(\theta_2^0, \theta_3^0)$, $i = 1, 2$, where t_1^* and t_2^* are the inner points of the design

$$\xi_\infty^* = \left(\begin{array}{ccc} t_1^* & t_2^* & \infty \\ 1/3 & 1/3 & 1/3 \end{array} \right),$$

can be expanded into a Taylor series in a vicinity of any point $(\hat{\theta}_2^0, \hat{\theta}_3^0)$ by degrees of $u = \theta_2^0 - \hat{\theta}_2^0$ and $v = \theta_3^0 - \hat{\theta}_3^0$.

Let us take $s_0 = 1$, $\eta_0 = 0.03$, $\hat{\theta}_1^0 = 1.25$, $\hat{\theta}_2^0 = 0.5$, and $\hat{\theta}_3^0 = 0.25$. These values are of practical importance (see Pirt (1975) or Block (1994)).

Consider the design $\zeta^* = \zeta_{\xi_\infty^*}^*$,

$$\zeta^* = \left(\begin{array}{ccc} x_1^* & x_2^* & x \\ 1/3 & 1/3 & 1/3 \end{array} \right),$$

where $c = s_0(\hat{\theta}_3^0 + v) + \eta_0$. Due to Theorem 8.3.1, the inner points of ξ_∞^0 can be expressed by x_1^* by the formula $t_i^* = t(x_i^*)$, where

$$t(x) = \left[(1+b)\ln\left(\frac{x}{\eta_0}\right) - b\ln\frac{c-x}{c-\eta_0}\right]/\theta_1^0 ,$$

where $b = \theta_2^0\theta_3^0/c$. The design ζ^* does not depend on θ_1^0. With $\theta_2^0 = \hat{\theta}_2^0 = 0.5$ and $\theta_3^0 = \hat{\theta}_3^0 = 0.25$, we find numerically that

$$x_{1(0)} := x_1^*(\hat{\theta}_2^0, \hat{\theta}_3^0) = 0.1618,$$

$$x_{2(0)} := x_2^*(\hat{\theta}_2^0, \hat{\theta}_3^0) = 0.2540,$$

Now with the help of recurrent formulas of Section 2.4, we find the coefficients of the expansion

$$x_{i<n>}(u,v) = \sum_{0 \le j+s \le n} x_{i(j,s)} u^j v^s,$$

$n = 1, 2, \ldots$, $x_{i<n>}(u,v) \to x_i^*(u + \hat{\theta}_2^0, v + \hat{\theta}_3^0)$, $n \to \infty$, $i = 1, 2$. These coefficients are given in Table 8.6. Inserting these coefficients into (8.27),

Table 8.6: Coefficients of Taylor expansions of x_1 and x_2 by degrees of u and v

u^i/v^j	0	1	2	3	4	5
0	.16176	.52428	−.01972	.02099	−.01715	−.03217
	.25403	.89405	−.00564	.00800	−.01390	.02515
1	−.03048	−.11980	−.00062	−.00467	.02332	
	−.01392	−.05670	−.00357	.00441	−.00598	
2	.03902	.15412	.00167	.00295		
	.01702	.06945	.00416	−.00561		
3	−.05351	−.21214	−.00320			
	−.02214	−.09045	−.00515			
4	.07782	.30940				
	.03047	.12451				
5	−.11876					
	−.04400					

we obtain the Taylor coefficients for the function

$$q(p) = \left[\det \bar{M}\left(\xi^*(\tilde{\Theta}_p), \tilde{\Theta}_p\right)\right]^{1/3},$$

where $p = (u, v)$ and $\tilde{\Theta}_p = (\hat{\theta}_2^0 + u, \hat{\theta}_3^0 + v)^T$, given in Table 8.7.

Table 8.7: Taylor coefficients of the function $(\det(\bar{M}(\zeta, \bar{\Theta}_\Delta)))^{1/3}$ by degrees of u and v

u^i/v^j	0	1	2	3
0	.00877	.07301	.16671	.04199
1	$-.02737$	$-.22778$	$-.51882$	
2	.06634	.55166		
3	$-.14698$			

Using coefficients from Table 8.6 and expanding the function $\ln y$, $y = x/\eta_0$, and $y = (c - x)/(c - \eta_0)$ into Taylor series we find the coefficients of the expansion

$$t_{i<n>}(u, v) = \sum_{0 \le i+j \le n} t_{i(i,j)} u^j v^s,$$

$n = 0, 1, 2, \ldots$, $t_{i<n>}(u, v) \to t_i^*(\hat{\theta}_2^0 + u, \hat{\theta}_3^0 + v)$, $n \to 0$, $i = 1, 2$. The coefficients are represented in Table 8.8.

Table 8.8: Taylor coefficients of the functions t_1 and t_2 by degrees of u and v

$\dfrac{u^i}{v^j}$	0	1	2	3	4	5
0	11.08559	20.57254	-35.41260	80.33176	-202.83386	540.81013
	16.40372	23.58760	-45.16761	113.39015	-316.69258	935.88713
1	7.14104	14.16692	-24.51964	54.62668	-132.94068	
	14.44336	18.39714	-38.54685	102.49576	-297.80900	
2	.34779	.18102	$-.50627$	1.64422		
	$-.30382$	$-.13480$	.48552	-1.76129		
3	$-.53855$	$-.29211$	.84036			
	.36261	.15841	$-.57558$			
4	.85658	.47484				
	$-.45962$	$-.19693$				
5	-1.39626					
	.61543					

Note that the support points of the locally D-optimal designs can be calculated on the basis of the Taylor expansions very quickly and even by hand. It is easy to check that the value $t_{i<n>}(u, v)$ with $n = 2$ can be used for a wide variety of the parameter values.

The obtained expansions can also be used for studying the robustness of the locally D-optimal design with respect to the miss-specification of the initial values.

Let us take the design

$$\xi_0 = \left(\begin{array}{ccc} t_{1(0)} & t_{2(0)} & \infty \\ 1/3 & 1/3 & 1/4 \end{array} \right),$$

which is locally D-optimal for $\Theta^0 = (0.25, 0.5, 0.25)^T$. Assume that the proper value of the vector Θ is equal to $\Theta_\Delta = (0.25 + \Delta_1, 0.5 + \Delta_2, 0.25 + \Delta_3)^T$, and $\Delta = (\Delta_1, \Delta_2, \Delta_3)^T$.

Let us consider the efficiency of design ξ_0 with respect to the locally D-optimal design for different $\Theta_\Delta \neq \Theta^0$:

$$I_\Delta = \left(\frac{\det M(\xi_0, \Theta_\Delta)}{\det M(\xi^*(\Theta_\Delta), \Theta_\Delta)} \right).$$

Using formula (8.27), we obtain

$$I_\Delta = \left[\frac{\det \bar{M}(\zeta_{\xi_0}(\Theta_\Delta), \tilde{\Theta}_\Delta)}{\det \bar{M}(\zeta^*(\tilde{\Theta}_\Delta), \tilde{\Theta}_\Delta)} \right]^{1/3},$$

where $\tilde{\Theta}_\Delta = (\hat{\theta}_2^0 + \Delta_2, \hat{\theta}_3^0 + \Delta_3)^T$,

$$\zeta_{\xi_0}(\Theta_\Delta) = \left(\begin{array}{ccc} \eta(t_{1(0)}, \Theta_\Delta) & \eta(t_{2(0)}, \Theta_\Delta) & c_\Delta \\ 1/3 & 1/3 & 1/3 \end{array} \right),$$

and $c_\Delta = s_0(\Delta_3 + \hat{\theta}_3^0) + \eta_0$.

Table 8.9: Efficiency of the design ξ_0

Δ_1	Δ_2	Δ_3	I
0	−0.1	−0.1	0.34
0	−0.1	0.1	0.98
0	0.1	−0.1	0.86
0	0.1	0.1	0.69

Δ_1	Δ_2	Δ_3	I
0.1	−0.1	−0.1	0.40
0.1	−0.1	0.1	0.20
0.1	0.1	−0.1	0.10
0.1	0.1	0.1	0.79
−0.1	−0.1	−0.1	0.69
−0.1	−0.1	0.1	0.14
−0.1	0.1	−0.1	0.45
−0.1	0.1	0.1	0.09

Δ_1	Δ_2	Δ_3	I
−0.05	0.05	0.05	0.44
−0.05	0.05	−0.05	0.76
−0.05	−0.05	−0.05	0.92
−0.05	−0.05	0.05	0.57
0.05	0.05	0.05	0.93
0.05	0.05	−0.05	0.57
0.05	−0.05	−0.05	0.28
0.05	−0.05	0.05	0.69

Table 8.10: Efficiency of the design ξ_0

Δ_1	Δ_2	Δ_3	I	Δ_1	Δ_2	Δ_3	I
0.01	−0.01	−0.01	0.954	−0.02	0.02	0.02	0.85
0.01	−0.01	0.01	0.987	−0.02	0.02	−0.02	0.95
0.01	0.01	−0.01	0.976	−0.02	−0.02	−0.02	0.99
0.01	0.01	0.01	0.997	−0.02	−0.02	0.02	0.91
−0.01	−0.01	−0.01	0.997	0.02	0.02	0.02	0.99
−0.01	−0.01	0.01	0.976	0.02	0.02	−0.02	0.91
−0.01	0.01	−0.01	0.988	0.02	−0.02	−0.02	0.82
−0.01	0.01	0.01	0.958	0.02	−0.02	0.02	0.95

Denote by $I_\Delta(n)$ the approximate value of I_Δ obtained by the substitution the design, $\zeta^*(\hat{\Theta}_\Delta)$ by its approximation

$$\zeta_{<n>}(\tilde{\Theta}_\Delta) = \begin{pmatrix} x_{1<m>} & x_{2<n>} & c_\Delta \\ 1/3 & 1/3 & 1/3 \end{pmatrix},$$

where $x_{i<n>} = x_{i<n>}(u, v)$ and $u = \Delta_2$, $v = \Delta_3$. Note that $I_\Delta(n)$ is very close to I_Δ if $n = 4$ and $\Delta_i \leq 0.05$, $i = 2, 3$.

The approximate values $I_\Delta \approx I_\Delta(4)$ are given in Tables 8.9 and 8.10. The results given in the tables allow one to make the following conclusions. Under moderate deviations $|\theta_{i(0)} - \hat{\theta}_{i(0)}| \leq 0.02$, $i = 1, 2, 3$, the locally D-optimal design constructed for the point $\hat{\Theta} = (\hat{\theta}_1^{(0)}, \hat{\theta}_2^{(0)}, \hat{\theta}_3^{(0)})^T$ remains very efficient. Under more serious deviations, the efficiency of this design quickly decreases.

For substantial deviations, it is preferable to use the maximin efficient designs. We will consider such designs in Section 8.7.

8.6   Locally E- and e_k-Optimal Designs

In this section, we present a theoretical and numerical study of the locally E- and e_k-optimal designs in the Monod model. The optimality criteria are defined in (8.11) and (8.12). Due to Theorem 8.2.3, it will be sufficient to study the optimal designs for the regression model $\beta^T K \varphi(x)$ on the interval $[\eta_0, \bar{c}]$, where the matrix K is defined by (8.22). It will be shown in Proposition 8.8.1 of the appendix that the functions $\varphi_1(x)$, $\varphi_2(x)$ and $\varphi_3(x)$ defined in (8.23) generate a Chebyshev system on the interval $[\eta_0, \bar{c}]$. Then it is well known (see Karlin and Studden (1966, Chap. 1)) that there exists a function $g(x)$,

$$g(x) = r_1\varphi_1(x) + r_2\varphi_2(x) + r_3\varphi_3(x), \tag{8.41}$$

with real coefficients r_1, r_2, and r_3 and (unique) points $\tilde{x}_1$, $\tilde{x}_2$, and $\tilde{x}_3$ satisfying $\eta_0 \leq \tilde{x}_1 < \tilde{x}_2 < \tilde{x}_3 \leq \bar{c}$ such that

$$g(\tilde{x}_i) = (-1)^{i+1}, \quad i = 1, 2, 3, \tag{8.42}$$

and such that the inequality

$$|g(x)| \leq 1. \tag{8.43}$$

holds for all $x \in [\eta_0, \bar{c}]$. Counting the number of possible zeros of the function $g'(x)$ shows that at least one of the points $\tilde{x}_i$ has to be a boundary point of the interval $[\eta_0, \bar{c}]$. Since $g(\eta_0) = 0$, it follows that $\tilde{x}_3 = \bar{c}$. The matrix

$$\tilde{F} = \left(e_i^T K \varphi(\tilde{x}_j) \right)_{i,j=1}^{3}$$

is nonsingular because the functions φ_1, φ_2, and φ_3 generate a Chebyshev system on the interval $[\eta_0, \bar{c}]$ and the matrix K is nonsingular. The following lemma gives the locally e_k- and E-optimal designs and will be proved in the appendix.

Lemma 8.6.1 *(1) If $\eta_0 > 0$ is sufficiently small, then the design*

$$\zeta_E = \begin{pmatrix} \tilde{x}_1 & \tilde{x}_2 & \bar{c} \\ \tilde{\omega}_1 & \tilde{\omega}_2 & \tilde{\omega}_3 \end{pmatrix} \tag{8.44}$$

is a locally E-optimal design in the regression model $\beta^T K \varphi(x)$ on the interval $[\eta_0, \bar{c}]$. Here, the support points are defined by condition (8.42) and (8.43) and the weights $\tilde{\omega}_i$ are given by

$$\tilde{\omega}_i = |\tilde{A}_i| \Big/ \sum_{j=1}^{3} |\tilde{A}_j|, \tag{8.45}$$

where $\tilde{A}_i = e_i^T \tilde{F}^{-1} r$, $i = 1, 2, 3$, and $r = (r_1, r_2, r_3)^T$ denotes the vector of coefficients of the function g defined in (8.41) – (8.43).

(2) For arbitrary $\bar{c} \leq c$, $\eta_0 > 0$, the design ζ_E defined by (8.44) is a locally E-optimal design for the regression model $\beta^T K \varphi(x)$ on the interval $[\eta_0, \bar{c}]$ if and only if the condition

$$r^T \bar{M}(\zeta_E) r = (r^T (K^T K)^{-1} r) \lambda_{\min}(K \bar{M}(\zeta_E) K^T)$$

is satisfied, where $r = (r_1, r_2, r_3)^T$ denotes the vector of coefficients of the function g defined in (8.41)–(8.43). In this case, the design ζ_E is the unique (locally) E-optimal design.

(3) If $\eta_0 > 0$ is sufficiently small, then the design

$$\zeta_{e_k} = \begin{pmatrix} \tilde{x}_1 & \tilde{x}_2 & \bar{c} \\ \tilde{\omega}_1(k) & \tilde{\omega}_2(k) & \tilde{\omega}_3(k) \end{pmatrix}.$$

is a locally e_k-optimal design in the regression model $\beta^T K \varphi(x)$ on the interval $[\eta_0, \bar{c}]$. Here, the support points are defined by condition (8.42) and (8.43) and the weights $\tilde{\omega}_i$ are given by

$$\tilde{\omega}_i(k) = |\tilde{A}_{ik}| \Big/ \sum_{j=1}^{3} |\tilde{A}_{jk}|, \quad k = 1, 2, 3,$$

with $\tilde{A}_{ij} = e_i^T \tilde{F}^{-1} e_j$ $(i, j = 1, 2, 3)$.

(4) For arbitrary $\eta_0 > 0$, the design ζ_{e_k} is a locally e_k-optimal design for the regression model $\beta^T K \varphi(x)$ on the interval $[\eta_0, \bar{c}]$ if and only if

$$\tilde{A}_{ik} \geq 0 , \quad i = 1, 2, 3 .$$

In this case, the design ζ_{e_k} is the unique (locally) e_k-optimal design.

Note that it follows from Theorem 8.2.3 that a design ζ is locally E-$(e_k$-)optimal for the regression model $\beta^T K \varphi(x)$ on the interval $[\eta_0, \bar{c}]$ if and only if the design ξ_ζ induced by the transformation (8.26) is locally E-(e_k)-optimal for Monod model (8.1) on the interval $[0, T]$, where $T = t(\bar{c})$. Therefore, if no constraints are imposed on the desired real-time operation, the locally e_k- and E-optimal designs yield designs with prohibitively long experiments. This fact was also observed empirically for the E-criterion by Vanrolleghem, Van Daele, and Dochain (1995).

For the numerical construction of the E-optimal design ζ_E, it is sufficient to maximize the function

$$Q(x_1, x_2) = \lambda_{\min}(K^T \bar{M}(\zeta) K)$$

on the set

$$U = \{(x_1, x_2)^T; \eta_0 \leq x_1 \leq x_2 \leq \bar{c}\} ,$$

where ζ is defined in (8.44) and (8.45). The design ζ_{e_k} can be constructed in a similar way and the optimality can be checked by the necessary and sufficient characterizations (2) and (4) in Lemma 8.6.1. We conclude this section with a comparison of locally D, E-, and e_2-optimal designs. The characteristics of the locally optimal designs can be found in Table 8.11, which shows the design (8.37) derived from the D-optimal design and the E- and e_2-optimal design

$$\xi_E^* = \begin{pmatrix} t_{1E}^* & t_{2E}^* & 2t_{2D}^* \\ w_1 & w_2 & w_3 \end{pmatrix}$$

$$\xi_{e_2}^* = \begin{pmatrix} t_{1E}^* & t_{2E}^* & 2t_{2D}^* \\ m_1 & m_2 & , m_3 \end{pmatrix}$$

on the interval $[0, 2t_{2D}^*]$. Note that ξ_E^* and $\xi_{e_2}^*$ have the same support points. In all cases considered in our numerical study, the necessary and sufficient conditions from parts (2) and (4) of Lemma 8.6.1 were fulfilled, which leads to the conjecture that the designs defined by (8.44) and (??) are in fact locally E- and e_k-optimal on any interval $[0, T]$ and for any initial condition $\eta_0 > 0$.

Table 8.11: Comparison of D-, E-, e_2-optimal designs for various values of $b = \vartheta_2 \vartheta_3$. The E- and $\xi_{e_2}^*$-optimal design have the same support $t_{1E}^*, t_{2E}^*, 2t_{2D}^*$ but different weights w_1, w_2, w_3 and m_1, m_2, m_3, respectively. The D-optimal design has equal masses at the points $t_{1D}^*, t_{2D}^*, 2t_{2D}^*$ and the efficiencies d_i and $\hat{d}_i$ are defined in (8.47), while $I(\xi)$ is defined by (8.46).

b	0.1	0.25	0.5	0.75	1	1.5	2
t_{1D}^*	2.92	3.29	3.95	4.62	5.30	6.67	8.04
t_{2D}^*	3.51	4.25	5.45	6.64	7.82	10.18	12.54
t_{1E}^*	2.65	2.94	3.47	4.03	4.59	5.73	6.88
t_{2E}^*	3.52	4.26	5.48	6.68	7.88	10.26	12.65
w_1	0.45	0.42	0.41	0.40	0.40	0.40	0.40
w_2	0.35	0.36	0.37	0.37	0.37	0.37	0.37
w_3	0.20	0.21	0.22	0.22	0.23	0.23	0.23
m_1	0.40	0.39	0.39	0.39	0.39	0.39	0.40
m_2	0.39	0.39	0.38	0.38	0.38	0.38	0.37
m_3	0.21	0.22	0.23	0.23	0.23	0.23	0.23
d_1	0.80	0.82	0.83	0.84	0.84	0.84	0.84
d_2	0.84	0.85	0.85	0.85	0.85	0.85	0.85
d_3	1.66	1.58	1.52	1.49	1.48	1.47	1.47
$\hat{d}_1$	0.82	0.83	0.84	0.84	0.84	0.84	0.84
$\hat{d}_2$	0.83	0.84	0.85	0.85	0.85	0.85	0.85
$\hat{d}_3$	1.60	1.50	1.47	1.46	1.46	1.46	1.46
$I(\xi_E^*)$	1.12	1.11	1.10	1.10	1.10	1.10	1.10
$I(\xi_{e_2}^*)$	1.11	1.10	1.10	1.10	1.10	1.10	1.10

The designs were first compared by their D-efficiencies

$$I(\xi) := \left(\frac{\det M(\xi)}{\det M(\xi_D^*)} \right)^{1/3} \tag{8.46}$$

and we did not observe substantial differences with respect to this criterion (see the last two columns in Table 8.11). For a more refined, comparison we calculated the asymptotic efficiencies

$$d_i = \frac{(M^{-1}(\xi_E^*))_{ii}}{(M^{-1}(\xi_D^*))_{ii}}, \quad i = 1, 2, 3,$$
$$\hat{d}_i = \frac{(M^{-1}(\xi_{e_2}^*))_{ii}}{(M^{-1}(\xi_D^*))_{ii}}, \quad i = 1, 2, 3, \tag{8.47}$$

for estimating the individual coefficients in the Monod model. We observe a very similar behavior of the E- and e_2-optimal designs, which provide more efficient estimates for the parameters ϑ_1 and ϑ_2 than the design ξ_D^* derived from the D-optimal. On the other hand, this design is more efficient for the estimation of the parameter ϑ_3. However, if improvement of accuracy in the estimation of the parameters ϑ_1 and ϑ_2 is considered more important, the E- and e_2-optimal design have some advantages.

8.7   Maximin Efficient Designs

As we have seen in the previous sections locally D- E-, and c-optimal designs for the Monod model are rather sensitive to the miss-specification of initial values for parameters. The notion of maximin efficient designs (see the discussion in Section 1.6) allows one to construct more robust designs if a set (say Ω) of possible values of parameters is known. A design will be called standardized maximin Φ-optimal (or, briefly, maximin efficient) if it maximizes

$$\Psi_\Omega(\xi) = \min_{\theta \in \Omega} \frac{\Phi(M(\xi, \theta))}{\Phi(M(\xi^*(v), \theta))}, \tag{8.48}$$

where $\xi^*(\theta)$ is a locally Φ-optimal design.

We will consider the case of D- and E-criteria,

$$\Phi_D(M(\xi, \theta)) = (\det M(\xi, \theta))^{1/m},$$

$$\Phi_E(M(\xi, \theta)) = \lambda_{\min}(M(\xi, \theta)),$$

where m is the number of parameters in the model under consideration; $m = 3$ in our case.

In this section, standardized maximin D- and E-optimal designs will be denoted by ξ_D^* and ξ_E^*, respectively.

The most important case for the choice of the set Ω in the maximin criterion arises if the experimenter is able to specify intervals for the location of each parameter θ_i; that is,

$$(\theta_1, \theta_2, \theta_3) \in \Omega = [z_{1,L}, z_{1,U}] \times [z_{2,L}, z_{2,U}] \times [z_{3,L}, z_{3,U}], \tag{8.49}$$

where $0 < z_{i,L} \leq z_{i,U} < \infty$ $(i = 1, 2, 3)$. In this section, we will compare standardized maximin Φ-optimal designs with uniform designs of the form

$$\xi_{\mathcal{U}(N), \bar{T}} = \begin{pmatrix} \frac{1}{N}\bar{T} & \cdots & \bar{T} \\ \frac{1}{N} & \cdots & \frac{1}{N} \end{pmatrix}, \tag{8.50}$$

which are commonly applied in microbiological models. As was shown in Dette, Melas, Pepelyshev, and Strigul (2005) with arbitrary $T \leq \infty$, standardize maximin D- and E-optimal designs for a compact set $\Omega \in [0, \infty) \times [0, \infty) \times [0, \infty)$ exist and are of the form

$$
\begin{pmatrix}
t_1 & \cdots & t_{n_1-1} & T \\
\omega_1 & \cdots & \omega_{n_1-1} & \omega_{n_1}
\end{pmatrix}.
$$

For constructing such designs, a numerical method introduced in the work cited above can be used. This method will be described in the following subsection.

8.7.1 A numerical procedure

In Section 8.7.2, we will calculate some standardized maximin efficient designs numerically and demonstrate that these designs have excellent efficiencies compared to locally optimal uniform designs. We will now briefly explain the algorithm used for these calculations. The algorithm is based on the following conjecture, which was satisfied in all examples in our numerical study. In this subsection, the function Φ denotes the D-,E-, or e_i-optimality criterion ($i = 1, 2, 3$) defined in Section 8.2.

Conjecture 8.4.1. *For any design ξ, the set*

$$
\Omega_0 = \Omega_0(\xi) = \left\{ \theta \,\middle|\, \theta = \arg\min_{\theta \in \Omega} \frac{\Phi(M(\xi, \theta))}{\Phi(M(\xi_\theta^*, \theta))} \right\}. \tag{8.51}
$$

is finite, say $\Omega_0 = \{\theta_{(1)}, \ldots, \theta_{(n_2)}\}$ ($n_2 \in \mathbb{N}$).

In all our considered examples, we observed that $n_2 \leq 4$, but a general bound could not be established formally. Now, consider the set

$$
\mathcal{U}_{n_1} = \Big\{ (u_1, \ldots, u_{2n_1}) = (t_1, \ldots, t_{n_1}, w_1, \ldots, w_{n_1}) \,\Big|
$$
$$
0 \leq t_1 < \cdots < t_{n_1} \leq T; w_i > 0, \textstyle\sum_{i=1}^{n_1} w_i = 1 \Big\} \tag{8.52}
$$

and note that each element of $\mathcal{U}_{n_1}$ defines a design with n_1 support points; that is,

$$
\xi_u = \begin{pmatrix}
t_1 & \cdots & t_{n_1-1} & t_{n_1} \\
w_1 & \cdots & w_{n_1-1} & w_{n_1}
\end{pmatrix}. \tag{8.53}
$$

It is proved in Dette, Melas, Pepelyshev, and Strigul (2005) that there exists an $n_1 \in \mathbb{N}$ and a $u \in \mathcal{U}_{n_1}$ such that the standardized maximin Φ-optimal design is given by $\xi^* = \xi_u$. We will now describe an iterative calculation of the standardized maximin Φ-optimal design observing that at least 3 support points are required. Thus, we set $n_1 = 3$ and choose an arbitrary (possibly locally Φ-optimal) starting design, say ξ_{u_0} with $u_0 \in \mathcal{U}_{n_1}$. We put

$s = 0$ and define

$$\bar{u}_{(s)} = \frac{\partial}{\partial u} \min_{\theta \in \Omega} \frac{\Phi(M(\xi_u, \theta))}{\Phi(M(\xi_\theta^*, \theta))} \bigg|_{u=u_{(s)}} = \frac{\partial}{\partial u} \min_{j=1,\dots,n_2} \frac{\Phi(M(\xi_u, \theta_{(j)}))}{\Phi(M(\xi_{\theta_{(j)}}^*, \theta_{(j)}))} \bigg|_{u=u_{(s)}}$$

$$= \min \left\{ \sum_{j=1}^{n_2} h_j \frac{\partial}{\partial u} \frac{\Phi(M(\xi_u, \theta_{(j)}))}{\Phi(M(\xi_{\theta_{(j)}}^*, \theta_{(j)}))} \bigg|_{u=u_{(s)}} \;\bigg|\; h_j \geq 0; \sum_{j=1}^{n_2} h_j = 1 \right\},$$

where we have used Conjecture 8.7.1 with $\Omega_0 = \Omega_0(\xi_u) = \{\theta_{(1)}, \dots, \theta_{(n_2)}\}$ and the formula for the derivative of the minimum. In the next step, we calculate

$$u_{(s+1)} = u_{(s+1)}(h_s) = (1 - h_s)u_{(s)} + h_s \bar{u}_{(s)}, \tag{8.54}$$

where the weight h_s maximizes the minimum Φ-efficiency,

$$\mathrm{eff}_\Phi(\xi, \theta) = \frac{\Phi(M(\xi, \theta))}{\Phi(M(\xi_\theta^*, \theta))} \tag{8.55}$$

among all designs of the form ξ_u with $u = u_{(s+1)}$ defined by (8.54); that is,

$$h_s = \arg\max \left\{ \min_{\theta \in \Omega} \mathrm{eff}_\Phi(\xi_{u_{(s+1)}(h)}, \theta) \mid 0 \leq h \leq 1 \right\}.$$

Obviously, we obtain

$$\min_{\theta \in \Omega} \mathrm{eff}_\Phi(\xi_{u_{(s+1)}}, \theta) \geq \min_{\theta \in \Omega} \mathrm{eff}_\Phi(\xi_{u_{(s)}}, \theta)$$

and in the case of equality, the design $\xi_{u_{(s+1)}}$ is standardized maximin Φ-optimal in the class of designs

$$\Xi_{n_1} = \left\{ \xi_u \mid u \in \mathcal{U}_{n_1} \right\}. \tag{8.56}$$

Otherwise, it follows by standard arguments that the sequence of designs $(\xi_{u_{(j)}})_{j \in \mathbb{N}_0}$ contains a weakly convergent subsequence with a limit, say $\xi_{n_1}^*$, that is a standardized maximin Φ-optimal in the class Ξ_{n_1}. Note that in all cases considered in our study, the sequence $(\xi_{u_{(j)}})_{j \in \mathbb{N}_0}$ was weakly convergent and it is usually not necessary to consider subsequences. We can now use the general equivalence theorem for standardized maximin Φ-optimality (see Dette, Haines, and Imhof (2003, Theorem 3.3)) to check if the design $\xi_{n_1}^*$ is standardized maximin Φ-optimal in the class of all approximate designs (for the standardized maximin D-optimality criterion, the corresponding equivalence theorem is stated in the appendix in Lemma 8.6.1. Otherwise, the procedure is continued with n_1 replaced by $n_1 + 1$ and an initial design in the class Ξ_{n_1+1} constructed as follows: We define

$$t^* = \arg\max_{t \in [0,T]} \min \left\{ \sum_{j=1}^{n_2} h_j \frac{\partial}{\partial \alpha} \frac{\Phi(M((1-\alpha)\xi_{n_1}^* + \alpha\xi_t, \theta_{(j)}))}{\Phi(M(\xi_{\theta_{(j)}}^*, \theta_{(j)}))} \bigg|_{\alpha=0+} \;\bigg|\; h_j$$

$$\geq 0; \sum_{j=1}^{n_2} h_j = 1 \right\}$$

where ξ_t denotes the Dirac-measure at the point t and

$$\alpha^* = \arg\max_{\alpha \in [0,1]} \min \left\{ \sum_{j=1}^{n_2} h_j \left. \frac{\Phi(M((1-\alpha)\xi_{n_1}^* + \alpha\xi_{t^*}, \theta_{(j)}))}{\Phi(M(\xi_{\theta_{(j)}}^*, \theta_{(j)}))} \right| h_j \right.$$

$$\geq 0; \ \sum_{j=1}^{n_2} h_j = 1 \Big\}.$$

The initial design $\xi_{u(0)}$ for the calculation of the standardized maximin Φ-optimal in the class Ξ_{n_1+1} is finally defined by the vector $u_{(0)} \in \mathcal{U}_{n_1+1}$, which is given by

$$u_{(0)} = (u_1^*, \dots, u_{n_1}^*, t^*, (1-\alpha^*)w_1^*, \dots (1-\alpha^*)w_{n_1}^*, \alpha^*),$$

where $u_1^*, \dots, u_{n_1}^*$ denote the support points of the design $\xi_{n_1}^*$ with corresponding weights $w_1^*, \dots w_{n_1}^*$. The first step of the procedure is now continued to obtain the standardized maximin Φ-optimal design in the class Ξ_{n_1+1}. If this design is not standardized maximin Φ-optimal in the class of all approximated designs, the procedure is repeated, increasing the number of support points by 1. The algorithm stops if the standardized maximin Φ-optimality of the calculated design has been confirmed by the Equivalence Theorem.

Note that the algorithm definitively terminates, because, as it is shown in Dette, Melas, Pepelyshev, and Strigul (2005), any standardized maximin D-, E-, or e_i-optimal design is supported at a finite number of points. Moreover, in our numerical study, all iterations usually stopped after a few steps and the standardized maximin Φ-optimal could quickly be identified using the described procedure.

8.7.2 A comparison of maximin and uniform designs

For the sake of brevity, we will restrict the calculation of standardized maximin optimal designs to a procedure that uses the optimal designs from the infinite design space $[0, \infty]$. As a consequence, we only have to tabulate designs for one design space, namely $[0, \infty]$. Moreover, the consideration of an infinite design space is justified by the following observations. First, it was demonstrated in Section 8.3 that efficient locally optimal designs on a finite design space can easily be obtained from the designs on an infinite design space using the following method. If

$$\xi_\theta^* = \begin{pmatrix} t_1^* & t_2^* & \infty \\ w_1^* & w_2^* & w_3^* \end{pmatrix} \tag{8.57}$$

denotes a locally D-, E-, or e_i-optimal design for the Monod model on the design space $[0, \infty]$ and the right boundary of the design space $[0, T]$ satisfies $T \geq 1.5t_2^*$, then the design

$$\tilde{\xi}_\theta^* = \begin{pmatrix} t_1^* & t_2^* & T \\ w_1^* & w_2^* & w_3^* \end{pmatrix} \tag{8.58}$$

on the finite design space has at least a Φ-efficiency of 0.98, where the Φ-efficiency is defined by (8.55). Similarly, it was observed in our numerical study that if

$$\xi^* = \left(\begin{array}{cccc} t_1^* & \cdots & t_{n-1}^* & \infty \\ w_1^* & \cdots & w_{n-1}^* & w_n^* \end{array} \right) \tag{8.59}$$

denotes a standardized maximin D-, E-, or e_i-optimal design for the Monod model on the design space $[0, \infty]$ and $T \geq 2t_{n-1}^*$, then the design

$$\tilde{\xi}^* = \left(\begin{array}{cccc} t_1^* & \cdots & t_{n-1}^* & T \\ w_1^* & \cdots & w_{n-1}^* & w_n^* \end{array} \right) \tag{8.60}$$

has at least maximin efficiency 0.98, where the maximin efficiency is defined by

$$\mathrm{eff}_{\Psi_\Omega}(\xi) = \frac{\Psi_\Omega(\xi)}{\sup_\eta \Psi_\Omega(\eta)} \tag{8.61}$$

and the robust optimality criterion $\Psi_\Omega(\xi)$ is given by (8.48). Second, we note that in microbiological studies, the length of the design interval $[0, T]$ can often be chosen by the experimenter.

In Tables 8.12 and 8.13, we present some standardized maximin optimal designs for various regions of the parameter space Ω, where the design interval is given by $[0, \infty]$. A typical vector of parameters observed in studies of microbial growth by $\eta_0 = 0.03$, $s_0 = 1$, $\theta_1 = 0.25$, $\theta_2 = 0.5$, and $\theta_3 = 0.25$ (see Pirt (1975) or Blok (1994)) and for an illustration of the robustness properties of the standardized maximin optimal designs, we took this point as the center of the set Ω required for the definition of the standardized optimality criteria. In Table 8.12, we display standardized maximin D-optimal designs for the Monod model on the set $[0, \infty]$, whereas Table 8.13 contains the corresponding standardized maximin E-optimal designs. It is interesting to note that in all cases, the standardized maximin optimal designs require at least four support points. Moreover, the number of support points is increasing with the size of the set Ω specified by the experimenter. This observation was also made by Dette and Biedermann (2003) for the Michaelis–Menten model. Note that the standardized maximin optimal designs are always supported at a finite number of points and that in all cases considered in our study, the optimal designs have at most six support points, including the right boundary point of the design space (see also Conjecture 8.4.1). As pointed out in the previous paragraph, implementable and very efficient designs of the form (8.60) can be derived from the standardized maximin optimal designs on the infinite designs space in the case $t_{n-1}^* < T$. In particular, compared to the designs (8.57) on the infinite design space $[0, \infty]$, these designs have at least an efficiency of 0.98, provided that the point t_{n-1}^* satisfies $2t_{n-1}^* < T$.

For this reason, we will now assume that the microbiological experiments can be carried out over a sufficiently long time T such that these strategies of design construction are applicable and compare the standardized maximin

Table 8.12: Standardized maximin D-optimal designs in Monod model for various regions $\Omega = [z_{1,L}, z_{1,U}] \times [z_{2,L}, z_{2,U}] \times [z_{3,L}, z_{3,U}]$.

Ω	t_1	t_2	t_3	t_4	t_5	t_6	w_1	w_2	w_3	w_4	w_5	w_6
[.24, .26] × [.47, .53] × [.24, .26]	10.93	15.83	17.32	∞			.325	.223	.124	.328		
[.23, .27] × [.45, .55] × [.24, .26]	10.51	14.68	18.60	∞			.270	.244	.194	.292		
[.23, .27] × [.45, .55] × [.23, .27]	10.40	14.50	18.53	∞			.264	.241	.203	.292		
[.23, .27] × [.43, .57] × [.24, .26]	10.46	14.48	18.47	∞			.262	.243	.202	.293		
[.23, .27] × [.43, .57] × [.23, .27]	10.23	14.06	16.84	19.41	∞		.246	.219	.105	.149	.281	
[.22, .28] × [.45, .55] × [.24, .26]	10.22	14.08	17.12	19.69	∞		.244	.222	.106	.148	.280	
[.22, .28] × [.45, .55] × [.22, .28]	9.78	13.50	16.69	20.18	∞		.204	.229	.136	.162	.269	
[.22, .28] × [.43, .57] × [.24, .26]	10.01	13.65	16.79	20.13	∞		.216	.226	.135	.154	.268	
[.22, .28] × [.43, .57] × [.22, .28]	9.81	13.42	16.77	20.35	∞		.207	.233	.134	.166	.261	
[.22, .28] × [.41, .59] × [.24, .26]	9.96	13.58	16.91	20.42	∞		.214	.230	.141	.153	.263	
[.22, .28] × [.41, .59] × [.22, .28]	9.76	13.34	16.86	20.56	∞		.205	.237	.136	.164	.258	
[.20, .30] × [.41, .59] × [.24, .26]	9.19	12.72	15.69	18.58	22.00	∞	.159	.235	.101	.098	.161	.246
[.20, .30] × [.41, .59] × [.20, .30]	8.56	12.10	15.32	19.16	23.33	∞	.147	.218	.103	.128	.163	.241
[.20, .30] × [.40, .60] × [.24, .26]	9.16	12.64	15.59	18.60	22.08	∞	.159	.232	.099	.102	.163	.245
[.20, .30] × [.40, .60] × [.20, .30]	8.51	11.98	15.16	19.10	23.67	∞	.147	.212	.102	.138	.167	.235

Table 8.13: Standardized maximin E-optimal designs in Monod model for various regions $\Omega = [z_{1,L}, z_{1,U}] \times [z_{2,L}, z_{2,U}] \times [z_{3,L}, z_{3,U}]$.

Ω	t_1	t_2	t_3	t_4	t_5	t_6	w_1	w_2	w_3	w_4	w_5	w_6
$[.24,.26] \times [.47,.53] \times [.24,.26]$	9.47	15.58	17.52	∞			.302	.257	.189	.252		
$[.23,.27] \times [.45,.55] \times [.24,.26]$	9.19	14.77	18.34	∞			.273	.258	.245	.223		
$[.23,.27] \times [.45,.55] \times [.23,.27]$	9.15	14.63	18.28	∞			.271	.256	.252	.222		
$[.23,.27] \times [.43,.57] \times [.24,.26]$	9.25	14.60	18.23	∞			.270	.257	.254	.220		
$[.23,.27] \times [.43,.57] \times [.23,.27]$	8.99	14.03	16.72	19.31	∞		.261	.204	.166	.161	.207	
$[.22,.28] \times [.45,.55] \times [.24,.26]$	9.03	14.02	16.75	19.31	∞		.255	.206	.156	.173	.209	
$[.22,.28] \times [.45,.55] \times [.22,.28]$	8.78	13.48	16.55	19.91	∞		.235	.197	.200	.197	.197	
$[.22,.28] \times [.43,.57] \times [.24,.26]$	9.22	13.88	16.91	2.11	∞		.234	.220	.197	.153	.195	
$[.22,.28] \times [.43,.57] \times [.23,.27]$	8.86	13.45	16.60	2.09	∞		.230	.208	.206	.169	.188	
$[.22,.28] \times [.43,.57] \times [.22,.28]$	9.02	13.71	16.89	2.31	∞		.246	.213	.201	.154	.186	
$[.22,.28] \times [.41,.59] \times [.24,.26]$	8.62	12.86	15.38	17.71	2.73	∞	.218	.174	.148	.147	.137	.177
$[.22,.28] \times [.41,.59] \times [.22,.28]$	8.52	12.60	15.51	18.59	22.22	∞	.202	.180	.171	.158	.129	.161
$[.20,.30] \times [.41,.59] \times [.24,.26]$	8.24	12.16	15.08	18.24	22.19	∞	.182	.202	.164	.149	.152	.151
$[.20,.30] \times [.41,.59] \times [.20,.30]$	8.49	12.57	15.46	18.56	22.29	∞	.203	.182	.168	.159	.131	.158
$[.20,.30] \times [.40,.60] \times [.20,.30]$	8.20	12.10	15.02	18.26	22.20	∞	.176	.204	.168	.152	.156	.144

optimal designs with some uniform designs, which provide an alternative design of experiment if there is only very vague prior information regarding the unknown parameter. All of our efficiency considerations are restricted to designs obtained from the optimal designs on an infinite design space $[0, \infty]$ by the procedure explained by (8.59)–(8.60). The efficiencies of the standardized maximin optimal designs on the interval $[0, T]$ are slightly larger, but the additional effort of calculating these designs for any interval $[0, T]$ under consideration is only justified if $T < 2t^*_{n-1}$.

To be precise, we consider the problem of designing an experiment for the Monod model with design space $[0, T] = [0, 40]$. For the uniform design, we chose the uniform distribution on 20 points in the interval $[0, 40]$ (i.e., is the design $\xi_{\mathcal{U}(20),40}$ defined in (8.50) for $N = 20$ and $\bar{T} = 40$). Note that it can be checked numerically that for $\eta_0 = 0.03$, $s_0 = 1$, $\theta_1 = 0.25$, $\theta_2 = 0.5$, and $\theta_3 = 0.25$, the locally D-optimal uniform design is the uniform distribution on the interval $[0, 32]$ and that $\xi_{\mathcal{U}(20),40}$ could be considered as an approximation to the locally D-optimal uniform design, which takes into account that the parameters required for the construction of the locally D-optimal uniform design have been misspecified. Moreover, we checked numerically that for the point $\eta_0 = 0.03$, $s_0 = 1$, $\theta_1 = 0.25$, $\theta_2 = 0.5$, and $\theta_3 = 0.25$, the uniform design $\xi_{\mathcal{U}(20),40}$ is only slightly less efficient compared to the locally D-optimal uniform design $\xi_{\mathcal{U}(20),32}$. The situation for the other criteria is similar. In Table 8.14, we compare this uniform design with the standardized maximin D-optimal designs derived from Table 8.12 and the procedure described by (8.59) and (8.60). The comparison is performed by considering the ratios

$$C_D(\xi, \theta) = \left(\frac{\det M(\xi, \theta)}{\det M(\xi_{\mathcal{U}(20),40}, \theta)} \right)^{\frac{1}{3}}, \tag{8.62}$$

$$C_i(\xi, \theta) = \frac{(e_i^T M^{-1}(\xi, \theta)e_i)^{-1}}{(e_i^T M^{-1}(\xi_{\mathcal{U}(20),40}, \theta)e_i)^{-1}}, \quad i = 1, 2, 3 \tag{8.63}$$

$$C_E(\xi, \theta) = \frac{\lambda_{\min} M(\xi, \theta)}{\lambda_{\min} M(\xi_{\mathcal{U}(20),40}, \theta)} \tag{8.64}$$

of the corresponding optimality criteria. Note that these ratios depends on the parameter θ and that for a given $\theta \in \Omega$, a larger value than 100% indicates that the design ξ is more efficient than the uniform design $\xi_{\mathcal{U}(20),40}$ with respect to the corresponding optimality criterion. For the sake of brevity Table 8.15 contains the maximum, minimum, and averaged values of these ratios, which are indicated by the symbols "max", "min", and "average", respectively. For example, in the column with the label C_D and "min", the reader finds the minimum ratio

$$\min_{\theta \in \Omega} C_D(\tilde{\xi}_D^*, \theta) = \min_{\theta \in \Omega} \left(\frac{\det M(\xi, \theta)}{\det M(\xi_{\mathcal{U}(20),40}, \theta)} \right)^{\frac{1}{3}}$$

taken over the set Ω, whereas in the column with the label C_D and "average" the corresponding integrated values with respect to the uniform distribution can be found; that is,

$$\int_\Omega C_D(\tilde{\xi}_D^*, \theta)\, d\theta$$

(recall that the design $\tilde{\xi}_D^*$ on the interval $[0, T]$ is obtained from the standardized maximin D-optimal design on the infinite design space in Table 8.12 by (8.59) and (8.60)).

We observe that the standardized maximin D-optimal design is always better than the uniform design $\xi_{\mathcal{U}(20),40}$, if the D-, E-, e_1-, and e_2-criterion are used for comparing competing designs, because the corresponding minimum values are larger than 100%. The improvement by using a standardized maximin optimal design instead of the uniform design with respect to these criteria can be substantial. For example, consider the set $\Omega = [0.20, 0.30] \times [0.40, 0.60] \times [0.20, 0.30]$ corresponding to a situation where only vague prior information regarding the unknown parameters in the Monod model is available. In this case, the minimum gain in D-efficiency by the standardized maximin D-optimal design is approximately 17%, the maximum is 43%, and the average, we have 25% improvement compared to the uniform design $\xi_{\mathcal{U}(20),40}$. The performance of the standardized maximin D-optimal design would be even better if the parameter space Ω could be specified more precisely. The advantages with respect to the E-, e_1-, and e_2-criterion are even larger. On the other hand, the uniform design $\xi_{\mathcal{U}(20),40}$ is more efficient for the estimation of the parameter θ_3. However, as pointed out in Section 8.3, the efficient estimation of θ_1 and θ_2 is usually more important for the Monod model, because in realistic situations (see Pirt (1975) or Blok (1994)) the parameter θ_3 can be estimated with much higher precision than the parameters θ_1 and θ_2. The situation for the E-optimality criterion is very similar. The standardized maximin E-optimal design should be preferred in all cases, except if the primary goal of the experiment is the parameter θ_3 and all other parameters are not of interest for the experimenter.

8.8   Appendix

We will begin by proving a Chebyshev property for a systems of functions, which will be crucial for a proof of the statements in Section 8.2–8.5. Recall that a system of functions $\psi_1(t), \ldots, \psi_m(t)$ is called a Chebyshev system (T-system) on an interval $[\alpha, \beta]$ if

$$\det\left(\psi_i(t_j)\right)_{i,j=1}^m > 0$$

for any $\alpha \le t_1 < \cdots < t_m \le \beta$ (see Karlin and Studden (1966, Chap.1)). The main property of a Chebyshev system is that any nontrivial linear

Table 8.14: Comparison of standardized maximin D-optimal design with uniform designs of the form (8.50) ($N = 20$, $\bar{T} = 40$) on the design space $[0, T] = [0, 40]$. The table shows the minimum, maximum and average values of the ratios defined by (8.62), (8.63) and (8.64) in %.

Ω	min					max					average				
	C_D	C_1	C_2	C_3	C_E	C_D	C_1	C_2	C_3	C_E	C_D	C_1	C_2	C_3	C_E
[.24,.26] × [.47,.53] × [.24,.26]	140	168	175	64	175	154	220	226	75	226	151	207	213	68	213
[.23,.27] × [.45,.55] × [.24,.26]	133	157	155	65	155	148	190	183	74	183	140	172	166	67	167
[.23,.27] × [.45,.55] × [.23,.27]	132	154	153	66	153	148	191	181	75	181	139	169	163	67	163
[.23,.27] × [.43,.57] × [.24,.26]	131	152	149	64	149	149	195	184	76	184	140	171	163	67	163
[.23,.27] × [.43,.57] × [.23,.27]	130	147	149	65	149	148	195	180	75	180	138	164	157	67	157
[.22,.28] × [.45,.55] × [.24,.26]	129	146	148	66	148	147	193	177	76	180	137	161	155	69	155
[.22,.28] × [.45,.55] × [.22,.28]	127	140	138	66	138	144	205	177	79	180	134	156	146	69	146
[.22,.28] × [.43,.57] × [.24,.26]	127	141	142	65	142	147	206	181	78	180	136	158	149	69	149
[.22,.28] × [.43,.57] × [.22,.28]	126	136	134	64	134	146	212	181	80	181	134	152	143	69	143
[.22,.28] × [.41,.59] × [.24,.26]	126	138	139	65	139	149	213	186	80	186	135	156	147	69	147
[.22,.28] × [.41,.59] × [.22,.28]	124	132	131	65	131	148	221	188	81	188	133	149	141	70	141
[.20,.30] × [.41,.59] × [.24,.26]	121	116	111	67	117	146	238	198	89	199	129	143	132	75	132
[.20,.30] × [.41,.59] × [.20,.30]	118	106	109	73	109	142	230	193	98	193	125	128	123	81	123
[.20,.30] × [.40,.60] × [.24,.26]	120	115	116	68	116	147	239	200	91	200	129	141	131	76	131
[.20,.30] × [.40,.60] × [.20,.30]	117	105	108	74	108	143	229	191	98	192	125	126	123	82	123

Table 8.15: Comparison of standardized maximin E-optimal design with uniform designs of the form (8.50) ($N = 20$, $\bar{T} = 40$) on the design space $[0, T] = [0, 40]$. The table shows the minimum, maximum and average values of the ratios defined by (8.62), (8.63) and (8.64) in %.

Ω	min					max					average				
	C_D	C_1	C_2	C_3	C_E	C_D	C_1	C_2	C_3	C_E	C_D	C_1	C_2	C_3	C_E
$[.24,.26] \times [.47,.53] \times [.24,.26]$	130	157	187	52	187	143	231	246	58	246	140	212	234	54	234
$[.23,.27] \times [.45,.55] \times [.24,.26]$	124	159	160	51	160	143	186	211	62	211	133	180	187	53	187
$[.23,.27] \times [.45,.55] \times [.23,.27]$	123	155	155	50	155	143	187	210	63	210	133	178	183	53	183
$[.23,.27] \times [.43,.57] \times [.24,.26]$	121	159	152	49	152	145	187	213	63	212	134	179	184	53	184
$[.23,.27] \times [.43,.57] \times [.23,.27]$	123	154	148	49	148	143	188	203	66	203	132	175	175	53	175
$[.22,.28] \times [.45,.55] \times [.23,.27]$	119	153	147	50	147	143	188	200	67	200	132	169	173	54	173
$[.22,.28] \times [.45,.55] \times [.24,.26]$	119	153	147	49	147	142	181	192	71	192	130	173	165	54	164
$[.22,.28] \times [.45,.55] \times [.22,.28]$	119	147	141	49	141	143	180	200	67	200	132	169	173	54	173
$[.22,.28] \times [.43,.57] \times [.24,.26]$	116	142	141	49	141	142	181	192	71	192	130	165	165	54	165
$[.22,.28] \times [.43,.57] \times [.22,.28]$	118	148	141	49	141	143	188	200	66	200	132	171	175	53	175
$[.22,.28] \times [.43,.57] \times [.24,.26]$	115	143	141	47	141	145	188	200	67	200	132	164	164	54	164
$[.22,.28] \times [.41,.59] \times [.24,.26]$	115	135	136	48	136	146	190	202	70	202	132	168	168	54	167
$[.22,.28] \times [.41,.59] \times [.22,.28]$	118	143	141	46	141	146	189	207	72	207	131	163	166	54	165
$[.22,.28] \times [.41,.59] \times [.24,.26]$	116	137	138	47	137	145	188	198	72	198	130	163	165	54	164
$[.23,.27] \times [.41,.59] \times [.23,.27]$	119	129	132	47	132	148	189	198	70	198	130	163	163	54	163
$[.20,.30] \times [.41,.59] \times [.24,.26]$	112	137	138	48	137	137	189	207	72	207	129	159	159	53	159
$[.20,.30] \times [.41,.59] \times [.20,.30]$	107	119	126	47	126	146	188	197	81	197	128	152	152	58	152
$[.20,.30] \times [.41,.59] \times [.24,.26]$	101	113	120	44	120	147	211	203	81	203	127	147	145	57	145
$[.20,.30] \times [.24,.26] \times [.20,.30]$	107	129	132	46	132	144	188	194	75	194	129	159	159	53	159
$[.20,.30] \times [.40,.60] \times [.24,.26]$	106	118	125	47	125	147	191	201	81	201	128	151	152	57	152
$[.20,.30] \times [.40,.60] \times [.20,.30]$	100	112	119	43	119	147	216	206	82	206	127	147	144	56	144

combination $\sum_{i=1}^{m} \alpha_i \psi_i(t)$ of the functions $\psi_1, \ldots, \psi_m$ has at most $m-1$ distinct roots in the interval $[\alpha, \beta]$.

Proposition 8.8.1

a) *The system of functions*

$$\left\{ \frac{1}{u}, \frac{1}{b_1 - u}, \frac{1}{(b_1 - u)(b_2 - u)} \right\}$$

is a Chebyshev system on any interval $[a, b] \subset (0, \infty]$ *whenever* $b_1, b_2 \geq b$.

b) *If a system of functions* $\psi_1(x), \ldots, \psi_m(x)$ *is a Chebyshev system on the interval* $[\eta_0, b]$, *then the system of functions*

$$\int_{\eta_0}^{u} \psi_1(x) \, dx, \ldots, \int_{\eta_0}^{u} \psi_m(x) \, dx$$

is also a Chebyshev system on any interval $[a, b] \subset (\eta_0, b]$.

Proof.

(a) Consider the determinant

$$J = \det \left(\psi_i(u_j) \right)_{i,j=1}^{3}, \ a \leq u_1 < u_2 < u_3 \leq b,$$

where $\psi_1(u) = 1/u$, $\psi_2(u) = 1/(b_1 - u)$, and $\psi_3(u) = 1/[(b_1 - u)(b_2 - u)]$. Let us multiply the i-th row by $u_i(b_1 - u_i)(b_2 - u_i)$, $i = 1, 2, 3$. By a linear transformation of the columns, the resulting determinant can be reduced to the Vandermonde determinant

$$\det(u_j^{i-1})_{i,j=1}^{3} = \prod_{j<i} (u_i - u_j) > 0.$$

Hence, $J > 0$ and the system $\psi_1(u), \psi_2(x), \psi_3(x)$ is a Chebyshev system on the interval $[a, b] \subset (0, \infty]$.

(b) Let us suppose that assertion (b) of Proposition 8.8.1 is not true. Then there exist real numbers $\alpha_1, \ldots, \alpha_m$, not all equal to zero, such that the linear combination

$$\sum_{i=1}^{m} \alpha_i \int_{\eta_0}^{u} \psi_i(x) \, dx =: q(u)$$

has at least m distinct zeros in the interval $[a, b] \subset (\eta_0, b]$. Moreover, we have $q(\eta_0) = 0$, and therefore the derivative

$$q'(u) = \sum_{i=1}^{m} \alpha_i \psi_i(u)$$

has m different roots in the interval $[\eta_0, b]$, which contradicts the Chebyshev property of the system $\{\psi_i(u)\}$.

∎

8.8.1 Proof of Lemma 8.2.1

The derivative of η is positive at the point $t = 0$, and by continuity, it must also be positive in a neighborhood of the origin. Moreover, the function $\bar{\eta}(t) \equiv c$ is obviously a solution of the differential equation (8.1). By the Implicit Function Theorem (see Gunning and Rossi (1965)), it follows that the differential equation (8.1) cannot have two differentiable solutions that coincide at a point. Thus, the function η should be less than c for any $t > 0$ and must be increasing. Consequently, there exists the limit $\lim_{t \to \infty} \eta(t)$. Now, (8.1) implies that the limit of the derivative of η also exists. Because η is a bounded function, it follows from (8.14) that

$$\lim_{t \to \infty} \eta'(t) = 0,$$

$$\lim_{t \to \infty} \eta(t) = c,$$

which completes the proof of Lemma 8.2.1.   ■

8.8.2 Proof of Theorem 8.2.1

Recall the definiton of the set $\Omega = \{\theta = (\vartheta_1, \vartheta_2, \vartheta_3)^T : \vartheta_i > 0, i = 1, 2, 3\}$ in Theorem 8.2.1; then the assertion of the theorem follows from results of Jennrich (1969) and the following lemma.

Lemma 8.8.1 *Let $\eta(t, \theta)$ denote the regression function determined by (8.1)–(8.3).*

(a) For any fixed vector $\theta \in \Omega$, there exist the derivatives

$$\frac{\partial \eta}{\partial \vartheta_i}(t, \theta), \quad \frac{\partial^2 \eta}{\partial \vartheta_i \partial \vartheta_j}(t, \theta), \quad i, j = 1, 2, 3.$$

(b) For any fixed $\theta^0 \in \Omega$, the function

$$g(\theta) = \sum_{j=1}^{n} (\eta(t_j, \theta) - \eta(t_j, \theta^0))^2$$

with $0 \leq t_1 < \cdots < t_n < \infty$ and $n \geq 3$ attains its minimum value (equal to zero) in the set Ω if and only if $\theta = \theta^0$.

(c) For any $\theta^0 \in \Omega$ and for any design ξ with more than three support points, it follows that

$$\det M(\xi, \theta^0) \neq 0.$$

Proof. Statement (a) is an immediate consequence of the identities (8.16) and (8.20). Let us suppose that condition (b) is not valid. Then, as $n \geq 3$, there exist two vectors $\theta_{(1)}$ and $\theta_{(2)}$ such that

$$\eta(t_i, \theta_{(1)}) = \eta(t_i, \theta_{(2)}), \quad i = 1, 2, 3. \tag{8.65}$$

Define $\eta_i = \eta(t_i, \theta_{(1)})$ $(i = 1, 2, 3)$ and consider the two functions

$$t_{(j)}(x) = t(x, \theta_{(j)}) = \eta^{-1}(x, \theta_{(j)}), \ \ j = 1, 2.$$

Due to (8.65), we have $t_{(1)}(\eta_i) = t_{(2)}(\eta_i) = t_i$ $(i = 1, 2, 3)$, and observing (8.16), we obtain

$$t_i = a_1(\theta_{(j)}) \int_{\eta_0}^{\eta_i} \frac{1}{u} \, du + a_2(\theta_{(j)}) \int_{\eta_0}^{\eta_i} \frac{1}{a_3(\theta_{(j)}) - u} \, du,$$

$i = 1, 2, 3$, $j = 1, 2$, where the constants $a_i(\theta)$ are given by

$$a_1(\theta) = \frac{1+b}{\vartheta_1} = \frac{1}{\vartheta_1} \frac{s_0 \vartheta_3 + \eta_0 + \vartheta_2 \vartheta_3}{s_0 \vartheta_3 + \eta_0},$$

$$a_2(\theta) = \frac{b}{\vartheta_1} = \frac{1}{\vartheta_1} \frac{\vartheta_2 \vartheta_3}{s_0 \vartheta_3 + \eta_0},$$

$$a_3(\theta) = c = s_0 \vartheta_3 + \eta_0.$$

It is easy to verify that the conditions $a_i(\theta_{(1)}) = a_i(\theta_{(2)})$, $i = 1, 2, 3$, imply $\theta_{(1)} = \theta_{(2)}$. Consequently, there exist two different vectors

$$a_{(j)} = (a_{1(j)}, a_{2(j)}, a_{3(j)}) = (a_1(\theta_{(j)}), a_2(\theta_{(j)}), a_3(\theta_{(j)}))^T, \ j = 1, 2,$$

such that the equations

$$a_{1(j)} \int_{\eta_0}^{\eta_i} \frac{1}{u} \, du + a_{2(j)} \int_{\eta_0}^{\eta_i} \frac{1}{a_{3(j)} - u} \, du = t_i,$$

are satisfied for all $i = 1, 2, 3$, and $j = 1, 2$. Subtracting the equalities for $j = 2$ from the equalities for $j = 1$, we obtain

$$\Delta_1 \int_{\eta_0}^{\eta_i} \frac{1}{u} \, du - a_{2(1)} \int_{\eta_0}^{\eta_i} \frac{\Delta_3}{(a_{3(1)} - u)(a_{3(2)} - u)} \, du$$
$$+ \Delta_2 \int_{\eta_0}^{\eta_i} \frac{1}{a_{3(1)} - u} \, du = 0 \qquad (8.66)$$

for some constants $\Delta_i = a_{i(1)} - a_{i(2)}$, $i = 1, 2, 3$. From Proposition 8.8.1, it follows that the functions

$$\psi_1(x) = \int_{\eta_0}^x \frac{du}{u}, \ \psi_2(x) = \int_{\eta_0}^x \frac{du}{b_1 - u}, \ \psi_3(x) = \int_{\eta_0}^x \frac{du}{(b_1 - u)(b_2 - u)}$$

form a Chebyshev system on an interval $(\eta_0, d]$ whenever $b_1, b_2 \geq d$. Inserting $d = \min\{a_{3(1)}, a_{3(2)}\}$, $b_1 = a_{3(1)}$, $b_2 = a_{3(2)}$, $\alpha_1 = \Delta_1$, $\alpha_2 = -a_{2(1)} \Delta_3$, and $\alpha_3 = \Delta_2$, we obtain from (8.66) that

$$\sum_{i=1}^3 \alpha_i \psi_i(\eta_j) = 0, \ j = 1, 2, 3,$$

where the coefficients α_i are not all equal to zero and we have, from Lemma 8.2.1, $\eta_i \leq \min\{a_{3(1)}, a_{3(2)}\}$ $(i = 1, 2, 3)$. This equality contradicts the main property of Chebyshev systems and the proof of assertion (b) is completed.

Finally, let us prove that assertion (c) is valid. From Proposition 8.8.1, it follows that the functions $\varphi_1(x)$, $\varphi_2(x)$, and $\varphi_3(x)$ defined in (8.23) generate a Chebyshev system on the interval $(0, \infty]$. If $n = 3$, a direct calculation shows shows that

$$M(\xi, \theta^0) = F^T W F,$$

where $W = \mathrm{diag}\{w_1, w_2, w_3\}$ and

$$F = \left(\frac{\partial \eta(t_i, \theta^0)}{\partial \vartheta_j} \right)^3_{i,j=1}.$$

Consequently, it follows that

$$\det M(\xi, \theta^0) = w_1 w_2 w_3 (\det F)^2, \qquad (8.67)$$

whereas due to (8.21), we have

$$\det F = \det K \cdot \det \left(\varphi_j(\eta_i) \right)^3_{i,j=1} \neq 0,$$

where we used the Chebyshev property of the system $\{\varphi_i(x)\}^3_{i=1}$. In the general case $n > 3$, let

$$\alpha = (i_1, i_2, i_3), \quad \xi_\alpha = \begin{pmatrix} t_{i_1} & t_{i_2} & t_{i_3} \\ w_{i_1} & w_{i_2} & w_{i_3} \end{pmatrix}, \quad 1 \leq i_1 < i_2 < i_3 \leq n$$

and denote by τ the set of all different multiindices α; then the Binet–Cauchy formula shows

$$\det M(\xi, \theta^0) = \sum_{\alpha \in \tau} \det M(\xi_\alpha, \theta^0)$$

and all terms in the sum on the right-hand side are positive. ∎

8.8.3   Proof of Lemma 8.3.1

It follows by a standard argument that the optimal weights (with respect to the D-criterion) are equal for any design with three support points. Consequently, it is enough to verify that

$$\frac{\partial}{\partial x_3} \det F(x_1, x_2, x_3) > 0$$

for $\eta_0 \leq x_1 < x_2 < x_3$, where $F(x_1, x_2, x_3) = \det(\varphi_i(x_j))^3_{i,j=1}$. To this end, we introduce the notation

$$L(x) = \frac{\partial}{\partial x} \det F(x_1, x_2, x),$$

$$G(x) = \det \begin{pmatrix} f_1(x_1) & f_2(x_1) & f_3(x_1) \\ f_1(x_2) & f_2(x_2) & f_3(x_2) \\ f_1(x) & f_2(x) & f_3(x) \end{pmatrix}, \qquad (8.68)$$

with $f_i(x) = \varphi_i(x)/v(x)$; then a direct computation shows that

$$0 \leq \det F(x_1, x_2, x) = v(x_1)v(x_2)v(x)G(x)$$

and

$$L(x) = v(x_1)v(x_2)v'(x)G(x) + v(x_1)v(x_2)v(x)G'(x) . \qquad (8.69)$$

Consequently, we obtain

$$L(x) = v(x_1)v(x_2)v(x)G'(x)$$

for $x = x_1, x_2$. Observing (8.23) and (8.68), it follows that the function $G(x)$ has zeros at $x = \eta_0, x_1, x_2$. Hence, there exist points u_1 and u_2 with $\eta_0 < u_1 < x_1 < u_2 < x_2$ and $G'(u_i) = 0$, $i = 1, 2$. Moreover, the functions $\{f_i'(x)\}$ form a Chebyshev system on the interval $[\eta_0, \bar{c}]$ and the functions $\{f_i(x)\}$ have the same property on the interval $(\eta_0, \bar{c}]$ by Proposition 8.8.1. Therefore, the functions G' and G have at most two zeros in the interval $(\eta_0, \bar{c}]$ and we obtain from $G(x) > 0$, $G'(x) > 0$ for $x > x_2$ that both terms on the right-hand side of (8.69) are positive for $x > x_2$. This implies $L(x) > 0$ for any $x > x_2$ and, consequently, the largest support point of the D-optimal three-point design must be attained at the boundary (i.e., $x_3 = \bar{c}$).   ∎

8.8.4   Proof of Lemma 8.3.2

Proof of assertion (1). Let $b > 0$ be an arbitrary fixed number and ζ be an arbitrary fixed design with at least $n \geq 3$ support points. Consider a transformation of the function of the equivalence theorem for D-optimality in the linear regression model $\beta^T \psi(x)$ defined by (8.33); that is,

$$d(x) = d(x, \zeta) = d_1(x) - d_2(x),$$

where

$$d_1(x) = d_1(x, \zeta) = \frac{1}{v^2(x)}\psi^T(x)\tilde{M}^{-1}(\zeta)\psi(x) - \frac{3b^2}{(1-x)^2},$$

$$d_2(x) = \frac{3}{v^2(x)} - \frac{3b^2}{(1-x)^2},$$

and $v(x)$ is given by (8.24) for $c = 1$. We define $g(x) = g_1(x) - g_2(x)$, with

$$g_1(x) = [x(1-x)^2 d_1'(x)]''',$$

$$g_2(x) = [x(1-x)^2 d_2'(x)]''',$$

and introduce

$$\begin{pmatrix} A & B & C \\ B & G & E \\ C & E & F = \tilde{M}^{-1}(\zeta) \end{pmatrix},$$

where $\tilde{M}(\zeta)$ is the information matrix of the design ζ in the linear regression model $\beta^T \psi(x)$. Assume for a moment that $F = 3b^2$. In this case, a direct computation gives

$$(1 - x)^3 g_1(x) = [-4Gx^2 + (10G + 2E)x - 6(E + G)],$$

$$x^5 g_2(x) = [144(1 + b)^2 - 36(2 + b)(1 + b)x]. \tag{8.70}$$

It is not difficult to see that all elements of the matrix $\tilde{M}^{-1}(\zeta)$ are positive and, in particular, $G > 0$ and $E > 0$. Moreover, for $x \in (0, 1)$, we have

$$g_2(x) > 0, \; g_1(x) < 0.$$

The first of these inequalities is obvious. The second inequality will be verified by considerating the roots $x_{(1)}$ and $x_{(2)}$ of the equation

$$-4Gx^2 + (10G + 2E)x - 6(E + G) = 0,$$

which are given by

$$x_{(1),(2)} = \frac{E + 5G \pm \sqrt{E^2 - 14EG + G^2}}{4G}.$$

Since E and G are positive, it follows that $x_{(1)}, x_{(2)} > 1$ and we have from (8.70) $g_1(x) < 0$ for $x \in (0, 1)$. Therefore, the function $g(x)$ is strictly negative for $x \in (0, 1)$ if the condition $F = 3b^2$ is valid.

Now, let

$$\zeta = \left(\begin{array}{cccc} x_1 & x_2 & \cdots & x_n \\ w_1 & w_2 & \cdots & w_n \end{array} \right), \; 0 \leq x_1 < \cdots < x_n \leq 1,$$

be a D-optimal design. From the proof of Lemma 8.3.1 and the Binet–Cauchy formula, it follows that $x_n = 1$. Notice that for $x \to 0$, we have $\psi(x) \to (0, 0, 0)^T$ and, consequently, the left boundary $x = 0$ of the design space is not a support point of a D-optimal design (i.e., $x_1 > 0$). By the Equivalence Theorem for D-optimality, we have

$$\tilde{d}(x_i) := v^2(x_i)d(x_i) = \psi^T(x_i)\tilde{M}^{-1}(\zeta)\psi(x_i) - 3 = 0 , \; i = 1, \ldots, n,$$

$$\tilde{d}'(x_i) = 0 , \quad i = 1, \ldots, n - 1.$$

$$\tag{8.71}$$

Let us assume that the design ζ contains $n > 3$ points. From (8.33), we obtain

$$\lim_{x \to 1} v^2(x)d(x) = 0$$

and it follows that $F = 3b^2$. Consequently, the arguments of the previous paragraph are applicable and we have from (8.71) that

$$d'(x) = 0$$

for $x = x_1, x_2, x_3$ and for some points $\tilde{x}_1 \in (x_1, x_2)$ and $\tilde{x}_2 \in (x_1, x_2)$. Thus, the function $d'(x)$ and the function $x(1-x)^2 d'(x)$ have at least five roots in the interval $(0, 1)$. Hence, $g(x)$, being the third derivative of the function, $x(1-x)^2 d'(x)$ has at least $5 - 3 = 2$ zeros in the interval $(0, 1)$. However, we proved above that this is impossible. Thus, the assumption $n \geq 4$ yields a contradiction and any D-optimal design for the regression model $\beta^T \psi(x)$ on the interval $[0, 1]$ is supported at exactly three points.

Finally, let us show that there exists a unique D-optimal design. The existence of a D-optimal design follows from continuity of the function $\psi(x)$ and compactness of the interval $[0, 1]$. Assume that $\zeta_{(1)}$ and $\zeta_{(2)}$ are two different D-optimal designs. Then it follows by a standard concavity argument (see Fedorov (1972)) that the design $\tilde{\zeta} := \zeta_{(1)}/2 + \zeta_{(2)}/2$ is also D-optimal. However, this is impossible because the design $\tilde{\zeta}$ contains more than three distinct points.

Proof of assertion (2). From the proof of assertion (1), it follows that the function

$$v^2(x)d(x)$$

has at most three maxima on the interval $[0, 1]$ and assertion (2) follows by a continuity argument. ∎

8.8.5 Proof of Lemma 8.6.1.

We will only present a proof of assertions (1) and (2). The remaining statements regarding the e_k-optimal designs are proved similarly. Let $h(x) = K\varphi(x)$ and $s = (K^T)^{-1}r$ and define

$$N(\zeta_E) = \int h(x)h^T(x) \, d\zeta_E(x) = K\bar{M}(\zeta_E)K^T; \qquad (8.72)$$

then it is easy to see that assumption (2) can be rewritten as

$$r^T \bar{M}(\zeta_E)r = s^T N(\zeta_E)s = (s^T s)\lambda_{\min}(N(\zeta_E)) . \qquad (8.73)$$

From (8.41–8.43), we obtain

$$s^T h(x_i) = (-1)^{i+1}, \quad i = 1, 2, 3, \qquad (8.74)$$

$$s^T h(x) \leq 1 \; \forall x \qquad (8.75)$$

and Elfvings theorem (Elfving (1952)) and (8.74) imply that the vector s is an eigenvector of the matrix $N(\zeta_E)$ (i.e., is $N(\zeta_E)s = \lambda s$). Observing (8.74) and (8.73), it follows that

$$1 = s^T N(\zeta_E)s = \lambda \cdot (s^T s) = \frac{\lambda}{\lambda_{\min}(N(\zeta_E))}$$

and from (8.75) we obtain the inequality

$$h^T(x)\left(\frac{ss^T}{s^Ts}\right)h(x) \le \frac{1}{s^Ts} = \lambda_{\min}(N(\zeta_E)) \; \forall x.$$

Part (2) of Lemma 8.6.1 now follows immediately from the corresponding equivalence theorems for E-criterion [see Pukelsheim (1993)]. The design is unique, because the function $g(x)$ defined in (8.41) is unique (see, e.g., Karlin and Studden (1966, Chap 1)).

To prove part (1) let us note that for $\eta_0 \to 0$ we have

$$\tilde{\varphi}^T(x) = v(x)\left(-\frac{\ln\frac{x}{\eta_0}}{\ln\eta_0}, \ln\frac{c-x}{c-\eta_0}, \frac{x-\eta_0}{c-x}\right)$$

$$\sim v(x)\left(1, \ln\frac{c-x}{c-\eta_0}, \frac{x-\eta_0}{c-x}\right) =: \tilde{\psi}(x)$$

and that any subset with two functions of

$$\left\{v(x), v(x)\ln\frac{c-x}{c-\eta_0}, v(x)\frac{x-\eta_0}{c-x}\right\}$$

generates a Chebyshev system on the interval $[\eta_0, \bar{c}]$. Define

$$\tilde{K} = \begin{pmatrix} \dfrac{1+b}{\vartheta_1}\ln\eta_0 & \dfrac{b}{\vartheta_1} & 0 \\[2mm] -\dfrac{b}{\vartheta_2}\ln\eta_0 & -\dfrac{b}{\vartheta_2} & 0 \\[2mm] -\dfrac{b\eta_0}{c\vartheta_3}\ln\eta_0 & -\dfrac{b\eta_0}{c\vartheta_3} & -\dfrac{b}{\vartheta_3} \end{pmatrix}.$$

and

$$\tilde{M}(\xi) := \int \tilde{\psi}(x)\tilde{\psi}^T(x) \, d\xi(x) .$$

A straightforward but tedious calculation shows that for sufficiently small η_0, the sign pattern of the matrix

$$(\tilde{K}^T)^{-1}(\tilde{M}(\xi))^{-1}\tilde{K}^{-1}$$

is of the form

$$\begin{pmatrix} + & - & + \\ - & + & - \\ + & - & + \end{pmatrix}$$

and, consequently, the matrix D has a simple eigenvalue for sufficiently small η_0 (see Gantmacher (1998)). It follows from general results on E-optimality [see Dette and Studden (1993)] that the E-optimal design for the vector of regression functions $\tilde{\psi}(x)$ is supported on the Chebyshev points. Part (1) of the lemma is now obtained by a continuity argument. ∎

Appendix

Remarks on Computer Calculation of Taylor Coefficients

Consider the problem of calculating Taylor coefficients for optimal design functions, defined in Section 2.1.

As, example, let us take the regression function

$$\eta(x, \Theta) = \theta_1 e^{-\theta_3 x} + \theta_2 e^{-\theta_4 x},$$

where $\theta_1, \theta_2 \neq 0$, $\theta_3, \theta_4 > 0$, $\theta_3 \neq \theta_4$, and $x \in [0, \infty)$.

Denote $u = (\theta_3 + \theta_4)/2$ and $z = (\theta_3 - \theta_4)/2$.

In Chapter 6 it was prove that a locally D-optimal design in this case is unique and is of the form $\{0, x_2^*, x_3^*, x_4^*; 1/4, 1/4, 1/4, 1/4\}$, where $x_i^* = \tilde{x}_i(z)/u$, $i = 2, 3, 4$, and $\tilde{x}_i(z) \to \tilde{x}_i(0)$ with $z \to 0$, where $2\tilde{x}_2(0)$, $2\tilde{x}_3(0)$, and $2\tilde{x}_4(0)$ are roots of the Laguerre polynomial of the third degree with parameter 0.

Thus, it will do to find Taylor coefficients of the function

$$x^*(z) = \arg \max \det \left(f(x_1) \vdots \ldots \vdots f(x_4) \right),$$

where

$$f(x_i) = \left(e^{-(1+z)x_i}, x_i e^{-(1+z)x_i}, e^{-(1-z)x_i}, x_i e^{-(1-z)x_i} \right)_{i=1}^{4}$$

in a vicinity of point $z = 0$.

In the given case, the goal function has peculiarity of fourth order. The problem can be solved with the help of the following computer program, which realizes formulas from Theorem 2.4.4 in Maple suit.

Program

1. $m := 4$;
   the number of points.

2. $f := (t)- > \mathbf{matrix}(1, m, [e^{-(1+u)t}, te^{-(1+u)t}, e^{-(1-u)t}, te^{-(1-u)t}]);$
   matrix line of basic functions

3. $p := \mathbf{vector}(m); p[1] = 0;$
   vector of optimal design points.

4. $M := \mathbf{matrix}(m, m);$
   for i **from** 1 **to** m **do**
   $\mathbf{copyinto}(f(p[i]), M, i, 1);$
   od;
   $detM := \mathbf{det}(M);$
   calculating the information matrix M and its determinant. Variable $detM$ is a symbolic expression of $u, p[2], p[3], p[4]$.

5. $x[0] := \mathbf{matrix}(m - 1, 1, [.4679111137, 1.652703644, 3.879385241]);$
   zero approximation.

6. $g := vector(m - 1);$
   for i **from** 1 **to** $m - 1$ **do**
   $g[i] := \mathbf{diff}(detM, p[i + 1]);$
   od;
   calculating of the left hand side of the basic equation $g(\cdot) = 0$.

7. $irr := 4;$
   the order of peculiarity.

8. $J := \mathbf{matrix}(m - 1, m - 1);$
   for i **from** 1 **to** $m - 1$ **do**
   for j **from** 1 **to** $m - 1$ **do**
   $J[i, j] := \mathbf{coeftayl}(\mathbf{subs}(\{\mathbf{seq}(p[l + 1] = x[0][l, 1], l = 1..m - 1)\}, \mathbf{diff}(g[i], p[j + 1])), u = 0, irr);$
   od;
   od;
   $invJ := \mathbf{inverse}(J);$
   Calculating of the matrix $J_{(irr)}$ and the inverse matrix.

9. $MAX := 10; h := \mathbf{matrix}(m - 1, 1);$
   the number of steps of the Algorithm 2.6.2

10. **for** c **from** 1 **to** MAX **do**
    for i **from** 1 **to** $m - 1$ **do**
    $p[i + 1] := \mathbf{sum}('x[k][i, 1] * u \wedge k',' k' = 0..c - 1);$
    od;
    for i **from** 1 **to** $m - 1$ **do**
    $h[i, 1] := \mathbf{coeftayl}(g[i], u = 0, c + irr);$
    od;
    $x[c] := \mathbf{evalm}(-invJ \& * h);$
    od;

recursion calculating the coefficients. At the step c optimal design points are replaced by the segment of the Taylor series. Calculating matrix row h. The coefficients at cth step are calculated by formula $-invJ * h$.

11. **print**(x).

Let us describe the performance of the program.
Matrices J and $invJ$ are of the form

$$J = \begin{pmatrix} -.47885 & .063589 & .007699 \\ .063589 & -.11423 & .017997 \\ .007699 & .017997 & -.03157 \end{pmatrix},$$

$$invJ = \begin{pmatrix} -2.3120 & -1.5116 & -1.4256 \\ -1.5116 & -10.606 & -6.4146 \\ -1.4256 & -6.4146 & -35.678 \end{pmatrix}.$$

At the first step, obtain

$$h = \begin{pmatrix} 0 \\ 0 \\ 0 \end{pmatrix}, \quad x[1] = \begin{pmatrix} 0 \\ 0 \\ 0 \end{pmatrix}.$$

At the second step, obtain

$$h = \begin{pmatrix} -.02456 \\ .003633 \\ .056620 \end{pmatrix}, \quad x[2] = \begin{pmatrix} .02919 \\ .36419 \\ 2.0066 \end{pmatrix}.$$

and so on.

Remarks

(i) The basic equation can be given otherwise, in an immediate way.

(ii) The zero approximation for design points can be found numerically.

(iii) Performing the program can be done substantially more quickly if it is possible to find the derivatives in block 4 analytically.

Thus, we have demonstrated that the computer realization of formulas from Theorem 2.4.4 is rather easy.

References

Atkinson, A.C., Donev, A.N. (1992). *Optimum Experimental Designs.* Clarendon Press, Oxford.

Atkinson, A.C., Demetrio, C.G.B., Zocchi, S.S. (1995). Optimum dose levels when males and females differ in response. *J. R. Statist. Soc., Ser. C*, **44**, 213–226.

Baranyi, J., Roberts, T.A. (1995). Mathematics of predictive food microbiology. *Int. J. Food Microbiol.*, **26**, 199–218.

Bechenbach, F., Bellman, R. (1961). *Inequalities.* Springer-Verlag. Berlin.

Becka, M., Bolt, H.M., Urfer, W. (1993) Statistical Evaluation of Toxicokinetic Data. *Environmetrics*, **4**, 311–322.

Becka, M., Urfer, W. (1996). Statistical aspects of inhalation toxicokinetics. *Environmental Ecolog. Statist.* **3**, 51–64.

Beverton, R.J.H., Holt, S.J. (1957). *On the Dynamics of Exploited Fish Populations.* Her Majesty's Stationary Office. London.

Bezeau, M., Endrenyi, L. (1986). Design of experiments for the precise estimation of dose-response parameters: The Hill equation. *J. Theor. Biol.*, **123**(4), 415–430.

Blok, J. (1994). Classification of biodegradability by growth kinetic parameters. *Ecotoxicol. Environ. Safety*, **27**, 294–305.

Blok, J., Struys, J. (1996). Measurement and validation of kinetic parameter values for prediction of biodegradation rates in sewage treatment. *Ecotoxicol. Environ. Safety*, **33**, 217–227.

Bock, J. (1998) *Bestimmung des Stichprobenumfangs.* Oldenbourg, Munchen.

Boer, E.P.J., Rasch, D., Hendrix, E.M.T. (2000). Locally optimal designs in non-linear regression: A case study for the Michaelis-Menten function. Balakrishnan, N., Melas, V.B., and Ermakov, S.M.(ed.). *Advances in Stochastic Simulation Methods.* Birkhauser, pp. 177–188.

Box, J.E.P. (1996). Scientific statistics, teaching, learning and the computer. *Proceedings in Computational Statistics* – 12th Symposium Held in Barcelona, Spain, 1996/COMPSTAT. Prat, A. (ed.). Physica-Verlag, Heidelberg, pp. 3–10.

Box, J.E.P., Draper, N.R. (1987). *Empirical Model Building and Response Surface*. John Wiley & Sons, New York.

Box, G.E.P., Lucas, H.L. (1959). Designs of experiments in nonlinear situations. *Biometrika* **46**, 77–90.

Box, J.E.P., Wilson, K.B. (1951). On the experimental attainment of optimum conditions, *J. R. Statist. Soc.* Series B, **13**, 1–38; discussion, 39–45.

Chaloner, K. (1989). Optimal Bayesian experimental design for estimation the turning point of a quadratic regression. *Commun. of Statist., Theory Methods*, **18**. 1385–1400.

Chaloner, K., Larntz, K. (1989). Optimal Bayesian designs applied to logistic regression experiments. *J. Statist. Plan. Inference*, **21**, 191–208.

Chang, F-C., Heiligers, B. (1996). *E*-optimal designs for polynomial regression without intercept. *J. Statist. Plann. Inference*, **55**, 371–387.

Chang, F.-C., Lin, G.-C. (1997). *D*-Optimal designs for weighted polynomial regression *J. Stat. Plan. Inference*, **62**, 317–331.

Chatterjee, S.K., Mandal, N.K. (1981). Response surface designs for estimating the optimal point. *Calcutta Statist. Assoc. Bull.*, **30**, 145–169.

Cheng, R.C.H., Kleijnen, J., Melas, V.B. (2000). Optimal design of experiments with simulation models of nearly saturated queues *J. Stat. Plan. Inference*, **86**, 19–26.

Cheng, R.C.H., Melas, V.B., Pepelyshev, A.N. (2000). Optimal design for evaluation of an extremum point. Eds. A.Atkinson, A., Bogacka, B. Zhigljavsky A. (eds.). *Optimum Design 2000*. Kluwer, Boston, pp. 15–24.

Chernoff H. (1953). Locally optimal designs for estimating parameters. *Ann. Math. Statist.* **24**, 586–602.

Cornish-Browden, A. (1979). *Fundamentals of Enzyme Kinetics*. Butterworth, London.

Cressie, N.A.C., Keightley, D.D. (1979). The underlying structure of a direct linear plot with applications to the analysis of hormone-receptor interactions. *J. Steroid Biochem.*, **11**, 1173–1180.

Cressie, N.A.C., Keightley, D.D. (1981). Analysing data from hormone-receptor assays. *Biometrics*, **37**, 235–249.

De Vore, R.A., Lorentz, G.G. (1993). *Constructive Approximation*. Springer-Verlag, New York.

Dette, H. (1993). A note on *E*-optimal designs for weighted polynomial regression. *Ann. Statist.*, **21**, 767–771.

Dette, H. (1996). Lower bounds for sciencies with applications. In: Brunner, E., Denker, M. (eds.). *Research Developments in Probability and Statistics: Festschrift zum 65-ten Geburtstag von M.L. Puri, 1996*. VSP, Utrecht, The Netherlands, 111–124.

Dette, H. (1997a). Designing experiments with respect to standardized optimality criteria. *J. Roy. Statist. Soc.*, Series B, **59**, 97–110.

Dette, H. (1997b). *E*-optimal designs for regression models with quantitative factors – a reasonable choice. *Can. J. Statist.*, **25**, 531–543.

Dette, H., Biedermann, S. (2003). Robust and efficient designs for the Michaelis-Menten model. *J. Am. Stat. Assoc.*, **98**, 679–686.

Dette, H., Haines, L., Imhof, L. (1999). Optimal designs for rational models and weighted polynomial regression *Ann. Statist.*, **27**(4), 1272–1293.

Dette, H., Haines, L., Imhof, L. (2003). Maximin and Bayesian optimal designs for regression models. http://www.ruhr-uni-bochum.de/mathematik3/preprint.htm

Dette, H., Haller, G. (1998). Optimal designs for the identification of the order of a Fourier regression. *Ann. Statist.*, **26**, 1496–1521.

Dette, H., Melas, V.B. (2001). *E*-optimal designs in Fourier regression models on a partial circle. http://www.ruhr-uni-bochum.de/mathematik3/preprint.htm

Dette, H., Melas, V.B. (2002). *E*-Optimal designs in Fourier regression models on a partial circle. *Math. Methods Statist.*, **11**(3), pp. 259–296.

Dette, H., Melas, V.B. (2003). Optimal designs for estimating individual coefficients in Fourier regression models. *Ann. Statist.*, **31**(5) 1669–1692.

Dette, H., Melas, V.B., Biederman, S. (2002). *D*-Optimal designs for trigonometric regression models on a partial circle - a functional-algebraic approach. *Statist. Probab. Let.*, **57**, 389-397.

Dette, H., Melas, V.B., Pepelyshev, A. (2000). Optimal designs for estimating individual coefficients in polynomial regression - a functional approach. http://www.ruhr-uni-bochum.de/mathematik3/preprint.htm

Dette, H., Melas, V.B., Pepelyshev, A. (2002). *D*-Optimal designs for trigonometric regression models on a partial circle. *Ann. Inst. Statist. Math.*, **54**(4), 945–959.

Dette, H., Melas, V.B., Pepelyshev, A. (2003).Optimal designs for a class of nonlinear regression models. http://www.ruhr-uni-bochum.de/mathematik3/preprint.htm

Dette, H., Melas, V.B., Pepelyshev, A. (2003). Standardized maximin *E*-optimal designs for the Michaelis-Menten model. *Statist. Sin.*, **13**, 1147–1163.

Dette, H., Melas, V.B., Pepelyshev, A. (2004a). Optimal designs for a class of nonlinear regression models.*Ann. Statist.*, **32**(3), 2142–2167.

Dette H., Melas, V.B., Pepelyshev, A.N. (2004b). Optimal designs for estimating individual coefficients in polynomial regression — a functional approach. *J. Statist. Plan. Inference*, **118**, 201–219.

Dette, H., Melas, V.B., Pepelyshev, A., (2004c). Optimal designs for 3D shape analysis with spherical harmonic descriptors. Preprint, Ruhr-Universitat Bochum. http://www.ruhr-uni-bochum.de/mathematik3/preprint.htm

Dette, H., Melas, V.B., Pepelyshev, A., Strigul, N. (2002). Efficient design of experiments in the Monod model. http://www.ruhr-uni-bochum.de/mathematik3/preprint.htm

Dette, H., Melas, V. B., Pepelyshev, A., Strigul, N. (2003). Efficient design of experiments in the Monod model. *J. Roy. Statist. Soc.*, Series B, **65**, 725-742.

Dette, H., Melas, V.B., Pepelyshev, A., Strigul, N. (2005). Design of experiments for the Monod model – robust and efficient designs. *J. Theor. Biol.* **234**, 537–550.

Dette, H., Melas, V.B., Strigul, N. (2005). Design of experiments for microbiological models. Berger, M.P.F., Wong, W.K. (eds.) *Applied Optimal Designs*. John Wiley & Sons, Chichester, pp. 137–180.

Dette, H., Melas, V.B., Wong, W.K. (2004a). Locally D-optimal designs for exponential regression. Preprint Ruhr-Universität Bochum. http://www.ruhr-uni- bochum.de/mathematik3/preprint.htm

Dette, H., Melas, V.B., Wong, W.K. (2004b). Optimal design for goodness-of-fit of the Michaelis–Menten enzyme kinetic function. Preprint Ruhr-Universität Bochum. http://www.ruhr-uni-bochum.de/mathematik3/preprint.htm

Dette, H., Studden, W.J. (1993). Geometry of E-optimality. *Ann. Statist.* **21**, 416–433.

Dette, H., Wong W.K. (1999). E-optimal designs for the Michaelis-Menten model. *Statist. Probab. Let.* **44**, 405–408.

Dudzinski, M.L., Mykytowycz, R. (1961). The eye lens as an indicator of age in the wild rabbit in Australia. *CSIRO Wildl. Res.*, **6**, 156–159.

Duggleby, R.G. (1979). Experimental designs for estimating the kinetic parameters for enzymecatalysed reactions. *J. Theor. Biol.*, **81**, 671–684.

Dunn, G. (1988). Optimal designs for drug, neurotransmitter and hormone receptor assays. *Statist. Med.*, **7**, 805–815.

Ehrenfeld, E. (1955). On the efficiency of experimental design, *Ann. Math. Statist.*, **26**, 247–255.

Elfving, G. (1952). Optimum allocation in linear regression theory. *Ann. Math. Statist.* **23**, 255- -262.

Ellis, T.G., Barbeau, D.S., Smets, B.F., Grady, C.P.L. (1996). Respirometric technique for determination of extant kinetic parameters describing biodegradation. *Water Environ. Res.*, **68**, 917–926.

Ermakov, S.M., Melas, V.B. (1995). *Design and Analysis of Simulation Experiments*. Kluwer Academic Publisher, London.

Ermakov, S.M. (ed.). (1983). *Mathematical Theory of Experimental Design*. Nauka, Moscow (in Russian).

Fedorov, V.V. (1972). *Theory of Optimal Experiments*. Academic Press, New York.

Fedorov, V.V., Hackl, P. (1997). *Model-Oriented Design of Experiments*. Springer-Verlag, New York.

Fedorov, V.V., Müller, W.C. (1997). Another view on optimal design for estimating the point of extremum in quadratic regression. *Metrika*, **46**, 147–157.

Ferenci, Th. (1999). "Growth of bacterial cultures" 50 years on: towards an uncertainty principle instead of constants in bacterial growth kinetics. *Res. Microbiol.*, **150**(7), 431–438.

Fisher, R. (1935). *The Design of Experiments.* Oliver Boud, London.

Ford, I., Silvey, S.D. (1980). A sequentially constructed design for estimating a nonlinear parametric function. *Biometrika,* **67**, 381-388.

Ford, I., Torsney, B., Wu, C.F.J. (1992). The use of a canonical form in the construction of locally optimal designs for non-linear problems. *J. R. Statist. Soc.,* Series B, **54**, 569–583.

Fu, W., Mathews, A.P. (1999). Lactic acid production from lactose by *Lactobacillus plantarum*: Kinetic model and effects of pH, substrate, and oxygen. *Biochem. Eng. J.,* **3**, 163–170.

Gantmacher, F.R. (1998). *The Theory of Matrices.* Chelsea, Providence. RI.

Goudar, C.T., Ellis, T.G. (2001). Explicit oxygen concentration expression for estimating extant biodegradation kinetics from respirometric experiments. *Biotechnol. Bioeng.,* **75**, 74–81.

Graybill, F.A. (1976). *Theory and Application of the Linear Model.* Wadsworth, Belmont CA.

Gunning, R.C., Rossi, H. (1965) *Analitical Functions of Several Complex Variables.* Prentice-Hall, Inc., NewYork.

Haines, L.M. (1992). Optimal design for inverse quadratic polynomials. *South African Statist. J.,* **26**, 25–41.

Haines, L. M. (1993). Optimal design for nonlinear regression models. *Commun. Statist. A,* **22**, 1613–1627.

Haines, L. M. (1995). A geometric approach to optimal design for one-parameter nonlinear models. *J. R. Stat. Soc.,* Series B, **57**, 575–598.

Han, C., Chaloner, K. (2003). D- and C-Optimal designs for exponential regression models used in viral dynamics and other applications. *J. Statist. Plan. Inference,* **115**, 585–601.

Hay, W.W., Meznarich, H.K., DiGiacomo, J.E., Hirst, K., Zerbe, G. (1988). Effects of insulin and glucose concentration on glucose utilization in fetal sheep. *Pediatr. Res.,* **23**, 381–387.

Hco, G., Schmuland, B., Wiens, D.P. (2001). Restricted minimax robust designs for misspesified regression models. *Can. J. Statist.,* **29**, 117–128.

He Z., Studden W.J., Sun D. (1996). Optimal designs for rational models. *Ann. Statist.,* **24**, 2128–2147.

Heiligers, B. (1991). *E-Optimal Polynomial Regression Designs.* Habilitationssrift, RWTH, Aahen.

Heiligers, B. (1994). *E*-Optimal designs in weighted polynomial regression. *Ann. Statist.,* **22**, 917–929.

Heiligers, B. (1998). *E*-Optimal designs in spline regression. *J. Statist. Plan. Inference,* **75**, 159–172.

Hill, P.D.H. (1978). A note on the equivalence of *D*-optimal design measures for three rival linear models *Biometrika,* **65**, 666–667.

Hoel, P. (1965). Minimax design in two-dimensional regression. *Ann. Math. Statist.*, **36**, 1097-1106.

Hoel, P.G., Levine, A., 1964. Optimal spacing and weighting in polynomial prediction. *Ann. Math. Statist.*, **35**, 1553-1560.

Holmberg, A. (1982). On the practical identifiability of microbial gowth models incorporating Michaelis-Menten type nonlinearities. *Math. Biosci.* **62**, 23–43.

Huang, M.-N.L., Chang, F.-C., Wong, W.K. (1995). *D*-optimal designs for polynomial regression without an intercept. *Statist. Sin.*, **5**, 441–458.

Imhof, L.A. (2001). Maximin designs for exponential growth models and heteroscedastic polynomial models. *Ann. Statist.*, **29**(2), 561-576.

Imhof, L., Krafft, O., Schaefer, M. (1998). *D*-Optimal designs for polynomial regression with weight function $w(x) = x/(1 + x)$. *Statist. Sin.*, **8**, 1271–74.

Imhof, L., Studden, W.J. (2001). *E*-optimal designs for rational models. *Ann. Statist.*, **29**(3).

Jennrich, R.I. (1969). Asymptotic properties of non-linear least squares estimators. *Ann. Math. Statist.* **40**, 633–643.

Johansen, S. (1984). *Functional Relations, Random Coefficients and Nonlinear Regression, with Application to Kinetic Data.* Lecture Notes in Statistics, No 22. Springer-Verlag, New York.

Karlin, S., Studden, W. (1966). *Tchebysheff Systems: With Application in Analysis and Statistics.* John Wiley & Sons, New York.

Kiefer, J. (1974). General equivalence theory for optimum designs (approximate theory). *Ann. Statist.* **2**, 849–879.

Kiefer, J. (1985). *Collected Papers.* Springer-Verlag, New York.

Kiefer, J., Wolfowitz, J. (1960). The equivalence of two extremum problems. *Can. J. Math.*, **14**, 363–366.

Kiefer, J.C., Wolfowitz, J. (1959). Optimum designs in regression problems. *Ann. Math. Statist.*, **30**, 271–294.

Kitsos, C.P., Titterington, D.M., Torsney, B. (1988). An optimal design problem in rhythmometry *Biometrics*, **44**, 657–671.

Knightes, C.D., Peters, C.A. (2000). Statistical analysis of nonlinear parameter estimation for Monod biodegradation kinetics using bivariate data. *Biotechnol. Bioeng.*, **69**, 160–170.

Kovrigin, A. B. (1980). Construction of E-optimal designs. *Vestnik Leningrad. Univ.*, **19**, 120, Abstract.

Kozlov, V.P. (2000). *Selected papers.* St. Petersburg University Publishers, St. Petersburg (in Russian).

Lancaster, H.O. (1969). *The Chi-Squared Distribution.* John Wiley & Sons, New York.

Lancaster, P. (1969). *Theory of Matrices.* Academic Press, New York.

Lau, T.S., Studden, W.J. (1985). Optimal designs for trigonometric and polynomial regression. *Ann. Statist.*, **13**, 383–394.

Lestrel, P. E. (1997). *Fourier Descriptors and Their Applications in Biology.* Cambridge University Press, Cambridge.

Lopez-Fidalgo, J., Rodriguez-Diaz, J.M. (2004). Elfving method for computing c-optimal designs in more than two dimensions. *Metrika* **59**, 235-244.

Mardia, K. (1972). *The Statistics of Directional Data.* Academic Press, New York.

McCool, J.I. (1979). Systematic and random errors in least squares estimation for circular contours. *Precision Eng.*, **1**, 215–220.

Melas, V.B. (1978). Optimal designs for exponential regression. *Math. Operationsforsh. Statist.*, **9**, 45–59.

Melas, V.B. (1982). A duality theorem and E-optimality. *Ind. Lab.*, **48**, 295–296 (translated from Russian).

Melas, V.B. (1995). Non-Chebyshev E-optimal experimental designs and decompositions of positive polynomial. I. *Vestnik Petersburg Univ.*, **28**(2), 31–35.

Melas, V. B. (1996). A study of E–optimal designs for polynomial regression. *Proceedings in computational statistics - 12th symposium held in Barcelona, Spain, 1996/COMPSTAT* Prat, A. (ed.). Physica-Verlag, Heidelberg, pp. 101–110.

Melas, V. B. (1998). Analytical theory of E-optimal designs for polynomial regression on a segment. *MODA-5 – Advances in Model-Oriented Data Analysis and Experimental Design* Atkinson, A.C., Pronzato, L., Wynn, H.P. (eds.). Physica-Verlag, Heidelberg, pp. 51–58.

Melas, V.B. (1999). *Locally Optimal Designs of Experiments.* Publishers of St. Petersburg State Technical University, St. Petersburg (in Russian).

Melas, V.B. (2000). Analytic theory of E-optimal designs for polynomial regression. *Advances in Stochastic Simulation Methods.* Balakrishnan, N., Melas, V.B. S. Ermakov (eds.). Birkhäuser, Boston, pp. 85–116.

Melas V.B. (2001). Analytical properties of locally D-optimal designs for rational models. Atkinson A.C., Hackel P., Müller W.J. (eds.) *MODA 6 — Advances in Model-Oriented Design and Analysis*, Physica-Verlag, Heidelberg, pp. 201–210.

Melas, V.B. (2004). On a functional approach to locally optimal designs. Atkinson A.C., Hackel P., Müller W.J. (eds.) *MODA-7 – Advances in Model Oriented Design and Analysis.* Physica-Verlag, Heidelberg, pp. 97–105.

Melas, V.B. (2005). On the functional approach to optimal designs for nonlinear models.*J. Statist. Plan. and Inference*, **132**, 93–116.

Melas, V.B., Krylova, L.A. (1998). E-Optimal designs for quadratic regression on arbitrary segments, *Vestnik Sankt-Peterburgskogo Univ.*, ser. 1, No. 15, 44–49 (in Russian).

Melas. V.B., Pepelyshev. A.N., Cheng. R.C.H. (2003). Designs for estimating an extremal point of quadratic regression models in a hyperball. *Metrika*, **58**, 193–208.

Merkel, W., Schwarz, A., Fritz, S., Reuss, M., Krauth, K. (1996). New strategies for estimating kinetic parameters in anaerobic wastewater treatment plants. *Water Sci. Technol.* **34**, 393–401.

Monod, J. (1949). The growth of bacterial cultures. *Ann. Rev. Microbiol.*, **3**, 371–393.

Müller, C.H. (1995). Maximin efficient designs for estimating nonlinear aspects in linear models. *J. Statist. Plan. Inference*, **44**, 117–132.

Müeller, W.G. (1998). *Collecting Spatial Data.* Physica-Verlag, Heidelberg.

Müller, Ch.H., Pázman A. (1998). Applications of necessary and sufficient conditions for maximum efficient design. *Metrika*, **48**, 1–19.

Ortiz, I., Rodziguez, C. (1998). *D*-optimal designs for weighted polynomial regression without any initial terms. Atkinson A.C., Pronzato, L., Wynn, H.P. (eds.) *MODA-5 – Advances in Model Oriented Design and Analysis.* Physica-Verlag, Heidelberg, pp. 67–74.

Ossenbruggen, P. J., Spanjers, H., Klapwik, A. (1996). Assessment of a two-step nitrification model for activated sludge. *Water Res.*, **30**, 939–953.

Pázman, A., Pronzato, L. (1992). Nonlinear experimental design based on the distribution of estimators. *J. Statist. Plan. Inference*, **33**, 385–402.

Pert, D.C. (1978). *The Bases of Cultivation of Micro-organisms and Cells.* Mir, Moscow.

Petrushev, P.P., Popov, V.A. (1987).*Rational Approximations of Real Functuons.* Cambridge Univ.Press.

Piquette, J.C. (2001). Method for obtaining corrective power-series solutions to algebraic and transcendental systems. *Utilitas Math.*, **59**, 3–26.

Pirt, S. J. (1975). *Principles of Microbe and Cell Cultivation.* John Wiley & Sons, New York.

Pólya, G., Szegö, G. (1971).*Aufgaben und Lehrsatze aus der Analysis.* **I**. Springer-Verlag, New York.

Pronzato, L., Walter E. (1985). Robust experimental design via stochastic approximation. *Math. Biosci.*, **75**, 103–120.

Pukelsheim, F. (1980). On linear regression designs which maximize information, *J. Statist. Plan. Inference*, **4**, 339–364.

Pukelsheim, F. (1993). *Optimal Designs of Experiments.* John Wiley & Sons, New York.

Pukelsheim, F., Rieder, S. (1992). Efficient rounding of approximate designs. *Biometrika*, **79**, 763-770.

Pukelsheim, F, Studden, W. J. (1993). *E*-Optimal design for polynomial regression *Ann. Statist.*, **21**, 401–415.

Pukelsheim, F., Torsney, B. (1991). Optimal designs for experimental designs on linearly independent support points. *Ann. Statist.*, **19**, 1614–1625.

Rao, S.R. (1973). *Linear Statistical Inference and Its Applications*, 2nd ed., John Wiley & Sons, New York.

Rasch D. (1990). Optimum experimental design in nonlinear regression. *Commun. Statist., Theor. Methods*, **19**, 4789–4806.

Rasch, D. (1996). "Replication-free" optimal design in regression analysis. *Proceedings in Computational Statistics – 12th Symposium Held in Barcelona, Spain, 1996/COMPSTAT*. Prat, A. (ed.)- Physica-Verlag, Heidelberg, pp. 403–409.

Rasch, D. (2003). Determining the optimal size of experiments and surveys in empirical research. *Psychol. Sci.*, **45**, 3–47.

Rasch, D., Herrendorfer, G., Bock, J., Victor, N., Guiard, V. (1996). *Verfahrensbibliothek Versuchsplanung und auswertung, Band I. [Collection of Procedures for Designing and Analysing Experiments, Vol. I]*. Oldenbourg, Munich.

Rasch, D., Herrendorfer, G., Bock, J., Victor, N., Guiard, V. (1998). *Verfahrensbibliothek Versuchsplanung und Auswertung, Band II. [Collection of Procedures for Designing and Analysing Experiments, Vol II]* Oldenbourg, Munich.

Ratkowsky, D.A. (1983). *Nonlinear Regression*. Dekker, New York.

Ratkowsky, D.A. (1990). *Handbook of Nonlinear Regression Models*. Dekker, New York.

Riccomagno, E., Schwabe, R.,Wynn, H.P. (1997). Lattice-based d-optimum design for Fourier regression *Ann. Statist.*, **25**, 2313–2327.

Rivlin, T.J. (1974). *Chebyshev Polynomials*. John Wiley & Sons, New York.

Sahm, M. (1998). Optimal designs for estimating individual coefficients in polynomial regression. Dissertation. Ruhr University Bochum.

Schirmer, M., Butler, B.J., Roy, J.W., Frind, E.O., Barker, J.F. (1999). A relative-least-squares technique to determine unique Monod kinetic parameters of BTEX compounds using batch experiments. *J. Contam. Hydrol.*, **37**, 69–86.

Schwabe, R. (1996). *Optimum Designs for Multi-Factor Models*. Springer-Verlag, New York.

Seber, G.A.J., Wild, C.J. (1989). *Nonlinear Regression*. John Wiley Sons, New York.

Silvey, S.D. (1980). *Optimal Design*. Chapman & Hall, London.

Silvey, S.D., Titterington, D.M. (1973). A geometric approach to optimal design theory. *Biometrica*, **60**, 21–32.

Sitter, R., Torsney, B. (1995). Optimal designs for binary response experiments with two design variables. *Statist. Sin.* **5**, 405–419.

Song, D., Wong, W.K. (1998). Optimal two point designs for the Michaelis-Menten model with heteroscedastic errors. *Commun. Statist. Theory Methods* **27**(6), 1503–1516.

Spruill, M.G. (1990). Good designs for testing the degree of a polynomial mean *Sankhyfa*, Series B, **52**, 67–74.

Stiegler, S.M. (1986). *The History of Statistics*. Harvard University Press.

Street, A.P., Street D.J. (1987). *Combinatorics of Experimental Design.* Oxford University Press, Oxford.

Studden, W.J. (1968). Optimal design on Tchebycheff points. *Ann. Math. Statist.*, **39**, 1435–1447.

Studden, W.J. (1980a). D_s-Optimal designs for polynomial regression using continued fractions *Ann. Statist.*, **8**, 1132–1141.

Studden, W.J. (1980b). On a problem of Chebyshev. *J. Approx. Theory*, **29**, 253–260.

Studden, W.J., Tsay, J.Y. (1976). Remez's procedure for finding optimal designs. *Ann. Statist.*, **4**, 1271–1279.

Szegö, G. (1959). *Orthogonal Polynomials.* American Mathematical Society, New York.

Szegö, G. (1975). *Orthogonal Polynomials.* American Mathematical Society, Providence, RI.

Titchmarsh, E.C. (1939). *The Theory of Functions*, Oxford University Press.

Van der Waerden, B.L. (1967). *Algebra.* Springer-Verlag, Berlin.

Vanrolleghem, P.A., Spanjers, H., Petersen, B., Ginestet, Ph., Takacs, I. (1999). Estimating (combinations of) activated sludge model no. 1 parameters and components by respirometry. *Water Sci. Technol.*, **39**, 195–214.

Vanrolleghem, P.A., Van Daele, M., Dochain, D. (1995). Practical identifiability of a biokinetic model of activated sludge respiration. *Water Res.* **29**, 2561–2570.

Versyck, K.J., Bernaerts, K., Geeraerd, A.H., Van Impe, J.F. (1999). Introducing optimal experimental design in predictive modeling: A motivating example. *Int. J. Food Microbiol.*, **51**, 39–51.

Vila, J.P. (1990). Exact experimental designs via stochastic optimization for nonlinear regression models. Compstat 1990, 291–296.

Weber, W.E., Liebig, H.P. (1981). Fitting response functions to observed data *Elektron. Datenverarbeit. Med. Biol.*, **12**, 88–92.

Wiens, D.P. (1992). Minimax designs for approximateky linear regression. *J. Statist. Plan. Inference*, **31**, 353–371.

Wiens, D.P. (1993). Designs for approximately linear regression. Maximazing the minimum coverage probability of confidence ellipsoids. *Can. J. Statist.*, **21**, 59–70.

Whittle, P. (1973) Some general points in the theory of optimal experimental design. *J. R. Statist. Soc., Series B*, **35**, 123-130

Wu, C.F.J. (1985). Efficient sequential designs with binary data. *J. Amer. Stat. Assoc.* **80**, 974–984.

Wynn, H.P. (1970). The sequential generation of D-optimum experimental designs. *Ann. Math. Statist.*, **41**, 1655–1664.

Young, J.C., Ehrlich, R. (1977). Fourier biometrics: harmonic amplitudes as multivariate Shape descriptors. *Syst. Zool.*, **26**, 336–342.

Index

Lecture Notes in Statistics

For information about Volumes 1 to 129, please contact Springer-Verlag

130: Bo-Cheng Wei, Exponential Family Nonlinear Models. ix, 240 pp., 1998.

131: Joel L. Horowitz, Semiparametric Methods in Econometrics. ix, 204 pp., 1998.

132: Douglas Nychka, Walter W. Piegorsch, and Lawrence H. Cox (Editors), Case Studies in Environmental Statistics. viii, 200 pp., 1998.

133: Dipak Dey, Peter Müller, and Debajyoti Sinha (Editors), Practical Nonparametric and Semiparametric Bayesian Statistics. xv, 408 pp., 1998.

134: Yu. A. Kutoyants, Statistical Inference For Spatial Poisson Processes. vii, 284 pp., 1998.

135: Christian P. Robert, Discretization and MCMC Convergence Assessment. x, 192 pp., 1998.

136: Gregory C. Reinsel, Raja P. Velu, Multivariate Reduced-Rank Regression. xiii, 272 pp., 1998.

137: V. Seshadri, The Inverse Gaussian Distribution: Statistical Theory and Applications. xii, 360 pp., 1998.

138: Peter Hellekalek and Gerhard Larcher (Editors), Random and Quasi-Random Point Sets. xi, 352 pp., 1998.

139: Roger B. Nelsen, An Introduction to Copulas. xi, 232 pp., 1999.

140: Constantine Gatsonis, Robert E. Kass, Bradley Carlin, Alicia Carriquiry, Andrew Gelman, Isabella Verdinelli, and Mike West (Editors), Case Studies in Bayesian Statistics, Volume IV. xvi, 456 pp., 1999.

141: Peter Müller and Brani Vidakovic (Editors), Bayesian Inference in Wavelet Based Models. xiii, 394 pp., 1999.

142: György Terdik, Bilinear Stochastic Models and Related Problems of Nonlinear Time Series Analysis: A Frequency Domain Approach. xi, 258 pp., 1999.

143: Russell Barton, Graphical Methods for the Design of Experiments. x, 208 pp., 1999.

144: L. Mark Berliner, Douglas Nychka, and Timothy Hoar (Editors), Case Studies in Statistics and the Atmospheric Sciences. x, 208 pp., 2000.

145: James H. Matis and Thomas R. Kiffe, Stochastic Population Models. viii, 220 pp., 2000.

146: Wim Schoutens, Stochastic Processes and Orthogonal Polynomials. xiv, 163 pp., 2000.

147: Jürgen Franke, Wolfgang Härdle, and Gerhard Stahl, Measuring Risk in Complex Stochastic Systems. xvi, 272 pp., 2000.

148: S.E. Ahmed and Nancy Reid, Empirical Bayes and Likelihood Inference. x, 200 pp., 2000.

149: D. Bosq, Linear Processes in Function Spaces: Theory and Applications. xv, 296 pp., 2000.

150: Tadeusz Caliński and Sanpei Kageyama, Block Designs: A Randomization Approach, Volume I: Analysis. ix, 313 pp., 2000.

151: Håkan Andersson and Tom Britton, Stochastic Epidemic Models and Their Statistical Analysis. ix, 152 pp., 2000.

152: David Ríos Insua and Fabrizio Ruggeri, Robust Bayesian Analysis. xiii, 435 pp., 2000.

153: Parimal Mukhopadhyay, Topics in Survey Sampling. x, 303 pp., 2000.

154: Regina Kaiser and Agustín Maravall, Measuring Business Cycles in Economic Time Series. vi, 190 pp., 2000.

155: Leon Willenborg and Ton de Waal, Elements of Statistical Disclosure Control. xvii, 289 pp., 2000.

156: Gordon Willmot and X. Sheldon Lin, Lundberg Approximations for Compound Distributions with Insurance Applications. xi, 272 pp., 2000.

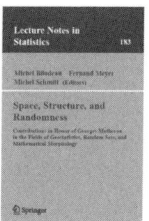

Space, Structure and Randomness
Contributions in Honor of Georges Matheron in the Fields of Geostatistics, Random Sets, and Mathematical Morphology

M. Bilodeau, F. Meyer and M. Schmitt (Editors)

This volume is divided in three sections on random sets, geostatistics and mathematical morphology. They reflect Georges Matheron's professional interests and his search for underlying unity.

2005. 416 p. (Lecture Notes in Statistics, Vol. 183) Softcover
ISBN 0-387-20331-1

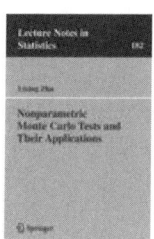

Nonparametric Monte Carlo Tests and Their Applications

L. Zhu

A fundamental issue in statistical analysis is testing the fit of a particular probability model to a set of observed data. Monte Carlo approximation to the null distribution of the test provides a convenient and powerful means of testing model fit. *Nonparametric Monte Carlo Tests and Their Applications* proposes a new Monte Carlo-based methodology to construct this type of approximation when the model is semistructured.

2005. 190 p. (Lecture Notes in Statistics, Vol.182) Softcover
ISBN 0-387-25038-7

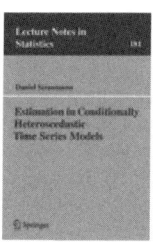

Estimation in Continually Heteroscedastic Time Series Models

D. Straumann

This monograph concentrates on mathematical statistical problems associated with fitting conditionally heteroscedastic time series models to data. This includes the classical statistical issues of consistency and limiting distribution of estimators. Particular attention is addressed to (quasi) maximum likelihood estimation and misspecified models, along with phenomena due to heavy-tailed innovations. The used methods are based on techniques applied to the analysis of stochastic recurrence equations.

2004. 228 p. (Lecture Notes in Statistics, Vol. 181) Softcover
ISBN 3-540-21135-7

Easy Ways to Order ▶   Call: Toll-Free 1-800-SPRINGER • E-mail: orders-ny@springer.sbm.com • Write: Springer, Dept. S8113, PO Box 2485, Secaucus, NJ 07096-2485 • Visit: Your local scientific bookstore or urge your librarian to order.